Catalytic Naphtha Reforming Process

Catalytic Naphtha Reforming Process

Soni O. Oyekan

CRC Press is an imprint of the
Taylor & Francis Group, an **informa** business

CRC Press
Taylor & Francis Group
6000 Broken Sound Parkway NW, Suite 300
Boca Raton, FL 33487-2742

First issued in paperback 2021

ISBN-13: 978-0-367-78096-8 (pbk)
ISBN-13: 978-1-138-03430-3 (hbk)

Visit the Taylor & Francis Web site at
http://www.taylorandfrancis.com

and the CRC Press Web site at
http://www.crcpress.com

Contents

Preface

It is with great pleasure that I offer this book as a product of my goal of conveying relevant information on key concepts, operations, and practices of catalytic naphtha reforming technologies and associated oil refining processes. This book will most likely be of use to those in academia and oil refining, petrochemicals, and allied businesses. I was motivated by several factors to write a book on the catalytic naphtha reforming process and technologies. Foremost is the need to review the collective technical and operational advancements with respect to efficient use of catalysts and catalytic reformers in oil refining.

The book on catalytic reforming by Donald Little published in 1985 and useful chapters in several other books provide fundamental and some practical aspects of catalytic reforming and light naphtha isomerization processes. Some of the chapters in separate books by James Gary and Glenn Handwerk; James Speight; and Bruce Gates, James Katzer, and G. C. A. Schuit provide relevant sections on catalytic reforming technologies. Books on catalytic naphtha reforming by George Antos, Abdullah Aitani, and Jose Parera provide comprehensive works with key basic and practical concepts of the catalytic reforming process and related sciences. The listed books were foundational for writing this book.

A second factor has been the need to incorporate key advancements from recent developments in catalytic reforming technologies and processes. Most of the recent innovations have been in continuous catalyst regenerative processes, and some of the enhancements are covered in this book. Some of the advancements in catalytic reforming have been driven by the need to provide high-octane reformate gasoline blendstock in current markets for unleaded and regulated environmentally friendly gasoline products. It is expected that this trend will continue as US and relevant global markets position to meet the projected Tier 3 Corporate Average Fuel Economy (CAFÉ) standard of 54.5 miles per gallon of gasoline for automobiles by 2025. In addition, as important as reformate production is, the higher demand for hydrogen production from catalytic reformers has re-emphasized interests in achieving reliable and efficient operations of the process units. Higher hydrogen demand is due to the increased requirements of oil refiners to produce ultralow-sulfur gasoline and diesel products. Furthermore, high-severity processing operations of catalytic reformers have led to increased organic chloride usage. Higher usage of organic chloride has exacerbated reliability challenges and heightened the need to establish cost-effective maintenance programs for piping and equipment, and for the other process units in oil refineries. Higher production of net hydrogen gas in catalytic reforming operations leads to increased use of chloride management strategies to protect oil refining equipment and process units from fouling and corrosion. The US Environmental Protection Agency (EPA) established stringent environmental regulations for the reduction of hydrogen chloride and toxic organic compounds (TOCs) in catalyst regeneration vent gases. Three major technologies are used for meeting the lower levels of hydrogen chloride and TOCs in regeneration vent gases. The caustic wash vent gas tower system, adsorbents, and Honeywell UOP's Chlorsorb are the available technologies for reducing hydrogen chloride and volatile organic compounds in regenerator vent gas. The challenges associated with use of these technologies are discussed.

Third, with the shift to more abundant, high-contaminant conventional crude oils and unconventional bitumen and shale oils, catalytic reforming process units have now developed another critical role in conjunction with crude distillation units as "gatekeepers" in the optimization of the crude slates that are processed in oil refineries. The highly sensitive platinum-containing catalysts used in catalytic reforming units are very susceptible to metal contaminant poisoning in naphtha feeds at fractional parts per million levels. Oil refiners with good analytical and continuous monitoring programs can use their assets beneficially by applying them for monitoring the quality of the naphtha feed to catalytic reformers. These resources could enable oil refiners to detect changes in naphtha feed qualities, especially with respect to catalyst contaminants. Oil refiners can then develop timely responses to naphtha feed quality changes in catalytic reformers. These contaminants could be due to contaminant contributions from one of the oils that is being processed in the crude slate. Appropriate adjustments can then be made to optimize the percentages of oil constituents in the crude slates in order to minimize or eliminate catalytic reformer and refinery reliability and productivity challenges associated with the constituent oils in the crude slate. Meticulous and continuous monitoring of key performance indicators of the crude distillation and catalytic reforming process units are required, and the benefits of timely, effective unit monitoring are covered in this book.

The factors listed motivated me to write this book to incorporate many of the current key challenges, process developments, effective operations, and monitoring of catalytic reformers in order to achieve optimized oil refining operations and desired refinery reliability and profitability. In addition, since I could have benefited from a brief introduction to the history of the oil and oil refining business at the start of my career in oil refining about four decades ago, I devoted the first chapter of this book to a brief history of the oil business and emphasized the pioneering business leadership of John D. Rockefeller in growing Standard Oil at the turn of the twentieth century. Standard Oil became the dominant global oil production, refining, and marketing company at the beginning of the twentieth century. The history of oil production can enhance our understanding of current geopolitical developments that have led to changes in crude oil production and supply, crude oil quality, crude oil prices, and their impacts on oil refinery operations, profitability, and the fiscal health of national and global economies. The basics of oil refining are covered in the second chapter, and several oil refinery processes and transportation fuel specifications are covered in the third chapter. Brief reviews of oil refining process units are also provided in the third chapter. The reviews help to emphasize the pivotal position of the catalytic reforming process unit in oil refineries for the processing of straight run naphtha and unsaturated naphthas from other downstream thermal and catalytic conversion units in oil refineries. Catalytic reforming technologies, catalysts, and processes are covered in detail in the six chapters that follow. The eighth chapter is devoted to coverage of key specially selected topics of vital interest to enable successful, reliable, and optimal operations of catalytic reforming process units and oil refineries. Due to several challenges that can arise due to inadequate turnaround planning, subpar execution of turnarounds, and safety problems during catalytic reformer startups, some best-practice concepts for turnaround maintenance are offered in Chapter 9. An overview of cost-effective management programs for precious metals is provided with great emphasis on platinum management. Platinum management principles discussed can also be extended to managing other platinum group metals, as well as silver, rhenium, and palladium.

One of the goals of writing this book is to encourage a more proactive approach in the planning, operations and maintenance of catalytic reforming units and oil refineries. A number of recommendations are provided in this book for enhancing the operations, reliability, and productivity of catalytic reformers and oil refineries. These recommendations can lead to cost-effective, safer, and optimal process unit operations and maximization of the profitability of oil refiners.

Acknowledgments

The contents, technical knowledge, and practical experiences covered in a book of this nature are developed via grounding in the principles of chemical engineering, catalysis, and oil refining processes. As a result, it is appropriate for me to start by acknowledging the training that I received from Professor Anthony Dent during my doctoral studies on reactor engineering and catalysis at Carnegie Mellon University. I am also grateful to Professors Csaba Horvath and Harding Bliss for the positive influences they had on me in my studies of reaction kinetics and thermodynamics, respectively, when I was at Yale University.

In the writing of a book of this scope, I have had to draw from course notes, technical presentations, and relevant data from my patents on the catalytic naphtha reforming process and other sources. The information that I share comes from about 40 years of technology development and training course notes that I shared with oil refining engineers and operators. Some of the older introductory information comes from the Petroleum Refining course that I taught at the New Jersey Institute of Technology, Newark, in 1983. Over the past 25 years, I taught courses on catalytic reforming, naphtha hydrotreating, paraffin isomerization, fluid catalytic cracking, aromatics isomerization, and sulfur guard bed systems to oil refining process technology personnel who worked with me at Sunoco, Amoco, and Marathon Petroleum Corporation.

My lecture notes were updated and used as part of a course that was offered in collaboration with Peter Kokayeff of Unocal and Dr. Stuart S. Shih of Mobil Oil for the American Institute of Chemical Engineers (AIChE) Continuing Education program between 1992 and 1998. The title of the course was Catalytic Processes in Petroleum Refining. Dr. Shih taught the hydrotreating processes section, the hydrocracking processing was taught by Kokayeff, and I taught the sections on catalytic reforming and fluid catalytic cracking processes.

A number of companies contributed to my development and experiences in research and development of oil refining technologies and operations of oil refining, and I gratefully acknowledge them. The companies are Exxon (now ExxonMobil), Engelhard (now BASF), DuPont, Sunoco, Amoco, and Marathon Petroleum Corporation.

For the continuous catalyst regenerative (CCR) process units and technologies, I acknowledge with special gratitude Axens and Honeywell UOP, for the opportunities to collaborate on several technical, revamp, and grassroots catalytic reforming projects with many of their personnel on CCR process-related challenges. In a number of sections, figures of process units from technical papers by Axens and Honeywell UOP have been used to enhance the quality of the description of the CCR processes. I thank the companies for granting me copyright permissions to use relevant images.

Lastly and foremost, I thank my wife, Priscilla Ann, for her love, support, and encouragement always, and especially during the time that I spent in writing this book. In addition, I am indebted to her for encouraging me to write with greater clarity and make the content of the book of higher value for a wide spectrum of persons such as business executives, financial analysts and planners, environmental engineers, oil refining technical personnel and operators, chemical plant operators, and persons in academic communities. I thank my wife also for the many useful suggestions that she offered with respect to the chapters in the book.

Author

Soni O. Oyekan, PhD, is president and CEO of Prafis Energy Solutions. He is a globally renowned expert in oil, gas, and oil refining. Dr. Oyekan was the corporate process technologist for Marathon Petroleum Corporation, BP/Amoco, and Sunoco separately, with responsibilities for naphtha processing, oil refining processes, and technology support over a span of 20 years. Prior to 1993, Dr. Oyekan conducted R&D studies at Exxon, Engelhard, and DuPont for 16 years, which resulted in his inventions for catalytic naphtha reforming. His key inventions are used globally to enhance the production of transportation fuels. Dr. Oyekan is the 2008 recipient of the AIChE William M. Grimes award for excellence in chemical engineering. He was awarded the NOBCChE Percy Julian award in 2009 for his outstanding contributions in oil refining.

The author, a resident of Richmond, Texas, is a Fellow of the American Institute of Chemical Engineers and a member of the AIChE Foundation Board of Trustees and of NOBCChE. He is a member of the Nigerian Society of Chemical Engineers, Sigma Xi, and Phi Kappa Phi societies.

Dr. Oyekan earned his BS degree (1970) in Engineering and Applied Sciences from Yale University, New Haven, Connecticut, and his MS (1972) and PhD (1977) degrees in Chemical Engineering from Carnegie Mellon University, Pittsburgh, Pennsylvania.

1 Introduction to Catalytic Reforming and the Oil Business

1.0 INTRODUCTION

Catalytic reforming is the process used in oil refineries for converting the low-value heavy naphtha fraction of crude oil to premium reformate, benzene, toluene and xylene (BTX), and hydrogen. As will be reviewed in detail, reformate constitutes about 30% of the gasoline that is produced. Hydrogen from catalytic reformers is used as coreactant in several hydroprocessing units in oil refineries. In some oil refineries, the catalytic reformer is the sole provider of a refinery's hydrogen. Additionally, benzene, toluene, and xylenes are used in the petrochemical industry in the production of other chemicals and solvents. Numerous technological advances have been made to improve catalytic reforming processes so as to provide increased productivity over much longer times between turnaround maintenance periods. Starting with a thermal naphtha cracking process, reforming technologies have evolved through fixed-bed semiregenerative catalytic reforming and fixed-bed cyclic regenerative reformers to the current high-performance, high-productivity continuous catalytic regenerative technologies. In the catalytic reforming process, as is the case for other catalytic processes, the catalyst during hydrocarbon processing loses activity with time and has to be regenerated. With the exception of continuous catalyst regenerative reforming processes, a reformer or a reactor shutdown is required to effect necessary catalyst regeneration and catalytic activity recovery for fixed-bed regenerative catalytic reformers.

In this book, some of the regulated gasoline quality requirements such as lower benzene concentrations and oxygenate blending that have some impact on catalytic reforming process operations are reviewed in appropriate chapters. Some of the expected regulated reductions of aromatics may further impact catalytic reforming units' utilization. The demand for BTX in the petrochemical industry and hydrogen for increased upgrading of heavy sour crude oil and unconventional bitumen-derived oils in oil refineries will continue to drive the need for more efficient and smart operations of catalytic reformers. Additionally, I have dubbed catalytic reformers "gate keepers" of oil refineries, as they are the first major catalytic conversion units after the crude distillation units. Catalytic reformers, with their highly sensitive platinum-containing catalysts, exhibit catalytic performance declines with the introduction of low concentrations of contaminants from the crude unit that are present in hydrotreated naphtha feeds. In the current era of the processing of a variety of crude oils and unconventional oils with highly variable qualities, changes in catalytic reforming operations are often relatable to changes in either one of the crude oils or crude slate that the oil refiner is processing.

1.1 EARLY HISTORY OF CRUDE OIL BUSINESS

There is a long history of the use of oil by people. Historians suggest that crude oil was used over 5000 years ago and that Noah applied heavy crude oil or tar to his ark to maintain the ark and protect it from water damage. Noah, the ark, and the flood are discussed in the book of Genesis of the Bible. It has also been reported that China used oil in laboratories as early as 200 BCE. There are some historical reports of the use of crude oil in China and Japan in the seventh century. Historians indicate that by 1100 CE, Christian crusaders were "setting petroleum alight" and frightening and overcoming their enemies in battles. We are also informed that early immigrants in North America gathered the "black goo" or oil and used it for medicinal purposes. The "black goo" was used as ointment and medicine for toothache, fever, burns, open sores, and rheumatism and for caulking canoes by the Seneca Indians.[1]

The vast global oil and gas businesses that we now participate in and enjoy owe some of their origins to a little-known Persian physician and chemist, Muhammad Ibn Zakariya Razi, who in the ninth century demonstrated that kerosene could be produced via distillation of oil to generate a clean product for lamps.[2] In the middle of the nineteenth century, a Polish chemist, Filip Neriusz Walter, and a Scottish chemist, James Young, were later credited as key contributors in improving the process for the production of kerosene from crude oil via distillation. By 1848, James Young had an operating oil refining company producing kerosene.[3] Ignacy Lukasiewics is credited by some with building the world's first oil refinery in 1856. His legendary achievements included the discovery of how to distill kerosene from crude oil, the invention of the modern kerosene lamp, the introduction of the first street lamp in Europe in 1853, and the construction of the world's first modern oil well in 1854.[4] About this time, there was significant oil production in Baku, and Meerzoeff is reported to have built the first modern Russian refinery at Baku by 1861.[5]

Abraham Pineo Gesner is recognized as one of the founders of the oil business in North America and Canada based on his three initial US patents on the production of kerosene from crude oils, which spurred extensive exploration and production of oil within the United States and Canada.[6] Kerosene was found to burn cleaner and cheaper than whale oil, and interest in the production of kerosene for lighting led to the beginnings of businesses based on oil. Oil and gas businesses have now grown exponentially into the global oil and gas business that we know.

After the startup of a plant by Abraham Pineo Gesner in 1854 for the production of kerosene for lamps, kerosene production reached 5000 gallons per day in the United States. The lure of instant wealth from the production of kerosene from crude oil led to increased exploration and production of oil. George Bissell, Edmund L. Drake, and William Smith were early innovators in developing technologies for the exploration and production of oil. By 1859, there were 34 startup companies, and they were producing a staggering 5 million gallons of kerosene annually. At the famous Pithole City, Pennsylvania, oil production reached 6000 barrels per day by 1865, and unfortunately, this was followed soon by diminishing oil production rates, which finally led to a deserted Pithole City by 1869.[1,7,8]

Major investors made their play for oil, and that group included the great entrepreneur, business genius, and oil magnate, John Davidson Rockefeller, who transformed the oil business through ownership of Standard Oil. By 1879, Standard Oil owned 90% of the oil refining capacity in the United States, and by 1891, it owned an overwhelming 25% of oil production, 90% of the oil refining capacity, and an extensive transportation system for moving oil and refined products in the United States and in the world. The most productive oil find known at that time occurred in 1901 in Spindletop, just outside of Beaumont, Texas, in the United States. The Spindletop oil production rate was estimated at 72,000 barrels per day, and money from Spindletop later fueled the growth of the Gulf Oil Company.[7]

1.2 POST-1900 HISTORY OF CRUDE OIL BUSINESS

As a result of the oil business boom in the United States, a number of companies such as Gulf Oil, Phillips, and Texaco Fuels Oil began to operate and compete with Standard Oil at the start of the twentieth century. With the invention of electricity by Thomas Edison, the kerosene business for lighting was severely challenged, and rescue came from Ford and the automobile industry. The automobile's internal combustion engine had to be powered with gasoline or petrol. New demand for gasoline led to the installation of more oil refining capacities, which led to an increased need for oil exploration and production. This was followed by expansion of infrastructure for storage, capital for pipelines, and railroads for movement of crude oil and distribution and marketing of finished products.[8–10]

John D. Rockefeller continued to lead and grow Standard Oil and the United States economy by his exceptional business acumen and practices. However, some of his business practices were unacceptable for his competitors. Essentially, Rockefeller's business strategies eliminated inefficient businesses as Rockefeller focused on acquisition of smaller competing companies. Rockefeller understood the benefits of vertical and horizontal integration and the economy of scale, as his leadership strategies for Standard Oil positioned his company as the leader in oil exploration and production, oil refining, petrochemicals, transportation and distribution, and retail marketing in the United States and worldwide.[11]

To compete effectively with Standard Oil, Royal Dutch Oil and Shell Transportation and Trading Company (Shell) amalgamated to form the Royal Dutch Shell Company in 1907. As the application of mechanization increased in the early 1900s, the demand for gasoline increased, as gasoline was used as fuel by automobiles and airplanes. Increased demand for gasoline was met with greater expansion of every aspect of the oil business from exploration and production to distribution and marketing of refined transportation fuels. Standard Oil continued to encounter increased distrust of its business practices, which in part had its origin in Standard Oil's business relationships with the railroad companies and its effort to eliminate competition in the business segments that Standard Oil was engaged in. Due to the highly aggressive business practices of Standard Oil and concerns about a possible oil business monopoly, the United States Supreme Court ruled in 1911 that Standard Trust or Standard

Oil Company had operated to monopolize and restrain trade and ordered that the company be dissolved into 34 smaller competing companies, significantly degrading the size and business clout of Standard Oil.[11,12]

A major global oil and energy–impacting event occurred in September 1960 in Baghdad, Iraq, with the formation of the Organization of the Petroleum Exporting Countries (OPEC) by five countries, namely Iran, Iraq, Kuwait, Saudi Arabia, and Venezuela. The OPEC group was later expanded to include Qatar (1961), Indonesia (1962), Libya (1962), the United Arab Emirates (1967), Algeria (1969), Nigeria (1971), Ecuador (1973), Gabon (1975), and Angola (2007). Gabon and Ecuador later terminated their memberships, while Indonesia suspended its membership in 2009 and reactivated it in 2016. The OPEC countries currently produce about 40% of global oil and have about 75% of the proved oil reserves.[11,12]

The 1973 oil embargo and subsequent events in the oil and gas industry continue to show the clout of OPEC in crude oil pricing and the politico-economic decisions of non-OPEC countries. Impacts of a number of OPEC decisions and drastic cyclical economic downturns for the oil and gas industry led to consolidations of oil and gas companies through mergers and acquisitions after 1995. Merger and acquisition activities led to the formation of super major oil companies such as ExxonMobil, British Petroleum, Royal Dutch Shell, Total S.A., and Chevron. Other major global companies based on assets and revenues include Saudi Aramco of Saudi Arabia, Sinopec, Chinese National Petroleum Corporation (CNPC) and PetroChina of China, Kuwait Petroleum Corporation of Kuwait, ENI of Italy, and Lukoil of Russia. There are other national oil-exporting companies that also feature prominently with respect to crude oil source and oil qualities and in disruption of crude oil supplies and pricing. National oil-exporting companies such as National Iranian Oil Company (NIOC), Petroleos de Venezuela, (PDVSA), Nigerian National Petroleum Corporation (NNPC), and Petroleos Mexicanos (PEMEX) are expected to continue to impact crude oil supply and pricing for a variety of factors. Some of the factors include the size of proved oil reserves for Iran and Venezuela and periodic oil production challenges in some of the countries such as Nigeria and Venezuela.

1.3 SOURCE OF CRUDE OIL

Crude or petroleum oil is a complex mixture of thousands of hydrocarbon compounds with broad ranges of molecular weights and boiling points. In addition to compounds that contain only carbon and hydrogen, there are also a variety of organic compounds containing separately sulfur, nitrogen, nickel, and vanadium, as well as other contaminant metals. Inorganic contaminant metals such as sulfur, nitrogen, nickel, vanadium, calcium, magnesium, and copper are referred to as heteroatoms, as they are different from the carbon and hydrogen atoms in hydrocarbons. The wide variety of thousands of hydrocarbon compounds provides the bases for the use of various fractions, components, and compounds of oil in energy generation and in the manufacture of products for transportation, heating, lubrication, chemicals, pharmaceuticals, plastics, automotive industries, agriculture, clothing, and packaging. As a result of the use of oil and oil products as feedstocks in so many chemically based manufacturing industries, oil is most essential for sustaining our manufacturing plants and driving national economies.

Oils are usually found in various reservoirs where they have accumulated over the past millions of years. During that time, conversion of dead organisms, zooplanktons and algae, have occurred to form complex hydrocarbon mixtures through intense compression and heating in geologic formations.(14,15) Crude oils are found in onshore and offshore deposits in many regions of the world. As per the 2015 ranking of countries with major proved oil reserves, Venezuela, Saudi Arabia, Canada, Iran, and Iraq are the top five with proved oil reserves for each country of at least 140 billion barrels.

Petroleum fossil oil is not a renewable resource and there have expectedly been some concerns with the rate of consumption of oil relative to the rate of frequent discoveries of sizeable oil finds for oil replacements. Oil reserves, oil daily production rates, and some technological advancements for producing unconventional oils such as shale and bitumen from oil sands are reviewed in some detail later in this chapter.

1.4 CRUDE OIL RESERVES

An estimate of proved oil reserves is a moving target, as major oil discoveries made in the last 15 years via application of innovative technologies for oil exploration and production have significantly changed the ranking of countries. In addition, advances in deepwater oil-production technologies have also greatly increased oil discoveries and production. A conservative estimate is that total global proved oil reserves are about 1.6 trillion barrels, with about 70% of that estimate in the OPEC countries as of the first quarter of 2016. OPEC countries such as Venezuela, Saudi Arabia, Iran, Iraq, and Kuwait dominate, with at least 100 billion barrels each in oil reserve, as shown in Table 1.1.(12,13)

As shown in Table 1.1, OPEC countries Venezuela, Iran, Iraq, Libya, and Qatar now own huge amounts of proved oil reserves due to major oil discoveries in the last 20 years. Data in Table 1.1 and Figure 1.1 show that Venezuela, which had 64.4 billion barrels of proved oil reserves in 1993 and substantially lower than the 261 billion barrels of Saudi Arabia, now owns 298 billion barrels of oil and ranks as the country with the highest proved oil reserve in the world. Saudi Arabia's huge oil reserves are centered in the giant Ghawar field, whose oil reserves are estimated to be in excess of 75 billion barrels. Venezuela owns huge extra-heavy oil reserves that are roughly estimated to be greater than 500 billion barrels in the country's Orinoco belt region. It is estimated that 220 billion barrels of those have been credited as part of the 298 billion barrels of proved oil reserve for Venezuela at this time. Due to capital and financial constraints, political issues, and the quality of the extra-heavy oil in the Orinoco Belt region, oil production has not been as high as it could be. Based on its massive proved oil reserves, Venezuela has the potential to achieve much higher production rates in future.

Non-OPEC countries such as Canada, Brazil, the United States, and Russia have also discovered huge reserves of unconventional oil and mostly bitumen from oil sands in Canada and shale oil in the United States. As shown in Table 1.1, Canada's proved oil reserve grew from 39.5 billion barrels in 1993 to 175.4 billion barrels in 2013. The United States has benefited from applications of advanced oil geophysical technologies for unlocking a number of discoveries of shale oil deposits, and that has boosted the US oil reserves from 30.2 billion barrels in 1993 to 44 billion barrels in 2013.(12,13)

TABLE 1.1
Countries with High Proved Oil Reserves, MMMB 1993 to 2013

Country	1993	2003	2013
Venezuela	64.4	77.2	297.7
Saudi Arabia	261.4	262.7	268.4
Canada	39.5	180.4	175.2
Iran	92.9	133.3	157.3
Iraq	100.0	116.0	150.0
Kuwait	96.5	99.0	101.5
UAE	98.1	97.6	97.8
Russia	n/a	79.0	93.0
Libya	22.8	39.1	48.5
Nigeria	21.0	35.3	37.3
USA	30.2	29.4	44.0
Kazakhstan	n/a	9.0	30.0
China	16.4	15.5	18.1
Qatar	3.2	27.0	25.2
Brazil	5.0	10.6	15.6

Source: Data taken from United States Energy Information Administration Reports. Giddens, P. H., *The Birth of the Oil Industry*, Porcupine Press, The Macmillan Company, New York, 1938; Williamson, H. F., Daum, A., Andreanoa, R., *The American Petroleum Industry, 1899–1959, The Age of Energy*, Northwestern University Press, Evanston, Illinois, p. 16.[10,11]
Note: MMMB is billions of barrels.

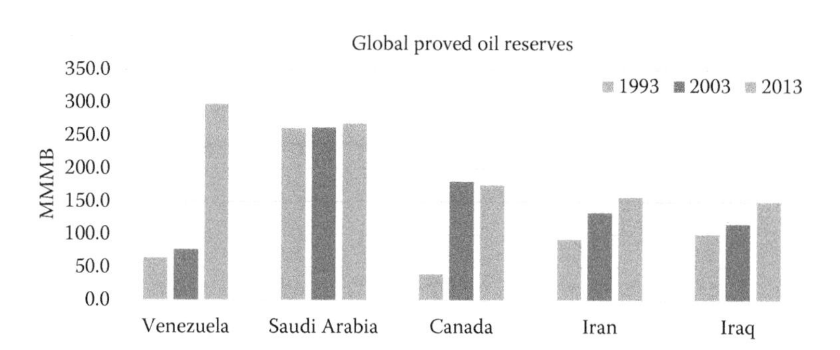

FIGURE 1.1 Top five countries' proved oil reserves for 1993 to 2013.
Note: MMMB is billions of barrels.

It is necessary to emphasize the ever-changing landscape and leaderboard of proved oil reserves with time. A recent independent assessment report of proven oil reserves as of 2016 by Rystad Energy, a Norwegian consulting company, using data not provided by oil-producing countries, now shows that the United States has moved ahead of Saudi Arabia, Russia, and Venezuela as the global leader in oil reserves, with

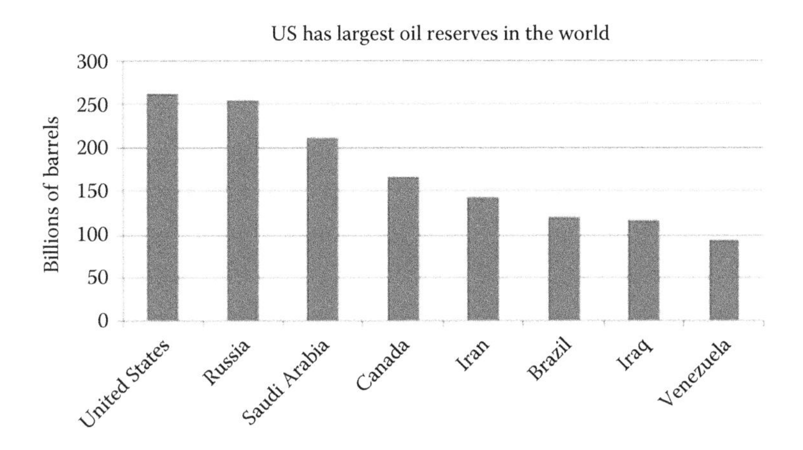

FIGURE 1.2 Proved oil reserves July 2016 by Rystad.
From Cunningham, N., US Has World's Largest Oil Reserve. Oil Price, 2016, http://oilprice.com/Energy/Energy-General/US-Has-Worlds-Largest-Oil-Reserves.html.(28)

over 260 billion barrels. Rystad Energy oil reserves data are as shown in Figure 1.2.(28) Rystad argued that there are no uniform ways to measure oil reserves and, as a result, oil reserves assessments differ grossly. It was suggested by Rystad that the United States holds over 260 billion barrels of oil reserves and more than 50% of that was located in shale formations.

1.5 INTRODUCTION TO OPPORTUNITY OILS AND UNCONVENTIONAL OILS

The goal of the review of the history of oil in this book is to share the author's appreciation of the rapid development of the oil business in the last 170 years. Additionally, items covered in this introductory section should provide appropriate background for a good understanding of the sources of crude oil, crude oil and liquid production, crude oil consumption rate, conventional and unconventional oil qualities, and oil prices. Opportunity oils are crude or unconventional oils and oil fractions that become available at "opportune" times and at favorable discounted prices. For opportunity crude oils, their prices are usually highly discounted relative to Western Texas Intermediate (WTI) and Brent crude oils. However, opportunity oils may vary widely with respect to the composition of their naphtha, distillate, gas oil, and residual oil fractions and heteroatom contaminant concentrations. Furthermore, the properties of naphtha, distillate, gas oil, and residual oil fractions from a crude distillation unit depend on the percentages of the constituent crude and unconventional oils in the crude slate. Since the ultimate properties of naphtha, distillate, gas oil, and residual oil feeds to downstream processing units are highly dependent on the oils in the crude slate, it is important to fully characterize the crude oils before purchasing and definitely before processing the oils in refineries. *Unconventional oils* in this book refers to bitumen-derived oils and tight or shale oils produced via nontraditional technologies such as surface mining, steam-assisted gravity drainage (SAGD) and hydraulic fracking.

Where applicable, unconventional oils should be reviewed in detail, as the properties of unconventional oils vary widely and are significantly different from the oils produced via conventional extraction technologies. Factors that can cause variability in the properties of unconventional oils are covered in some detail in a subsequent chapter in this book. Factors that are relevant in the ranking of countries with respect to oil reserves, daily oil production, and global oil consumption rates are provided.

1.6 GLOBAL OIL PRODUCTION

Crude oil production rates vary over time due to a multitude of factors. Contributory factors include production and supply disruptions as a result of vandalism of oil production assets, direct attacks and kidnappings of oil production personnel, regional conflicts, political factors, and OPEC-imposed production constraints on its members. OPEC-imposed constraints could impact the oil production rates of OPEC and non-OPEC countries as well as prices of crude oils. The oil glut of 2015 led to the collapse of oil prices, with negative consequences for the development and production of unconventional oils, as their costs of production are typically higher than those of conventional crude oils. The top-ranked daily oil-producing country has been Saudi Arabia and, as shown in Table 1.2, Saudi Arabia maintained its position between 2003 and 2013. Saudi Arabia's production rate has been a dominant factor with respect to some of the oil production decisions made by OPEC. The daily production rates of petroleum oil for the top 15 countries are listed in Table 1.2. Daily crude oil production rates of the top five countries for the years of 2003, 2008, and 2013 are as shown in Figure 1.3.[12,13]

Oil production ranking for countries differs considerably, however, when it is based on total petroleum and other hydrocarbon liquids, as shown in Table 1.3. Total

TABLE 1.2
Daily Crude Oil Production of Countries in MMBPD

Country	2003	2008	2013
Saudi Arabia	10.1	10.7	11.5
Russia Federation	8.6	10.0	10.8
USA	7.4	6.8	10.0
Canada	3.0	3.2	4.0
Iran	4.0	4.4	3.6
Iraq	1.3	2.4	3.1
Venezuela	2.9	3.2	2.6
Kuwait	2.4	2.8	3.1
Nigeria	2.2	2.2	2.3
Mexico	3.8	3.2	2.9
Kazakhstan	1.1	1.5	1.8
Norway	3.3	2.5	1.8
Libya	1.5	1.7	1.0
Angola	0.9	1.8	1.8
Algeria	1.8	1.8	1.7

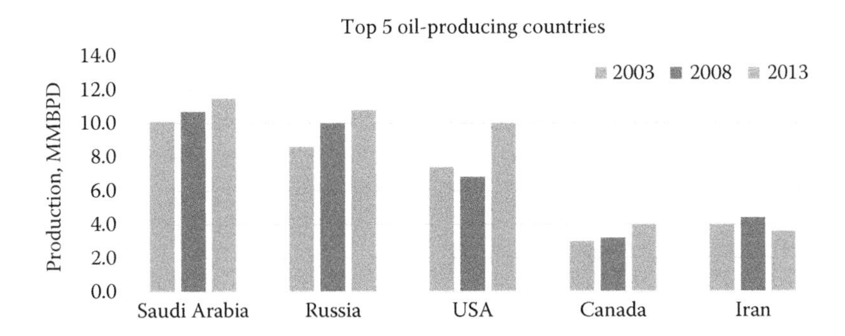

FIGURE 1.3 Daily oil production rates in millions of barrels of oil per day (MMBPD) of top five countries.

petroleum and other hydrocarbon liquids as defined consist of crude oils, condensates, tight oil, extra-heavy crude oil, and bitumen produced by a country. Tight oil refers to oils that are produced from shale formations. Special unconventional technology referred to as hydraulic fracking is required for successful production of shale or tight oil. Bitumen is oil derived from oil sands and produced via surface mining and steam-assisted gravity drainage technologies. Bitumen as used in Table 1.3 represents the sum of all bitumen-derived oils, including diluted bitumen (Dilbit) and upgraded bitumen, and is referred to as a syncrude oil.

Diluted bitumen is bitumen that has a diluent such as water, condensate, naphtha, or crude oil added to permit facile transportation of the bitumen. An upgraded

TABLE 1.3
Total Petroleum and Other Liquids Production MMBPD

Country	2006	2010	2014
USA	8.3	9.7	14.1
Saudi Arabia	11.1	10.9	11.6
Russia	9.7	10.3	10.9
China	3.9	4.4	4.6
Canada	3.3	3.4	4.4
UAE	3.0	2.8	3.5
Iran	4.2	4.2	3.5
Iraq	2.0	2.4	3.4
Brazil	2.2	2.7	3.0
Mexico	3.7	3.0	2.8
Kuwait	2.7	2.5	2.8
Venezuela	2.8	2.6	2.7
Nigeria	2.4	2.5	2.4
Qatar	1.3	1.8	2.1
Norway	2.8	2.1	1.9

Note: MMBPD is millions of barrels of oil per day.

bitumen oil is bitumen that has been upgraded via thermal and catalytic processes in order to improve the qualities of the resultant oil and render it suitable for processing in oil refineries. Synthetic crude oils are usually higher priced relative to the diluted bitumen oil or Dilbit, which is an oil sands–derived oil that has not been subjected to significant thermal and/or catalytic upgrading.(16,17) Details of unconventional oil production technologies are covered later in this section. Other liquids included in the data in Table 1.3 are natural gas plant liquid (NGPL), biofuels, gas-to-liquids (GTL), and coal-to-liquid (CTL) products.(12)

Projected increases in the production of natural gas liquids and light tight oil would most likely maintain the position of the United States as the top producer of total petroleum and other liquids in the world through 2040, as shown in Figure 1.4.(12)

Total oil liquid production for the United States shows that light tight oil and natural gas liquid production would increase significantly after 2015 and would be much higher than conventional oil rates.

2015 total liquid and gas production for the United States, Russia, and Saudi Arabia in Figure 1.5 shows that the United States maintained its position as the top producer of total liquids and gas.

Unconventional oils represent a growing segment of oils from oil sands, shale rock, and deepwater that have to be extracted via the application of innovative, nontraditional drilling and extraction technologies. There are a variety of technologies used, and the major ones are surface mining and steam-assisted gravity drainage for oil sands and fracking for light tight oil from shale rock formations, as indicated previously.

Steam-assisted gravity drainage extraction is applied in addition to the extensively used surface mining for extracting bitumen from oil sands. Current extractive production of bitumen via surface mining is energy intensive, as only 10% of the

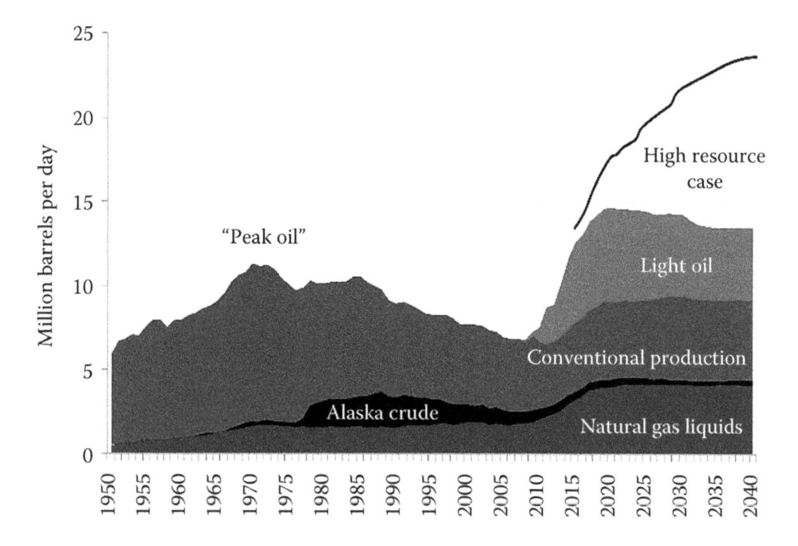

FIGURE 1.4 US total liquids through 2040.
From US Energy Information Administration [EIA] Reports; Lance, R., The Benefits of US Crude Exports, IPAA Annual Meeting. November 13, 2014. www.conocophillips.com/files/resource.(12,27)

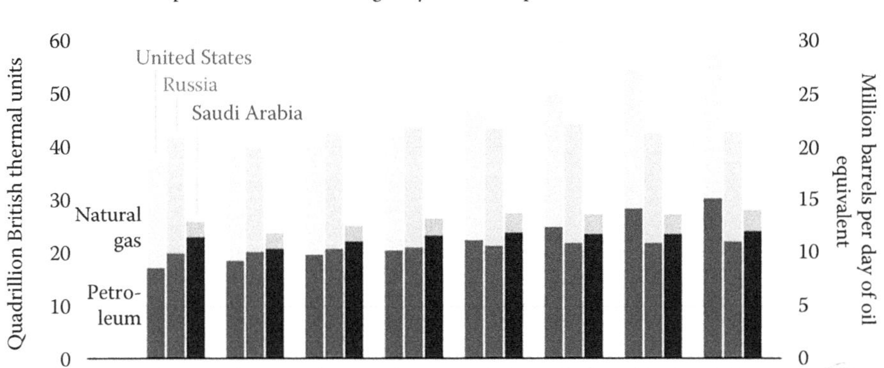

FIGURE 1.5 Total liquids and gas production 2008–2015.
Note: Lower bar represents total hydrocarbon liquids and upper bar total gas. Units are in million barrels per day of equivalent oil and quadrillion British thermal units. (From US Energy Information Administration [EIA] Reports.[12])

FIGURE 1.6 Surface mining for recovery of bitumen.

sand, clay, and water is bitumen. A surface mining site is as shown in Figure 1.6, and a schematic of the steam-assisted gravity drainage technology is shown in Figure 1.7. In both SAGD and surface mining extractive processes, significant amounts of energy from natural gas use and for separation of oil from sands are required to produce a barrel of bitumen from oil sands.[17]

The need for innovative energy-efficient technology for extracting bitumen directly from oil sands is great. Surface mining extraction is currently in use in Canada and Mexico for the production of bitumen.

Light oils consist of tight and shale oils that are produced essentially from impermeable shale rock type formations. Shale is fine-grained sedimentary rock that is easily broken into parallel layers. Shale rock formations can contain large amounts of oil and/or natural gas. A major extraction challenge is that the hydrocarbons in the shale formation do not flow easily and must be dislodged for accessibility and recovery. In order to extract the oil efficiently and profitably, two production

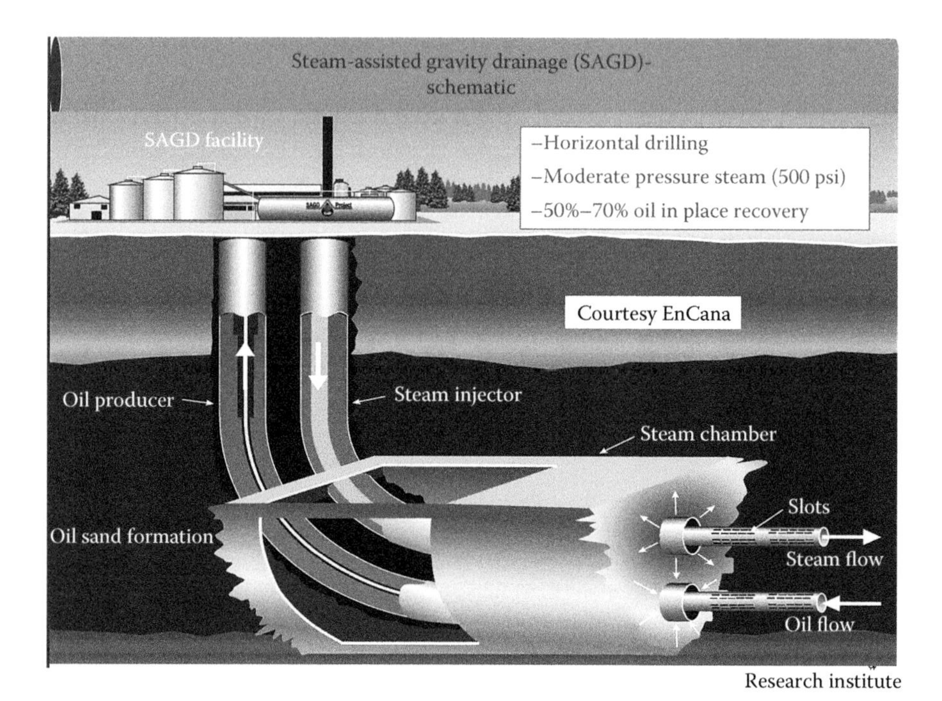

FIGURE 1.7 Steam-assisted gravity drainage recovery of bitumen from oil sands.

technologies are used. The technologies are referred to as horizontal drilling and hydraulic fracking. Instead of the typical vertical drilling with possibly a J-type bend used for conventional oils, horizontal drilling involves an initial drilling in the vertical direction for a short distance, followed by drilling horizontally and possibly for a few miles to access oil deposits. To loosen oil deposits spread over a wide area of sedimentary rock, water and chemicals are added at high pressure during oil production to free the oil in the deposits. The application of hydraulic fracking has led to record production of shale or light tight oil and shale gas in the United States. Unfortunately, the 2015 collapse of global oil prices moderated the rate of production of light tight oil. Similarly, applications of improved technologies for exploration and production have led to oil finds in deepwater off the coasts of West Africa and South America and in the Gulf of Mexico.

1.7 OIL CONSUMPTION

The US economy required the use of 20 million barrels per day in 1993, and this has gradually been reduced to 18.9 million barrels per day by 2013 due to energy conservation, automotive fuel efficiency, and environmental regulatory programs. Energy conservation programs include increased use of renewable transportation fuels, natural gas, and renewable fuels for power generation. There have been similar modest declines in oil consumption rates for the Organization for Economic Cooperation and Development (OECD) countries, notably for Germany, the United

TABLE 1.4
Daily Oil Consumption Rates in MMBPD

Country	2003	2008	2013
USA	20.0	19.5	18.9
China	5.8	8.3	10.8
Japan	5.5	4.4	4.6
India	2.5	3.2	3.7
Russian Federation	2.7	2.9	3.3
Saudi Arabia	1.7	2.4	3.2
Brazil	2.0	2.4	3.0
Canada	2.2	2.3	2.3
Germany	2.7	2.5	2.4
UK	1.7	1.6	1.5
France	2.0	1.9	1.7
Italy	1.9	1.7	1.3
Spain	1.9	1.7	1.3
Mexico	1.9	2.1	2.0

Note: The oil consumption rates are in millions of barrels per day or MMBPD.

Kingdom, Italy, France, and Spain, as these countries have implemented energy conservation programs. Daily oil consumption rates for selected years, 2003, 2008, and 2013, are provided in Table 1.4. Oil demand growth for non-OECD countries such as China and India is on the rise, though tempered slightly by slower economic growth rates in the past two decades.[13]

Oil demand declines for the United States and other OECD countries were partially offset by drastic demand increases of over 80% for China and over 45% for India between 2003 and 2013 due to increased oil refining for the production of transportation fuels and other refined products.

1.8 CRUDE OIL AND OTHER HYDROCARBON LIQUID QUALITIES

There are many types of crude oils, and they are classified broadly as sweet light, heavy sour, extra-heavy sour, and heavy blended synthetic crude oils. Sweet and sour are descriptive terms for crude oils that are based on the amount of organic sulfur in the crude oils. Sweet crude oils are those with sulfur concentrations that are less than 0.5 wt. %. Crude oils that contain greater than 0.5 wt. % sulfur are referred to as sour oils. Heavy sour crude oils are oils with American Petroleum Institute (API) gravity of less than 20 and concentrations of sulfur compounds that are greater than 0.5 wt. %. Extra-heavy oils are typically those with API gravities that are less than 10 API. More specifically, the World Energy Council (WEC) defined extra-heavy oil as crude oil having a gravity of less than 10° API and a reservoir viscosity of no more than 10,000 centipoises. When reservoir viscosity measurements are not available, extra-heavy oil is considered by the WEC to have a lower limit of 4 API.

Note:

Gravity of oils is determined and reported as degrees API.
Specific gravity is defined as the density of an oil relative to the density of water, where density of water is defined approximately as 1 gram/cubic centimeter.
Degrees API = (141.5/Specific gravity) − 131.5.

Based on the definitions, Louisiana Light Sweet (LLS), West Texas intermediate, and Brent are light sweet crude oils; Arabian Light is a sour crude oil; and both Maya and Tia Juana heavy are heavy sour crude oils. The profitability of refining a crude oil should be fully assessed, and the assessment should incorporate expected costs associated with processing the crude oil and relevant related asset maintenance costs. Due to the broad properties of crude oils, a current crude oil assay database is highly recommended. Pertinent crude oil properties such as sulfur, API gravity, total acid neutralization number (TAN), concentrations of contaminant metals, and concentrations of oil fractions on distillation should also be considered in estimating the value and price of a crude oil relative to the price of the benchmark crude oil used for setting its price. Typical benchmark crude oils include and are not limited to Brent, West Texas intermediate, and Dubai Fateh. Another benchmark for crude oil pricing that is favored by OPEC is what is referred to as the OPEC basket of oils. The OPEC basket of oils price is based on the prices of a collection of seven crude oils from Algeria, Saudi Arabia, Indonesia, Nigeria, Dubai, Venezuela, and the Mexican isthmus.

It is worth noting that oil refiners that have the necessary bottom-of-the-barrel upgrading processing assets in their refineries are in enviable positions to take advantage of lower-cost crude and synthetic oils, as they can reliably and profitably process a wide variety of purchased crude and synthetic oils. Selected oil quality data for a small number of crude oils used in oil refineries are provided in Table 1.5.

General crude oil qualities should be taken as merely representative of analyzed samples of the crude oils, and it should be understood that oil qualities at the time of purchase could vary depending on the specific areas from which the oils are produced, the degree of upgrading of the oils, chemicals used during production, and flow enhancers for oil transportation. As indicated previously, it could be beneficial and profitable for the oil refiner to update its database for crude oil assays as often as feasible so as to fully take advantage of possible discounts for gross variations in crude oil qualities when the oils are available.

Bitumen from oil sands is extra-heavy oil and usually has gravity that is less than 10 API. As discussed, a variety of extraction technologies are applied for extracting bitumen, and two of the most favored ones are surface mining and steam-assisted gravity drainage.[16,18] Two of the largest deposits of oil sands are in the Orinoco Belt of Venezuela and Alberta, Canada. Bitumen is a highly viscous liquid that does not flow easily and can be transported in pipelines only after the addition of condensate diluent to produce Dilbit or the addition of a synthetic oil to produce Synbit. Since bitumen usually contains high amounts of contaminants, it has to be upgraded and blended with other oils so as to reduce its sulfur, nitrogen, nickel, vanadium, and acidic contents and render it suitable for profitable oil refining (see Figure 1.8). The qualities of the Hamaca from Venezuela and Athabasca bitumen from Canada, as

TABLE 1.5
Selected Properties for Crude Oils

Crude Oil	LLS	WTI	WTS	AL	BR	TIA/H	MAYA
Gravity, API	36.1	40.8	34.1	33.4	38.3	12.3	22.2
Sulfur, wt. %	0.45	0.34	1.64	1.77	0.40	2.80	3.30
Pour point, C	−37.0	−29.0	−46.0	−54.0	−42.0	−16.0	−36.0
Viscosity @100F	4.3	4.9	4.6	8.4	3.9	8.6	102.0
V, wppm	1.2	1.6	6.4	13.5	6.0	386	314.0
Ni, wppm	7.1	1.6	3.7	3.3	1.0	38.5	52.0
CCR, wt. %	1.1	1.1	3.3	3.6	2.1	11.2	12.0
TAN, mgKOH/g	0.58	0.10	0.1	0.00	0.10	3.90	0.28

Notes:

1. LLS is Light Louisiana Sweet; WTI is West Texas Intermediate; WTS is West Texas Sour; AL is Arab Light; BR is Brent; TIA/H is Tia Juana Heavy. Maya is from Mexico.
2. Kinematic viscosity is at 100 F and the units are centistokes.
3. The crude oil characteristics are taken from a variety of sources. US Energy Information Administration (EIA) Reports; BP Statistical Review of World Energy, June 2014; Oyekan, S. O., Torrisi, S., Opportunities and Challenges in Transportation Fuels Production, Paper Presented at the AIChE Regional Process Technology Conference, Galveston, TX, October 2, 2009.[12–16]
4. CCR is Conradson Carbon and is a measure of the concentration of asphaltenes and how the heavy the oil is.
5. TAN is total acid neutralization number and the units are in mg of KOH per gram of the oil, where KOH is potassium hydroxide.
6. Viscosity is kinematic viscosity at 100°F.
7. Pour point is in degrees centigrade.

well as the upgraded Hamaca oil and Western Canadian Select (WCS), are given in Table 1.6.[16,19,20] Western Canadian Select is a heavy blended crude oil comprised of sweet synthetic (upgraded) oils, other crude oils, and condensate.

It is beneficial for oil refiners to have good and current knowledge of the qualities of the crude oils that they are purchasing and available for processing in their refineries so as to anticipate potential operational and reliability challenges while taking advantage of opportunities to maximize refinery profitability. Also, since upgraded bitumen is usually sold in synthetic mixtures containing other crude oils and condensate, the quality of a given bitumen-derived oil is likely to vary substantially, and an up-to-date assay database program should be utilized for profitable processing of such oils in combination with the other crude oils in refineries.

A discussion of crude oil would be grossly incomplete without discussing the impact of the significant increase in the production of shale or light tight oil in the United States on crude oil prices. It is important to discuss tight oils with respect to the challenges that should be expected in processing significant percentages of shale oil in crude slates. Though organic sulfur, contaminant metals, and acid concentrations of tight oils are usually similar to those of sweet crude oils, the high paraffinic content and percentages of their oil fractions make processing of light

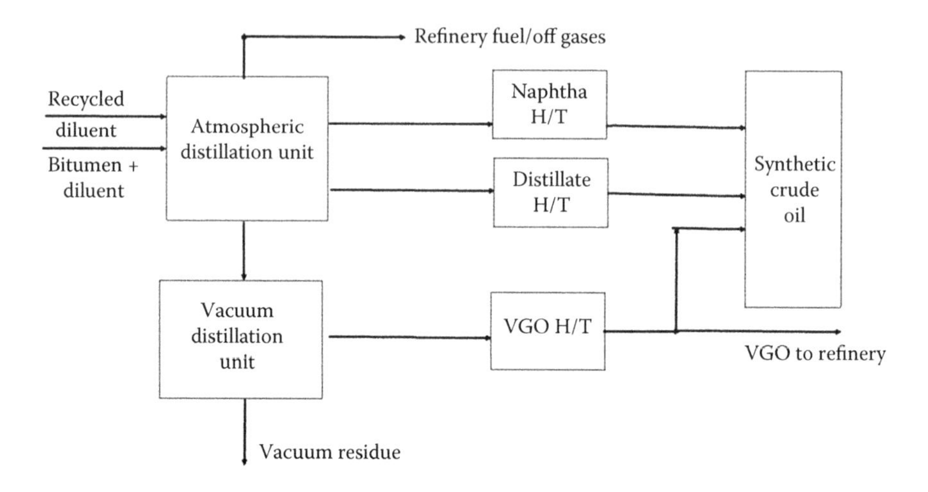

FIGURE 1.8 A typical bitumen upgrader showing use of diluent and generation of synthetic crude oil.
Note: Recycled diluent could be a naphtha or light oil. Naphtha H/T is the naphtha hydrotreater. VGO is vacuum gas oil. (From Gray, M. R., *Tutorial on Upgrading of Sands Bitumen*, University of Alberta, Canada.[17])

TABLE 1.6
Bitumens and Their Upgraded and Blended Oils

Crude Oils	Hamaca Bitumen	Hamaca	Athabasca Bitumen	WCS
Gravity, API	8.4	26	7.7–9.0	19–22
Sulfur, wt. %	3.8	1.55	4.4–5.1	2.8–3.5
Ni, wppm	115	42	69–85	46–59
V, wppm	388	152	81–218	115–140
TAN, mg KOH/g	2.8	0.7	>5.4	<1.0

tight oils more challenging than the processing of conventional oils. However, due to favorable prices relative to WTI and Brent and reliable supply, tight oils are highly desirable, cost-efficient oils for some United States oil refiners. A number of US oil refiners are installing additional crude distillation units for efficient and profitable refining of higher percentages of light tight oils in their refineries' crude slates. The properties of tight oils are quite different from those of other crude oils, as shown by the properties of the Bakken and Eagle Ford oils in Table 1.7.[21,22] Bakken and Eagle Ford oils are cost-effective replacements for imported light sweet crude oils. Tight oils had effectively replaced most of the imported light crude oils, especially Bonny Light from Nigeria, in oil refineries in the United States by 2016.

Since light tight oils are usually priced lower relative to benchmark WTI and Brent, they are desirable cost-effective feedstocks for refineries. However, challenges that have been identified in processing light tight oils include the negative impact of its high paraffinic contents, as this causes significant precipitation of some

TABLE 1.7
Tight Oils Compared to Conventional Crude Oils

Crude Oil	Eagle Ford	Bakken	LLS	WTI	AL	Brent	Bonny
Gravity, API	45.7	40	36.1	40.8	33.4	38.3	32.9
Sulfur, wt. %	0.04	0.2	0.45	0.34	1.77	0.40	0.2
Pour Point, C	−27.4	N/a	−37.0	−29.0	−54.0	−42.0	−14.4
V, wppm	0.1	0.1	1.2	1.6	13.5	6.0	0.4
Ni, wppm	0.1	0.4	7.1	1.6	3.3	1.0	4.3
CCR, wt. %	0.2	0.8	1.1	1.1	3.6	2.1	1.4
TAN, mgKOH/g	0.1	0.1	0.58	0.10	0.00	0.10	0.3

of the asphaltenes of heavier crude oils in the crude unit. Significant asphaltene precipitation leads to increased reliability issues in atmospheric crude units and associated equipment. Furthermore, refiners who had upgraded their refining assets to process heavy crude oils may find that their bottom-of-the-barrel conversion units are underutilized. A number of oil refining technologies and catalyst providers suggest that installation of additional atmospheric distillation units, reconfiguring of refinery oil processing schemes, and use of the fluid catalytic cracker (FCC) for processing could be beneficial for maximizing profitability from processing high percentages of shale oil in refineries.[22–26]

Another major factor worth emphasizing with respect to crude oil qualities are the chemicals that are used during crude oil extraction, production, and transportation, as the chemicals could produce metals and chlorides that are either contaminants or corrosion agents in downstream processing of the crude oil fractions after the atmospheric and vacuum distillation units. Examples include the use of antifoaming agents containing silicon and condensate and naphtha used as flow improvers for crude oils and unconventional oils. Chemicals added during production and for transportation could undergo thermal and catalytic reactions to generate contaminant metals that can poison precious and base metals containing catalysts in oil refining processes. Poisoning by added contaminant metals other than the sulfur, nitrogen, iron, nickel, and vanadium typically inherent in crude oil could lead to poor catalytic performances of process units and underutilization of oil refining assets due to process units' reliability challenges.

Another class of seemingly cost-effective oil feedstocks for oil refiners are those typically characterized as opportunity crude oils and oil fractions. Detailed characterization of the oils should be conducted and is recommended before they are processed in high percentages in crude slates for crude units and process units in oil refineries. Documented cases of silicon and other poisoning events of the naphtha hydrotreater and catalytic reforming units are discussed in subsequent chapters.

REFERENCES

1. My Channel Evolution of the Oil Industry, Documentary on the History of American Oil and Petroleum, October 2013, http://youtube/TheBestFilmArchives.

2. Bilkadi, Z., The Oil Weapons. *Saudi Aramco World*, 46(1), 20–27, 1995.
3. Russell, L. S., *A Heritage of Light Lamps and Lighting in the Early Canadian Home*, University of Toronto Press, Toronto, Canada, 2003.
4. Frank, A. F., *Oil Empire and Visions of Prosperity in Austrian Galicia*, Harvard University Press, Cambridge, 2005.
5. Matveichuk, A. A., *Intersection of Oil Parallels Historical Essays*, Russian Oil and Gas Institute, Moscow, 2004.
6. Murray, T. J., Dr. Abraham Gesner: The Father of the Petroleum Industry, *J. R. Soc. Med.*, 86(1), 43–44, 1993.
7. Owen, E. W., *Trek of the Oil Finders*, American Association of Petroleum Geologists, Tulsa, Oklahoma, p. 12, 1975.
8. McKain, D. L., Bernard, A. L., *Where It All Began, The Story of the People and Places Where the Oil Industry Began, West Virginia and Southeastern Ohio*, D. L. McKain, Parkersburg, West Virginia, 1994.
9. Gordon, J. S., *10 Moments That Made American Business*, American Heritage, February/March 2007.
10. Giddens, P. H., *The Birth of the Oil Industry*, Porcupine Press, The Macmillan Company, New York, 1938.
11. Williamson, H. F., Daum, A., Andreanoa, R., *The American Petroleum Industry, 1899–1959, The Age of Energy*, Northwestern University Press, Evanston, Illinois, p. 16.
12. U.S. Energy Information Administration (EIA) Reports, https://www.eia.gov/.
13. BP Statistical Review of World Energy World Petroleum Congress, Moscow, June 2014.
14. Leverett, F., Lecture # 2, "The Global Energy Industry, Geopolitical and Geo-Economics of Global Energy", 2007, MIT Open Courseware, http://www.ocn.mit.edu
15. Speight, J. G., *The Chemistry and Technology of Petroleum*, 2nd edition, Marcel Decker, New York, 1991.
16. Oyekan, S. O., Torrisi, S., Opportunities and Challenges in Transportation Fuels Production, *Paper Presented at the AIChE Regional Process Technology Conference*, Galveston, TX, October 2, 2009.
17. Gray, M. R., *Tutorial on Upgrading of Sands Bitumen*, University of Alberta, Canada, 2015.
18. Stratiev, D., Dinkov, R., Petkov, K., Stanulov, K., Evaluation of Crude Oil Quality, *Petroleum & Coal*, 52(1), 35–43, 2010.
19. Mai, A., Bryan, J., Goodarzi, N., Kantas A., Insights into Non-Thermal Recovery of Heavy Oil, *Paper Presented at World Heavy Oil Conference*, Calgary, Alberta, Canada, 2006.
20. Dusseault, M. B., Comparing Venezuelan and Canadian Heavy Oil and Tar Sands, *Paper presented at the Petroleum Society's Canadian International Petroleum Conference*, Calgary, Alberta, Canada, June 12–14, 2001.
21. Hill, D., North Dakota Refining Capacity Study, Final Technical Report, DOE Award No. DE-FE000516, 2011.
22. Bryden, K., Federspiel, M., Habib, E. T., Schiller, R., Processing Tight Oils in FCC: Issues, Opportunities and Flexible Catalytic Systems. W. R. Grace Catalagram No. 114, 2014.
23. US. Energy Information Administration, Technical Options for Processing Additional Light Tight Oil Volumes within the United States, 2015.
24. Benoit, B., Zurlo, J., *Overcoming the Challenges of Tight/Shale Oil Refining*, GE Water & Process Technologies, Processing Shale Feed, 2014, www.eptq.com.
25. Olsen, T., Working with Tight Oil, *AIChE CEP*, 35–39, April 2015.
26. Jeff Koebel, AFPM Q&A Technology Conference 2013 Answer Book, Dallas, October 7–9, 2013.
27. Lance, R., The Benefits of US Crude Exports, IPAA Annual Meeting. November 13, 2014. www.conocophillips.com/files/resource
28. Cunningham, N., US Has World's Largest Oil Reserve. Oil Price, July 2016, http://oilprice.com/Energy/Energy-General/US-Has-Worlds-Largest-Oil-Reserves.html

2 Basics of Crude Oil Refining

2.0 INTRODUCTION TO BASICS OF OIL REFINING

A good working knowledge of the basic chemistry and operations of crude oil refining should enable engineers, chemists, technologists, and oil refining executives to better manage their oil refining assets. The basic chemical composition, chemicals, and some of the key contaminants in crude oils that could negatively impact oil refining processes and equipment are covered in this chapter. Most of the capital expansion projects in oil refineries usually have as a primary goal additional bottom-of-the-barrel upgrading technology assets for enhancing the profitability of oil refining companies. In recent years, some of the asset upgrading has been in response to ensuring that refining companies can produce transportation fuels such as gasoline, diesel, and bunker fuels from heavy sour crude and unconventional oils.

A simplified configuration of a refinery is given in Figure 2.1. Crude oil is processed to produce refined oil products such as liquefied petroleum gas (LPG), propylene, gasoline, kerosene, diesel, heating oil, sulfur, asphalt, and coke. Oil refineries also produce feedstocks such as naphtha, propylene, benzene, and xylenes for the petrochemical industry. The major operating cost for oil refiners is that required for acquiring crude oils, and that cost can represent 75% to 80% of a refinery's operating budget. As a consequence, bottom-of-the-barrel upgrading technologies such as delayed coking units are usually key processing units in grassroots oil refineries. Many refineries have also been revamped to add coking process units to expand the oil refiner's capabilities to process cheaper heavy sour-type crude oils and reduce crude oil cost. Since most of the crude oils available are more of the medium-sour to heavy-sour types, oil refiners with adequate bottom-of-the-barrel processing technologies have significant competitive advantages.[1] The chemistry of crude oils, oil refining economics, refinery configurations, and process units are covered in this chapter. A simplified refinery configuration is provided in Figure 2.1 in order to familiarize the reader with terms that are used in this book.

2.1 CRUDE OIL COMPOSITION

Crude oil consists of a variety of complex hydrocarbons and some small amounts of hydrocarbon compounds containing elements such as sulfur, nitrogen, nickel, vanadium, iron, copper, and so on. Though the crude oils are different, as discussed in Chapter 1, carbon and hydrogen are the predominant elements, which typically vary within fairly narrow limits for the oils, as shown in Table 2.1.[2,3] Due to the variety of complex hydrocarbons in crude oils, the boiling ranges for the compounds are quite broad, and this property enables the application of fractional distillation to separate

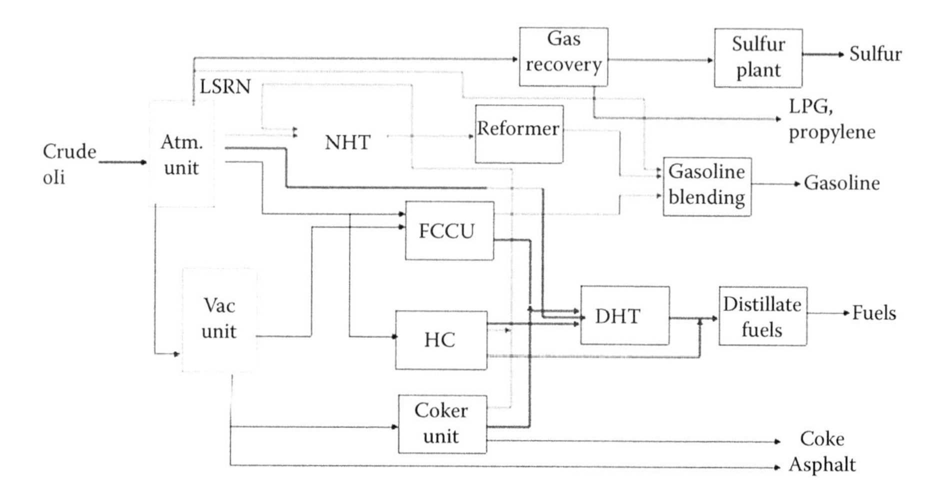

FIGURE 2.1 Simplified oil refinery configuration.
Note: Atm. unit is atmospheric crude distillation unit; LSRN is light straight-run naphtha; LPG is liquefied petroleum gas; DHT is distillate hydrotreater; FCCU is fluid catalytic cracking unit; HC is hydrocracker; NHT is naphtha hydrotreater; Vac unit is vacuum distillation unit.

crude oils into useful fractions based on the boiling-point ranges of the hydrocarbons in crude oils. Speight provides detailed information on a variety of physical, thermal, and chemical properties of crude oils in his book.[(2)]

Specific gravity, kinematic viscosity, sulfur, Conradson carbon, aniline points, thermal conductivity, specific heat, and many more properties and correlations where applicable for crude oils are covered appropriately by Speight.[(2)]

Hydrocarbons in crude oils vary in complexity from methane with one carbon atom to complex compounds with over 90 carbon atoms. Because crude oils contain thousands of compounds, crude oils are usually converted or processed in groups of compounds within a designated boiling point range, which are referred to as oil fractions. Oil fractions contain similar hydrocarbon compounds that vary in complexity and molecular weights progressively from the lighter to heavier fractions of crude oil. The main hydrocarbon types present in crude oils are paraffins, naphthenes, and aromatics. The paraffin compounds are usually further classified

TABLE 2.1
Crude Oil Composition

Element	Wt. %
C	84–89
H	11–14
S	0.1–6.0
N	0.1–2.0
Misc. Elements	~0.1

as straight-chain paraffins called normal paraffins and branched paraffins called iso-paraffins.

2.1.1 Paraffins

These are saturated single-chain or branched-chain hydrocarbons with a formula defined as C_nH_{2n+2} and as shown in Figure 2.2. Paraffins are also referred to as alkanes.

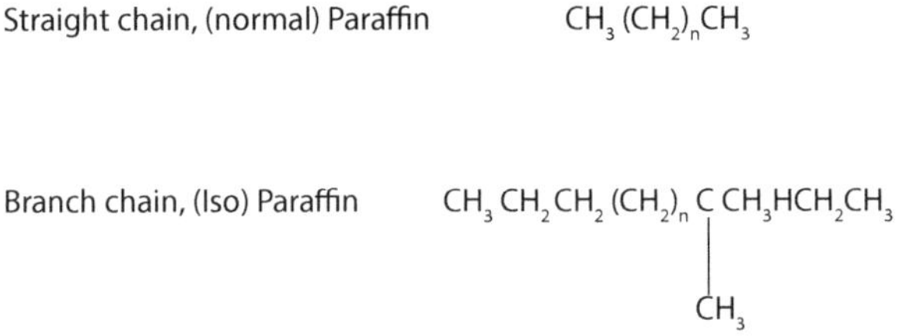

Straight chain, (normal) Paraffin $CH_3(CH_2)_nCH_3$

Branch chain, (Iso) Paraffin $CH_3CH_2CH_2(CH_2)_nC(CH_3)CH_3HCH_2CH_3$

FIGURE 2.2 Normal and branched paraffins.

From organic chemistry, it is known that the paraffins represented by the formula C_nH_{2n+2} can be generated with the first carbon number giving us methane, and the sequence can be used to get subsequent compounds such as ethane, propane, butane, pentane, hexane, heptane, and so on.

Data taken from Speight and incorporated in this book list the number of other compounds associated with a compound of a specific carbon number that also have the same molecular weight as and different molecular structure from the straight-chain compound. Hydrocarbon compounds with the same molecular weight and different molecular structures are called isomers. Isomers are usually differentiated by their physical and chemical properties. A number of isomers for the alkane or paraffin series of compounds are listed in Table 2.2.

TABLE 2.2
Hydrocarbon Compounds and Isomers

Carbon Atoms	No. Isomers
4	2
5	3
6	5
7	9
8	18
9	35
12	355
18	60,253

Source: Adapted from Speight, J. G., *The Chemistry and Technology of Petroleum*, 2nd edition, Marcel Dekker, New York. 1991.[2]

Some of the prominent paraffin compounds with special significance in the production of gasoline blending components are butane, pentane, hexane, heptane, octane, and iso-octane or trimethyl-pentane. For those experienced in oil refining processes, they are reminded of butane as a component of liquefied petroleum gas or LPG and as a blending component for gasoline in the winter months due to reduced Reid vapor pressure specifications for gasoline in the United States. They are also reminded of butane as the feed to the butane isomerization unit (Butamer) to generate isobutane, which then serves as a feed to the alkylation process unit. Butane isomerization and alkylation processes are discussed in Sections 2.6 and 2.10 of this book. Octane numbers of light naphtha are increased or upgraded in light naphtha isomerization process (Penex) units, and this is reviewed later in this chapter.

Heptane and iso-octane or trimethyl-pentane are significant for a number of other factors. Octane numbers are used as a measure of the antiknock quality of gasoline. Knock in automotive engines refers to abnormal ignition or autoignition of gasoline in the engine of automobiles caused by low-octane quality gasoline. The higher the octane of gasoline, the less the autoignition. Per the octane ratings established for compounds, significantly branched paraffin compounds usually have high octane numbers relative to their linear straight-chain isomers. Heptane and trimethyl pentane are used as references in the determination of octane numbers of compounds, blending fractions, and gasoline. Trimethyl pentane (iso-octane) and heptane are assigned octane numbers of 100 and 0, respectively. Heptane and some naphthenes such as methyl cyclopentane and cyclohexane are used as model compound feeds in catalytic studies for the naphtha reforming process. The application of paraffins and especially heptane as model compound feeds in fundamental catalytic research is reviewed in Chapter 3.

2.1.2 Naphthenes

Naphthenes are saturated hydrocarbons containing one or more rings, each of which may have one or more paraffinic side chains. They are also referred to as cycloparaffins. The side chains could be an alkyl represented by a formula of C_nH_{2n+1}. Naphthenes are more easily converted to aromatics in naphtha reforming reactions. Cyclohexane and methyl cyclopentane are suitable as model compound (Figure 2.3) feeds in fundamental studies of catalytic reforming catalysts. The compounds permit specific assessments of the contributions of the dehydrogenation/hydrogenation and acidic isomerization functionalities in naphtha reforming.

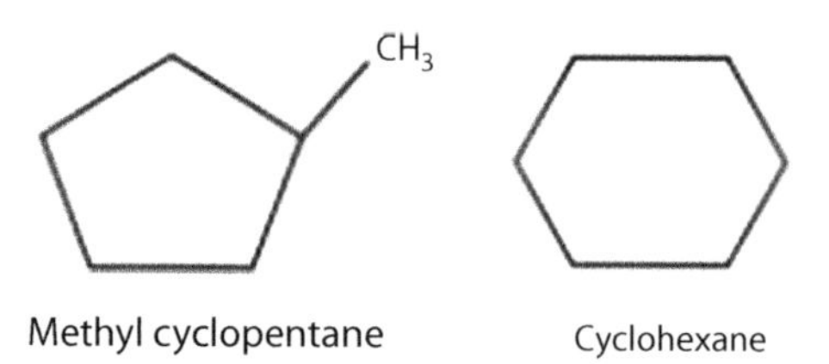

FIGURE 2.3 Methyl cyclopentane and cyclohexane.

2.1.3 Aromatics

Aromatics are hydrocarbons containing one or more benzene rings or aromatic nuclei. Representatives of these compounds include benzene, ethylbenzene, propyl-benzene, ortho-xylene, meta-xylene, and para-xylene in the naphtha boiling range. They also include naphthalene, phenanthrene, and numerous compounds of naphthalene and phenanthrene that have various multiringed compounds such as paraffins and naphthenes as part of the molecular structure. Aromatics have high octane numbers, and single-ring aromatics and alkyl benzenes are usually excellent gasoline blending compounds. Multiring aromatics in the naphtha feed with boiling points that are greater than 380°F are usually coke precursors in naphtha reforming over platinum-containing catalysts.

Some of the more commonly discussed aromatic compounds are shown in Figures 2.4 and 2.5.

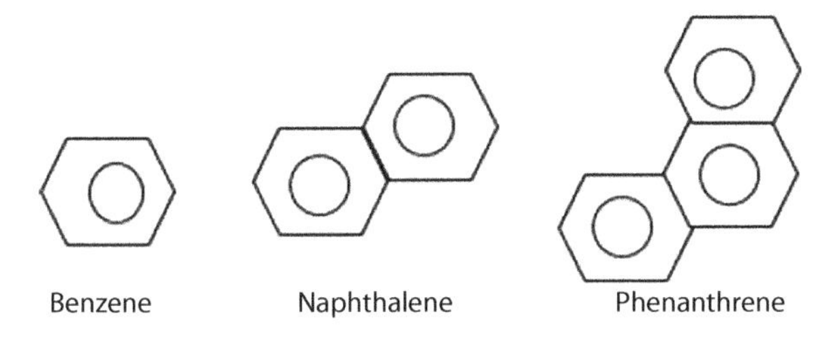

FIGURE 2.4 Benzene, naphthalene phenanthrene.

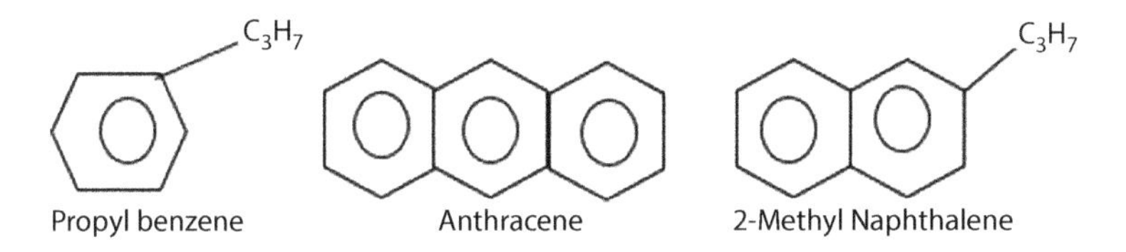

FIGURE 2.5 Propyl benzene, anthracene, 2-methyl naphthalene.

2.1.4 Olefins

Olefins are unsaturated hydrocarbons with a single double bond in their molecular structures. They are also referred to as alkenes and have a molecular formula that can be represented generally for that class of hydrocarbons as C_nH_{2n}. Olefins are typically not known to exist in conventional crude oils. Speight discussed a crude oil from Pennsylvania in the United States that was determined to contain up to 3 wt. % of olefins in the crude oil and in the distillate fraction of that crude.(2) The concentrations of the olefins were confirmed via spectroscopic and chemical analytical methods.

Hydrocrackers, fluid catalytic cracking units, delayed coking units, and visbreakers downstream of the atmospheric and vacuum distillation towers usually produce naphthas with significant concentrations of olefins. As an extension, olefins could be present in synthetic crude oils produced via the use of upgraded bitumen that has been blended with crude oils and crude oil fractions to produce marketable unconventional oils. It is recommended that good assays of such blended synthetic oils and other crude oils be conducted to fully characterize them before processing in oil refineries.

Another class of unsaturated hydrocarbons that could be found in crude oil fractions and especially from downstream processing units after the crude atmospheric and vacuum distillation units are dienes and acetylenes. Hydrocarbon compounds classified as dienes are characterized by having two double bonds in their molecular structure. The predominant ones include propadiene, butadienes, and pentadienes, which could be present in appreciable concentrations in the olefin feed to the alkylation process unit. Diene compounds are sometimes removed in special catalytic process units upstream of the alkylation unit.

The other unsaturated compound type is the alkynes that are characterized by having a triple bond between two of the carbon atoms in their structures. The most popular compound of this class is acetylene, which is typically used in welding during the construction of plants and during repairs and maintenance of piping and equipment. Repairs to steel and other metal equipment such as reactors or regenerators requiring the use of a welding torch should be conducted with the greatest care after adequate preparations and inspections to ensure total inerting of vessels and piping have been achieved before initiating any welding work.

In summary, the major hydrocarbon compounds in crude oils and especially those that could be present in naphtha fractions have been reviewed. Paraffins, iso-paraffins, naphthenes, and aromatics are the hydrocarbons that are typically found in straight-run naphtha. Olefins, dienes, and acetylenes in naphtha are products from hydrocrackers, fluid catalytic crackers, delayed coking, visbreakers, and other thermal cracking process units.

2.1.5 Heteroatom Compounds in Oils

The elemental composition of crude oils in Table 2.1 shows that sulfur, nitrogen, nickel, and vanadium (metal content) are present and the amounts are quite significant. Sulfur can be as high as 6 wt. % of some crudes, and total metals could be as high as 0.1 wt. %. Heteroatoms negatively impact the performances of catalysts in hydroprocessing units. It is important and useful to have a good understanding of the nature of the heteroatoms and how they impact catalyst performance in hydroprocessing units.

The metal contaminant content is defined as the sum of the nickel, vanadium, and other metals such as copper and iron in a crude or oil fraction. When data are available only for nickel and vanadium, metal contents have been used for the sum of nickel and vanadium, and the differences should be noted and understood. Sulfur, nitrogen, and metal contents of crude oils vary greatly, as shown for highly paraffinic crude oils from Libya, Algeria, Nigeria, and Indonesia, in which the sulfur and metal contents are about 0.2 wt. % and 3 wppm, respectively.[6] Arabian Light and Kuwait crudes usually contain sulfur and metals in the range of 1.5–4.0 wt. % and 10–30 wppm, respectively; the heavier Venezuelan Jobo, Mexican Maya, and Canadian Cold Lake contain 2–4.5 wt. % sulfur and 100–500 wppm metal contents (Table 2.3).[7] It is also necessary to be aware of oil fractions in which nickel, vanadium, iron, copper, sulfur, and nitrogen compounds occur in appreciable concentrations in order to fully

TABLE 2.3
Properties of Crude Oils

Crude Oil	Arabian L	Arabian H	Kuwait	Jobo	Cold Lake	Maya
Gravity, API	34.4	28.2	32	8.8	11.1	22.2
Sulfur, wt. %	1.70	2.80	2.5	4.00	4.30	3.30
Ni, wppm	3.0	16.0	7.0	117.0	63.0	52.0
V, wppm	11.0	53.0	26.0	440.0	180.0	314.0
Metals, wppm	14.0	69.0	33	557.0	243.0	366.0
CCR, wt. %	3.3	7.6	5.4	13.8	11.9	12.0

determine the degree of hydrotreating required to meet product specifications and protect catalysts in conversion units downstream of the hydrotreaters. Applying appropriate catalytic processes and capture technologies for contaminants makes it possible for refiners to meet their goals of producing transportation fuels that meet specifications and environmental air quality regulations. Additionally, oil refiners are challenged in the United States, as they have to meet EPA Tier 3 regulations of lower specifications of 10 wppm sulfur in gasoline and 20 wppm sulfur for diesel as well as reduced sulfur oxide (SO_x) and nitrogen oxide (NO_x) emissions in oil refineries.

Crude oils that are processed in a refinery typically contain many undesirable impurities, such as sand, drilling mud, polymers, corrosion byproducts, and inorganic salts. The concentrations of inorganic salts in crude oils vary depending on the sources of the crude oils, and the salts are typically chlorides of sodium, calcium, and magnesium. When a crude slate or mixture of a number of crude oils is processed in a refinery, the salt content of the crude oil slate can vary greatly. As reviewed in Section 2.4, it is necessary to remove undesirable impurities, especially salts and water, from crude oil in the desalter so as to protect the atmospheric and vacuum distillation units and downstream oil catalytic processing units and equipment.

An unusual contaminant found in some of the crude oils is mercury. Mercury data shown in Table 2.4 were extracted from a larger dataset for crude oils supplied to the United States by some of the countries in Africa, Asia, Europe, and South America. Based on the 2004 data, the concentrations of mercury are higher in some Asian crude oils and especially for those from Thailand and Vietnam. The highest average concentration of 593.1 ppm was determined for a crude oil from Thailand. However, crude oils imported from Thailand and Vietnam were 0.003% and 0.15%, respectively, of the total volume imported by the United States in 2004.[(4)] A paper presented at the 2012 American Fuels and Petrochemical Manufacturers (AFPM) provided suggestions for analytical methods for measuring mercury in equipment and how to better manage crude oil mercury.[(5)]

2.1.6 Sulfur Compounds in Oils

The sulfur compounds in crude oil are mainly organo sulfurs with minor concentrations of inorganic sulfur compounds. These sulfur compounds as well as iron scales are introduced into hydroprocessing units from piping and equipment. It

TABLE 2.4
Mercury in US Crude Oil Supply in 2004

Country	Volume, MMBPD	% of US Supply	Avg. Hg, ppm
Algeria	78.70	1.310	13.30
Angola	112.00	1.860	1.60
Nigeria	394.60	6.570	1.80
Thailand	0.19	0.003	593.10
Viet Nam	9.30	0.150	65.50
Canada	591.50	9.845	2.10
Norway	52.40	0.870	19.50
Saudi Arabia	547.20	9.110	0.90
Mexico	585.00	9.740	1.30
Argentina	21.50	0.360	16.10
Venezuela	474.50	7.900	4.20

is known that the higher the density of the crude oil, the higher the concentration of sulfur.(3) Within a crude oil, sulfur concentrations and types vary with respect to the carbon number ranges of the oil fractions. Sulfur compounds also increase in complexity in the higher-boiling oil fractions such as distillate, gas oil, and residual oil relative to those in the naphtha fraction. Thus, the sulfur content of a crude oil is usually distributed unevenly between the oil fractions, with the bottom fractions containing higher concentrations of sulfur. To emphasize this point, sulfur content data are provided for Kuwait crude oil fractions in Table 2.5.(8)

Groups of sulfur compounds and the corresponding oil fractions in which they are present are shown in Figure 2.6. Depending on the final boiling point of a naphtha cut, mercaptans, sulfides, disulfides, thiophenes, and benzothiophenes can be found in the naphtha fractions. Dibenzothiophene sulfur compounds are found in kerosene and distillate fractions and benzonaphthothiophenes in gas oil fractions.

TABLE 2.5
Sulfur Content of Kuwait Crude Oil Fractions

Crude Oil Fraction	Boiling Range, oF	Sulfur, wt. %
Naphtha	C5–300	0.02
Kerosene	300–450	0.18
Heating Oil	450–650	1.23
Heavy Heating Oil	650–700	2.37
Heavy Gas Oil	660–1020	2.91
Residue	<700	4.22
	<1050	5.12

Source: Data taken from Schuman S. C., Shalit, H., Hydrodesulfurization, *Catal. Rev. Sci. & Engr.*, 4, 245–318, 1971.(8)

Compound	Nomenclature	Crude oil fraction
RSH	Mercaptans (Thiols)	Naphtha
RSR'	Sulfides	Naphtha
RSSR'	Disulfides	Naphtha
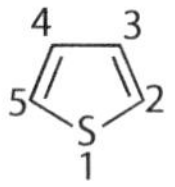	Thiophene	Naphtha
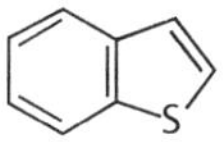	Benzothiophene	Naphtha/kerosene
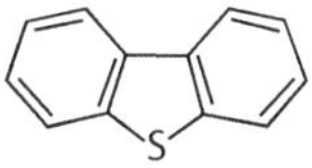	Dibenzothiophene	Distillate
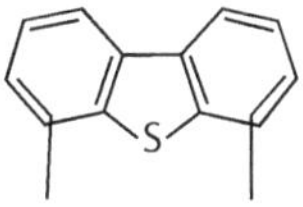	4, 6-Dimethyl dibenzothiophene	Distillate
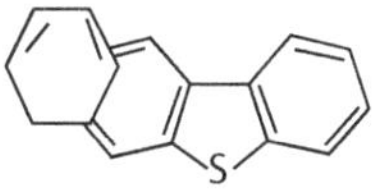	Benzonaphthothiophenes	Gas oil

FIGURE 2.6 Organic sulfur compounds in crude oil.

2.1.7 Nitrogen Compounds in Oils

Organic nitrogen compounds, though low in concentrations and typically in the 0.1–1.0 wt. % concentration range, have significant deleterious effects on the processing of crude oil fractions and the reliability of equipment in oil refineries. Nitrogen compounds in some of the feedstocks to catalytic hydrotreating and hydrocarbon conversion units can moderate the performance or poison the alumina support, precious metals, and catalytic cracking catalysts. In addition, ammonia generated during hydrogenation of organic nitrogen compounds reacts with hydrogen chloride and hydrogen sulfide in the product separation sections to produce ammonium salts. Ammonium salts deposited on downstream equipment and piping lead to fouling and corrosion of refinery piping, vessels, and equipment. In sections of this book, oil refinery reliability-enhancing methodologies and catalyst management systems are reviewed and recommendations provided for adequate nitrogen monitoring and management.

Nitrogen compounds are generally classified as basic and nonbasic. Speight described a method for differentiating basic and nonbasic nitrogen based on whether the organic nitrogen can be titrated with perchloric acid in a 50:50 solution mixture with glacial acetic acid and benzene. Based on this type of analysis, it has been determined that basic nitrogen is about 30% of the total nitrogen in crude oils.[2] Benzoquinoline is another important basic nitrogen compound in crude oils. Nonbasic organic nitrogen compounds are as shown in Figure 2.7, and basic organic nitrogen compounds are as listed in Figure 2.8.

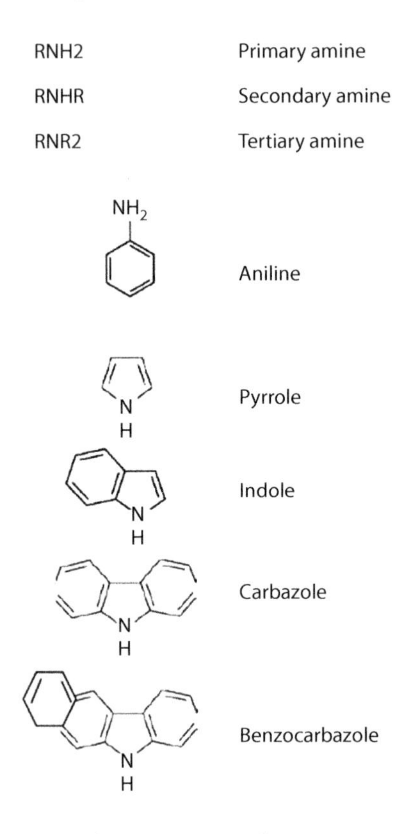

FIGURE 2.7 Nonbasic organo-nitrogen compounds.

2.1.8 Asphaltenes

Asphaltenes are high-molecular-weight hydrocarbons that are found in crude oil and present mostly in the heavy fractions of crude oils. The compounds consist of carbon, hydrogen, nitrogen, oxygen, and sulfur, as well as trace amounts of nickel and vanadium. They usually do not have a chemical molecular formula as is the case for alkanes, alkenes, alkynes, and naphthenes. They are hydrogen-deficient compounds, as their carbon-to-hydrogen stoichiometric ratio is about 1:1.2 for some asphaltenes, and that is also dependent on the source of the asphaltenes. It should be noted that the 1:1.2 ratio is close to the 1:1 that is typically used to represent the high hydrogen deficiency of catalytic coke. Asphaltenes are complex compounds with a number of aromatic rings, pyridine, thiophenes, and alkyl chains embedded in their structure, as shown in Figure 2.9.[9,10]

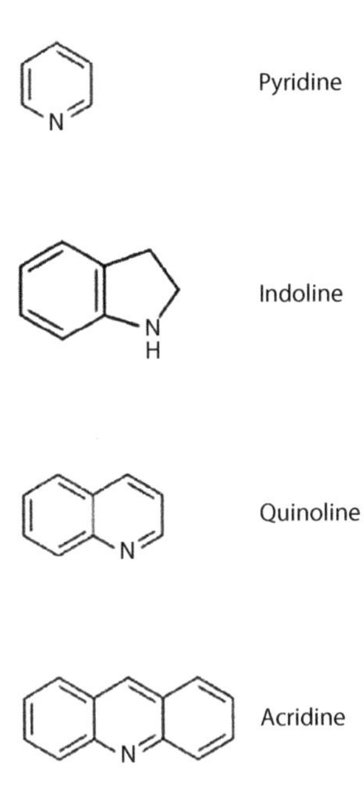

FIGURE 2.8 Basic organo-nitrogen compounds.

Asphaltene compounds are significant proportions of crude oil residues, though they are not truly soluble in most crude oils. Asphaltene particulates are held in suspension by a material called resins. These smaller suspending aggregates are soluble in crude oils. Resins and other aromatics present in micelles, a form of molecular aggregation of asphaltenes, resins, and other aromatics, stabilize the suspension in oil.[11] As a result of these aggregated composites, asphaltenes impart

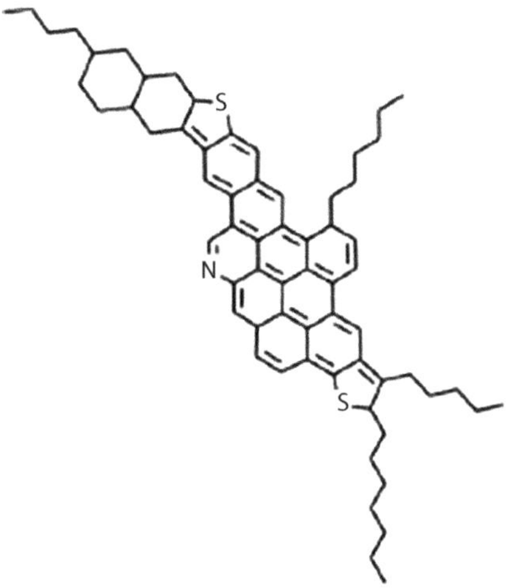

FIGURE 2.9 An asphaltene compound.

high viscosity to crude oils and negatively impact crude oil production. Asphaltenes negatively impact crude oil utilization due to segregation in storage tanks that are not well mixed. Asphaltenes are known to contribute substantially to the causes of fouling in the heat exchangers of the crude oil distillation preheat train. Some have suggested that asphaltenes present within micelles can be broken down or destabilized by reaction with paraffins under high temperatures. It has been suggested that after the micelle is weakened, freed polar asphaltenes agglomerate and are transported to heat exchanger tube walls, where they stick and foul the tubes.[12]

2.1.9 Porphyrins

Porphyrins are a group of various heterocyclic organic compounds that are composed of four modified pyrrole compounds that are interconnected, as shown in Figures 2.10 and 2.11. They are found in the heavier fractions of crude oils such as in oil residues and asphalt. Porphin is the name of the porphyrin with embedded heteroatoms, as shown in Figure 2.10. Porphyrins in crude oils are present as nickel and vanadyl

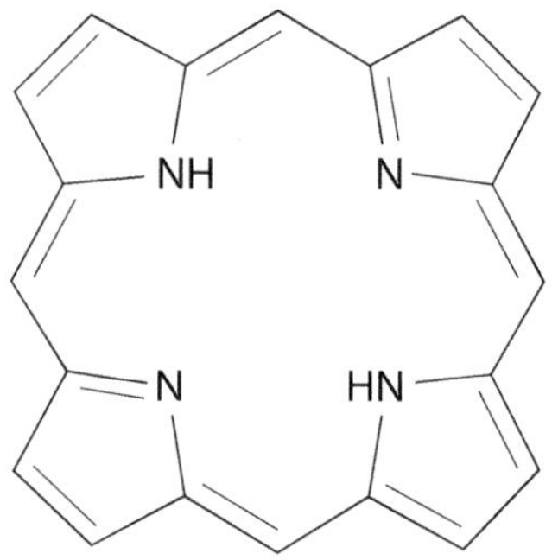

NH N N HN

FIGURE 2.10 A porphyrin with nitrogen atoms.

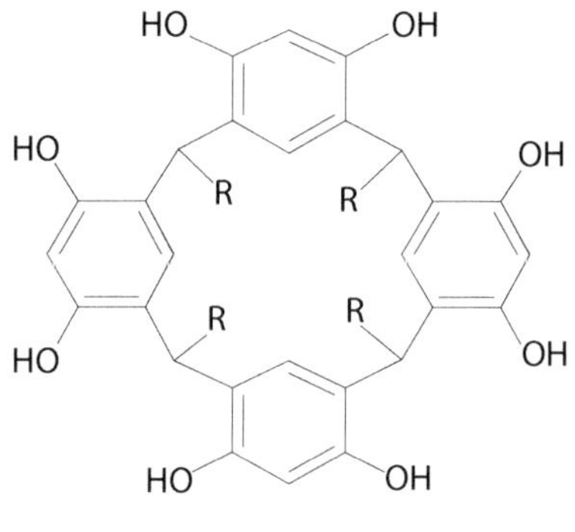

HO OH HO OH R R R R HO OH HO OH

FIGURE 2.11 A porphyrin with hydroxyl groups in its structure.

porphyrins with nickel or vanadium in the porphyrin structure. Porphyrin compounds vary greatly in molecular structures and, as in the case of asphaltenes, are not represented by a chemical formula.

2.2 OIL REFINING ECONOMICS

Oil refining involves converting a crude oil or slate to produce petroleum-derived products such as refinery fuel gas, liquefied petroleum gas, diesel, jet fuel, kerosene, fuel oil, lubricating oil, greases, waxes, petrochemical feedstocks, asphalt, sulfur, and coke. The overall economics, profitability, and viability of a refinery are highly dependent on a number of major factors. The factors are crude oil costs; the refinery configuration of assets; operating costs for the refining of oil to the product slate the refinery is targeting; and the cost of labor, maintenance, and utilities for meeting stringent product qualities and environmental compliance by the refinery. Cost factors listed are applicable on a large scale to a company with several refineries and the need to maintain the profitability of each and all the refineries, as typically accomplished by leading global oil refining companies. Some of the enabling operational and reliability factors that maintain the high profitability of those oil refining and integrated companies in the top echelon annually are covered in subsequent chapters of this book.

2.2.1 Oil Refining Profitability

Crude oil acquisition cost is the predominant operating cost in oil refining, as it could be as much as 80% of the operating cost, as indicated in a previous section. As a result, economic indicators for gross oil refining profitability are determined mainly by calculating differences between the revenues from selected petroleum products derived for a specific crude oil or from crude oils in general and the average cost of crude oils. The resultant differences as calculated by the preceding general definition are called refining margins. A specific definition of the refining margin for a specific crude oil is typically calculated, and that is the crack spread for that crude oil. Due to differences in petroleum product prices by region within a country, the calculated crack spreads are usually appropriate for each relevant specific region. The cost of transportation and regionally mandated product qualities are accounted for as well, and these lead to differences in oil refining crack spreads for retail and marketing regions.

Crack spreads are essentially gross indicators of potential profitability, as other factors such as refinery utilization and the costs of chemicals, catalysts, labor, and refinery optimization and reliability programs contribute to the net profitability of a refinery. The following crack spread definitions are used as deemed appropriate for planning by oil refiners to determine gross profitability per barrel of a crude oil.

- The 3-2-1 crack spread is defined as the difference between the sum of revenues for 2 barrels of gasoline and 1 barrel distillate and the cost of 3 barrels of crude oil.

- The 6-3-2-1 crack spread is defined as the difference between the sum of revenues for 3 barrels of gasoline, 2 barrels of distillate, and 1 barrel of residual fuel oil and the cost of 6 barrels of crude oil.
- The 5-3-2 crack spread is defined as the difference between the sum of revenues for 3 barrels of gasoline and 2 barrels of distillate and the cost of 5 barrels of crude oil.
- The 2-1-1 crack spread is defined as the difference between the sum of revenues for 1 barrel of gasoline and 1 barrel of distillate and the cost of 2 barrels of crude oil.[14]
- Other specific crack spreads have also been used to home in on the specific gross profitability of a refinery based on its expected product mix from a crude oil or crude slate. The 4-2-1-1, which represents the conversion of 4 barrels of crude oil to 2 barrels of gasoline, 1 barrel of diesel and 1 barrel of residual oil, is also used by oil refiners.

The most commonly used crack spread is the 3-2-1, which probably reflects the fact that, on average, oil refiners typically target about 60% conversion of a barrel of oil to gasoline in the refining of crude oil. Crude oil prices are set by supply and demand balance in the global marketplace. They are also set by the oil assays and the value of the expected product slate from the crude oil. Crude oils yield different distributions of oil fractions, as shown in Table 2.6 and Figure 2.12. Light sweet crude oils are oils with high American Petroleum Institute gravities and low sulfurs, and they tend to yield higher percentages of desired gasoline and distillate products, as shown in Figure 2.13. In addition, the low sulfur quality and generally low total acidities make sweet crude oils desirable for most refineries. Heavy sour crude oils tend to yield lower gasoline and more residual oil. Due to the high API gravity and low sulfur, the prices of light sweet crude oils are usually higher than those of the heavier sour crude oils. The prices are usually provided by the New York Mercantile Exchange (NYMEX) futures. Crude oil prices are usually priced relative to West

TABLE 2.6
Comparison of Distillation Yields from Crude Oils

	Condensate	Light L S	Light H S	Heavy H S	Very Heavy	Synthetic
API	42	39	31	27	21	32
C3/C4	1.5	0.5	0.8	0.3	0.3	0.3
Gasoline	78.5	37.0	33.0	19.0	19.0	21.0
Distillate	12.0	24.0	23.0	39.0	19.0	39.0
FCC Feed	8.0	30.0	32.0	26.0	33.0	39.0
Resid.	0.0	8.0	8.0	14.0	28.0	0.0

Source: Adapted from Natural Resources Canada, Government of Canada.[14]

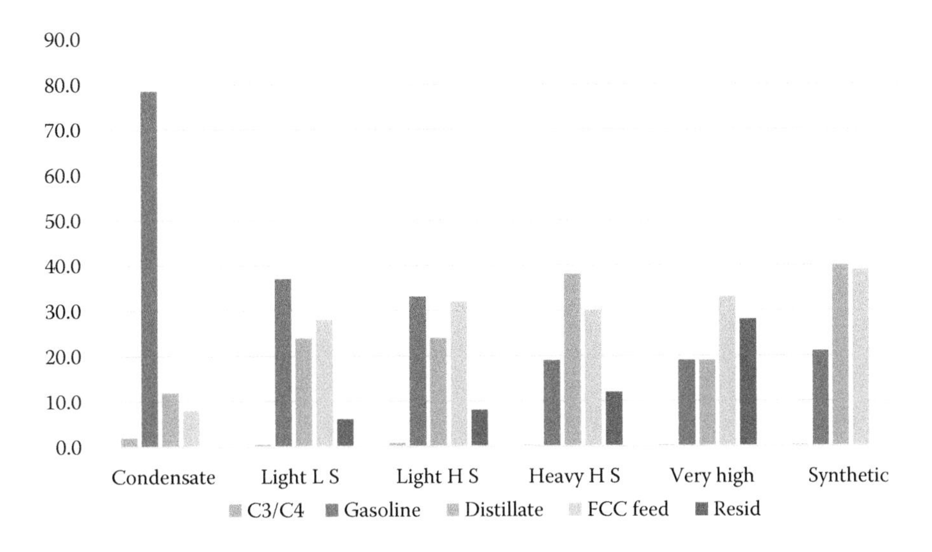

FIGURE 2.12 Comparison of refinery yields by crude type on distillation.

Texas Intermediate and Brent as benchmark crude oils. Future crude oil prices are typically used in the estimation of the crude oil crack spread.

The crude oils in Table 2.6 are defined as follows: Light LS is light low sulfur, Light HS is light high sulfur, Heavy HS is heavy high sulfur, and Synthetic oil is upgraded bitumen from oil sands with oil blended into it.

Publicized prices for crude oil, gasoline, and distillates (heating fuel oil) from NYMEX can be used in calculating crude oil crack spreads. In the United States, crack spreads that are greater than $9.00 usually indicate good potential for enhanced profitability for every barrel of crude oil processed. The 3-2-1 crack spread for the Gulf Coast of the United States is as provided in Figure 2.13. As shown in Figure 2.13, gross crack spread for Light Louisiana Sweet was higher than $5 per barrel, with the exception of the period after the third week of October, when it dipped below $5 and was negative after the first week of November in 2012.[15]

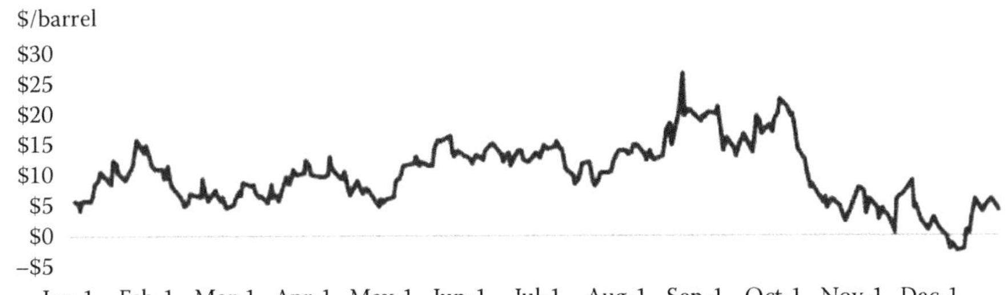

FIGURE 2.13 US Gulf Coast LLS 3-2-1 crack spread in 2012.

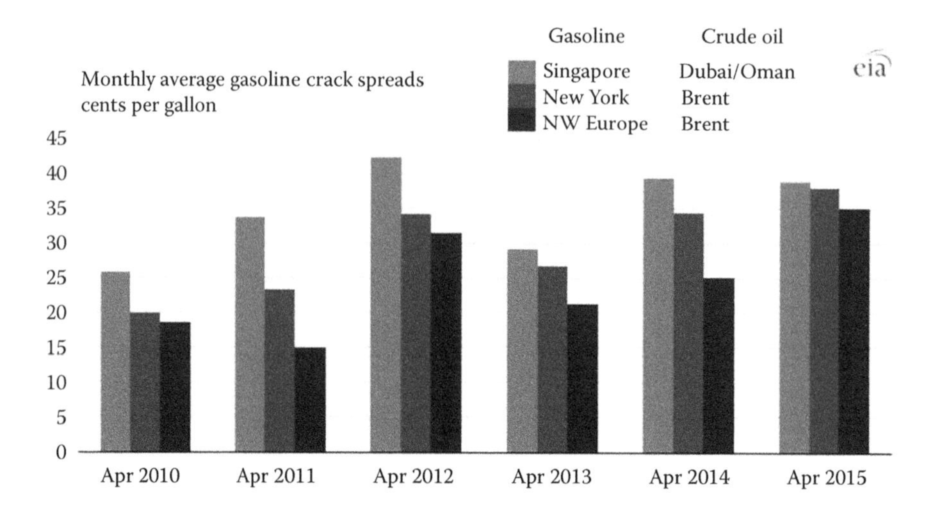

FIGURE 2.14 Gasoline/crude oil crack spread 2010–2015.
Adapted from US Energy Information Administration, Low Crude Oil Prices, Increased Gasoline Demand, Lead to High Refiner Margin, May 15, 2015.[16]

Crack spreads usually vary unpredictably, as shown in Figure 2.14 for the single gasoline product case where the crack spread is based on the difference between the price of a barrel of gasoline and a barrel of crude oil for the Singapore, New York, and Northwest Europe markets. The Singapore crude oil crack spread is based on Dubai/Oman crude oil, whereas those of New York and Northwest Europe are based on Brent crude oil. Crack spreads for the Singapore market for gasoline from Dubai crude oil are higher than those based on Brent crude oil for the New York and Northwest Europe markets. There are no obvious relationships between the magnitude of the crack spreads and the prices of the Brent and Dubai/Oman oils. Singapore crack spreads ranged from 25 cents per gallon or $10.50 per barrel in April 2010 to 42 cents per gallon or $17.64 per barrel in April 2012. Northwest Europe and New York crack spreads were much higher in April 2015 during a time of low crude oil prices relative to earlier years.[16] Additionally, the seasonal switch for gasoline from winter to summer grades may have led to an increase in the crack spread. Summer gasoline grades' Reid vapor pressures (RVPs) are usually regulated to be much lower than those allowed for gasolines in the winter months.

Since more expensive gasoline blending components displace butane, which is used in the winter months, the prices of gasoline in the United States increase in the summer months, as was reflected in the higher crack spreads in 2012.

2.3 REFINERY CONFIGURATIONS

Oil processing and marketing drivers for high profitability in oil refining and oil-refined products encompass a broad spectrum of factors and inputs. Oil-refined product marketing, which is critically important to the competitiveness of an oil refining or integrated oil company, is not covered in detail in this book, as the focus is on the profitability of oil refining operations. Crude oil and crude oil qualities were covered in Section 2.1, and the overall conclusion is that high profitability can be driven by high regional crack spreads, and, in many cases, the crack spreads are based on specific crude oils such as WTI, Brent, or LLS. In addition, cracking and delayed coking process units within refinery configurations can contribute greatly to achieving good refining profitability, as they provide oil refiners with the flexibility to refine a wide range of conventional, unconventional, light sweet, medium sour, and heavy sour crude oils. Refinery flexibility can also enable an oil refiner to refine more abundant, less expensive heavy sour crude oil and thereby enhance the refiner's profitability. Refinery configurations and capital-intensive process units such as the fluid catalytic cracker, delayed coker, and hydrocracker provide significant advantages in operations for oil refiners. These process units are reviewed in briefly in this chapter.

For grassroots refineries and depending on the availability of capital, it is recommended and often the practice that fundamental feasibility analyses and studies be conducted adequately to enable the refiner to generate an appropriate scope for the planned refinery. The scope should be forward looking and could incorporate an assessment of current and future expansion plans for the refinery. At this crucial planning stage, strategies with respect to refinery capacity, single and multiple trains based on crude and vacuum distillation units, and complexity in terms of number and size of process units should be fully considered. Adequate utilities, pipelines, and ports for the acquisition of crude oils and oil fractions, adequate tank farms for crude oils and oil fractions, hydrogen and utility supply, and regional availability of maintenance and engineering service companies should be factored into the initial considerations. Some of the considerations should include getting enough land to permit future expansion and upgrading of the refinery from either hydroskimming or cracking to a coking plant and for significant expansion of crude oil distillation capacity. It is most important to complete a thorough front end engineering design (FEED) and select and work with a success-oriented, cost-effective engineering procurement and construction company (EPC). Similar project planning features should be applied to other process units' revamps and asset expansions for the refinery. Smaller oil refining companies that may not have the necessary technical and engineering expertise and personnel to conduct the type of comprehensive strategy and feasibility studies suggested should consider using engineering and technical services consulting companies, as their services are cost effective.[17]

Oil refineries are distinguishable based on their configurations and degree of complexity, which in turn are based on processing capabilities and the expected yields of transportation and other oil-refined products for a given barrel of crude oil. Refinery configuration and processing capabilities enable refiners to process a wide variety of crude oil types. Typically, distillation of a crude oil yields essentially refinery fuel gas, liquefied petroleum gas, naphtha, distillate, vacuum gas oil, and residual oil. The naphtha fraction as defined in a broad sense and for simplicity in this book includes butane and lower carbon hydrocarbons such as pentane and hexane, which are referred to as light naphtha, and C6 to C12 hydrocarbons, which are classified as the heavy naphtha fraction. The boiling-point range of a naphtha fraction is considered to be between 65 and 400°F. The boiling-point ranges of subsequent fractions such as distillate, gas oil, and residual oils are dependent on the optimal cuts that a specific refiner has established for its operations and profitability. Typically, the approximate final boiling points of naphtha, distillate, and vacuum gas oil fractions are defined at about 400, 660, and 1010°F, respectively, and, as stated, these final boiling points may also vary considerably and are dependent on the production strategies of the oil refiner. The boiling range of the residual fraction is simply determined as those hydrocarbons with boiling points greater than 1100°F. A simplified list of initial products and fractions from the crude and vacuum distillation units is as shown in Table 2.7.

The range of products that can be produced from an oil refinery is highly dependent on the configuration and complexity of the refinery. There are four general classes of refinery configurations that are based on the degree of complexity and variety of processing units within the refineries. The four classifications of refinery complexity are topping, hydroskimming, cracking, and coking refineries. As indicated, increased complexities enable oil refiners to process a wide variety of crude oils. Oil refiners operate and cater to various niche, regional, and global markets and therefore usually plan at the grassroots refinery conception stages to meet the demands of their specific markets.

A term that is applied for comparing the complexities of refineries is the Nelson complexity index (NCI), which is based on the crude distillation capacity and the downstream upgrading process units' capacities. The Nelson complexity index, otherwise simply referred to as the Nelson index, provides what is termed a pure cost-based index of an oil refinery, as it gives a relative measure of the construction cost

TABLE 2.7
Complexity Factors for Key Oil Refining Process Units

Unit	Factor
Distillation Capacity	1.0
Vacuum Distillation	2.0
Thermal Processes	5.0
Fluid Coking	6.0
Delayed Coking	6.0
Visbreaking	2.5
Catalytic Reforming	5.0
Catalytic Cracking	6.0
Catalytic Hydrocracking	6.0
Catalytic Hydrotreating	3.0
Alkylation/Polymerization	10.0
Aromatics/Isomerization	15.0
Asphalt	1.5
Others	6.0

Source: Johnston, D., *Oil Gas J.*, 94(12), March, 1996.[19]

of a refinery based on the crude distillation and downstream upgrading capacities of that refinery.[18] The complexity index for a refinery is calculated by summing up the individual upgrading units' capacities by their complexity factors and dividing that by the crude distillation capacity. Essentially, it requires calculating a ratio based on the capacity of the downstream upgrading unit relative to the capacity of the crude unit and multiplying by the complexity factor. On this basis, the crude unit is assigned the number 1 as its complexity factor. The Nelson index basically compares the costs of various upgrading units such as a fluid catalytic cracking unit or a catalytic reformer to the cost of a crude distillation unit. Computation of the index provides a reasonable comparison of the relative cost of a refinery based on the added cost of various upgrading units and the relative upgrading capacities.[19–21] Adding all the derived complexity indices of the upgrading units to that of the crude unit leads to the Nelson complexity index for that refinery. More complex refineries in terms of the number and size of available upgrading process units would have higher complexity indices. Key units that are required in a refinery for conversion of crude oil to high-value refined products include catalytic reformers, fluid catalytic crackers, hydrocrackers, butane isomerization units, alkylation units, delayed or fluid coking units, light naphtha isomerization units, visbreaking units, and solvent deasphalting units. The Nelson complexity index is based on the processing units that are inside the battery limits (ISBL) of the refinery. Other items such as electric power, tank farms for crude oils, intermediate oil fractions and products, equipment for product blending, and waste treatment and disposal units are considered outside the battery limits (OSBL) of specific oil fraction upgrading units and are not considered in the determination of the complexity index of a refinery.

In addition, the equivalent distillation capacity (EDC) for a refinery can be calculated by multiplying the crude distillation capacity by that refinery's calculated Nelson complexity index. Equivalent distillation capacity is sometimes used as a gauge of the flexibility and processing capability of a refinery. Based on the Nelson indices, most of the oil refineries in the United States are in the top tier in the list of the most complex and profitable refineries in the world. For example, the Phillips 66 company indicated over 20 years ago that its US refineries ranged in Nelson complexity index from 7.0 at its Ferndale refinery in Washington, because of the fluid catalytic cracker, alkylation unit, and hydrotreating units, to 14.1 for its Los Angeles refinery in California, as that refinery has a fluid catalytic cracker, alkylation, hydrocracker, catalytic reformer, and coking units.[21] Similarly, there are also a number of highly complex grassroots refineries located in other countries. Efficient and reliable operation of complex oil refineries provide significant flexibility and profitability for the oil refiners who own such refining assets.

Nelson complexity indices are used in comparing the oil upgrading capabilities of individual refineries; a group of refineries belonging to an oil refiner; and oil refineries in a region, countries, and key global marketing areas. The complexity index for the Mina Abdulla refinery in Kuwait is estimated as shown in Table 2.8. Using the same complexity indices for the process units, the complexity index for oil refineries in Japan was estimated at 6.45.[19] The summary of the complexity indices for various regions of the world is listed in Table 2.9.[19–21] Other complexity metrics such as those developed by Solomon Associates are also widely used in the oil refining business as a gauge of the complexities of oil refineries.

TABLE 2.8
Estimated Complexity Index for the Mina Abdulla Refinery

Refining Process	Capacity, MBPD	Capacity, % of Crude Capacity	Process Complexity Index	Complexity Index
Atmospheric Distillation	242	100.0	1	1.00
Vacuum Distillation	134	55.4	2	1.11
Thermal Operations	59	24.4	6	1.46
Catalytic Cracking	0		6	0.00
Catalytic Reforming	0		5	0.00
Catalytic Hydrocracking	36	14.9	6	0.89
Catalytic Hydrorefining	94	38.8	3	1.17
Catalytic Hydrotreating	78	32.2	2	0.64
Alkylation/Polymerization	0		10	0.00
Aromatics/Isomerization	0		15	0.00
Lubes	0		10	0.00
Asphalt	0		1.5	0.00
Hydrogen, MMSCFD	184	76.0	1	0.76
Oxygenates (MTBE/TAME)	0		10	0.00
Estimated Nelson Complexity Index				7.03
Equivalent Distillation Capacity, MBSD				1701.00

Source: Johnston, D., *Oil Gas J.*, 94(12), March, 1996.[19]
MMSCF: millions of standard cubic feet

2.3.1 Topping Refineries

Topping refineries are simply crude oil fractionation plants that generate oil fractions that are sold to the oil refining and petrochemical companies that have the necessary production assets to process the oil fractions further for greater profitability. Products of topping refineries typically consist of refinery fuel gas, liquefied petroleum gas, light and heavy naphtha, distillate, atmospheric gas oil, atmospheric residual oil, and small-scale asphalt production. A simplified depiction of a topping refinery is shown in Figure 2.15.

Topping refineries are applicable and useful in countries and regions of a country with limited oil refining capacities, as topping units are used to generate lower-quality gasoline, diesel, and kerosene for local utilization. The disposition of oil fractions from a topping refinery is as shown in Table 2.10. With current global initiatives for environmental protection and the need to meet more stringent 10 ppm sulfur, higher gasoline octanes, and higher cetane indices and ultralow sulfur levels for diesel, topping refineries are not equipped to produce such higher-quality products. Prior to 2010 and the period of production of significant amounts of tight oil, the oil refining business in the United States had experienced major rationalization of oil refining assets of low capacities, and that led to the elimination of many of the oil

TABLE 2.9
Worldwide Refinery Complexity

Region	No. of Refineries	Distillation Capacity, MBPD/CD	Catalytic Cracking, MBPD/CD	Catalytic Reforming, MBPD/CD	Nelson Complexity Index	Average Refinery Size, MBPD/CD
Middle East	49	6147.42	309.44	635.97	4.2	125.46
Latin America	80	6384.90	1319.65	402.36	4.7	79.81
Africa	44	2806.97	173.58	331.35	3.3	63.79
Europe	116	14,511.76	2215.01	2275.86	6.5	125.10
Asia	135	14,675.29	2209.11	1664.27	4.9	108.71
CIS	56	10,060.95	526.57	1247.59	3.8	179.66
Other	32	2671.69	321.33	366.50	5.3	83.49
Canada	23	1848.45	379.00	349.50	7.1	80.37
USA	169	15,354.14	5,283.45	3623.19	9.5	90.85
World Total	705	74,461.58	12,737.14	10,896.59	5.5	104.1

Source: Johnston, D., *Oil Gas J.*, 94(12), March, 1996.; Jechura, J., Colorado School of Mines, Petroleum Refining Overview inside.mines.edu/~jjechura/Refining/01_Introduction.pdf.[19,20]

Notes:

1. Refinery Nelson Complexity = Sum of {(Unit capacity/CDU capacity) × Nelson Index)} for all units on refinery. Fabusuyi, T., Downstream Sector: Crude Oil Refining.[22]
2. CIS is Commonwealth of Independent States or the Russian Commonwealth.
3. The capacities, MBPD, are in thousands of barrel of oil per calendar day. The data set is taken from the *Oil and Gas Journal*. Johnston, D., *Oil Gas J.*, March, 1996.[19]

refineries with capacities of less than 50 thousand barrels per day (MBPD). However, recently in the United States, with the increased production of tight oil and some incompatibility challenges in the coprocessing of tight oils with heavier sour and more naphthenic crude oils, topping units are becoming more profitable options for processing significant percentages of tight oils in the crude slates of refineries.

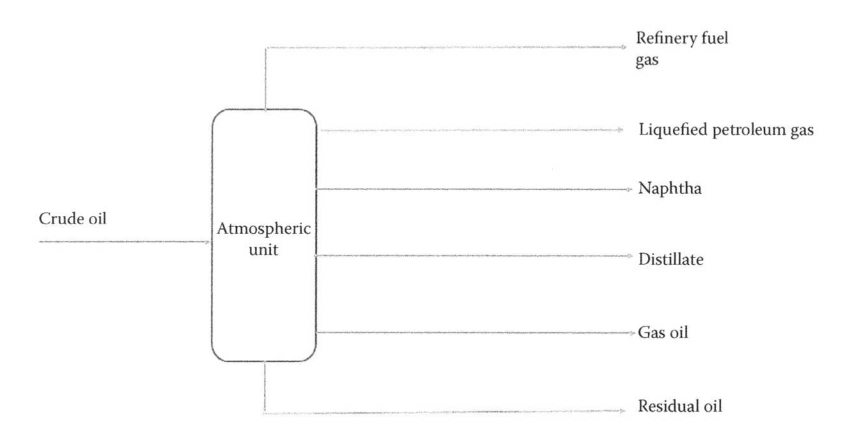

FIGURE 2.15 A simplified topping refinery showing key oil fraction products.

TABLE 2.10
Disposition and Use of Crude Oil Fractions

Oil Fraction/Gas	Disposition/Use	Boiling Range, F
Refinery Fuel Gas	Fuel	
Liquefied Petroleum Gas	Fuel, Petrochemicals Gasoline (Butane)	<65
Naphtha	Petrochemicals, Gasoline	65–400
Distillate	Diesel, Jet Fuel Kerosene	400–660
Vacuum Gas Oil	Fuel Oil, Lube Oil Gasoline	660–1010
Residual Oil	Fuel Oil, Lube Oil, Asphalt	>1010

Dedicated crude units can be used to fractionate tight oils, and the oil fractions can then be coprocessed with other oil fractions from conventional heavier sour crude oils in downstream processing units such as the catalytic reformer, hydrocracker, and fluid catalytic cracker and delayed coking units. This separate distillation of tight oils and conventional heavy sour crude oil slates eliminates asphaltene precipitation and reliability challenges that can occur when tight oils and other heavier oils are codistilled in the same crude units.

Another application is the use of topping refineries to generate crude oil fractions that could be exported, thereby circumventing the US limitation on exportation of whole crude oils. Fortunately, this has not been a prevalent practice among most United States oil refiners. The December 2015 decision by the US Congress to remove restrictions on US crude oil exports should also eliminate the economic incentives to build topping units solely for use in indirect exporting of crude oils as oil fractions.[23]

A word of caution is necessary with respect to the operation of topping oil refineries, and that is that the same level of safety and environmental management of air, water, and oil contaminants should be maintained as required in the more complex oil refineries. This caution is extended with greater emphasis to the operation of "rogue" topping oil refineries with open crude distillation units where operators and other persons are exposed to copious amounts of air and oil carcinogen pollutants continuously and that also lead to polluted environments and reduced lifetimes of the affected persons.

2.3.2 Hydroskimming Refineries

A hydroskimming refinery is a little more complex than a topping refinery. Though it has the usual crude oil atmospheric distillation unit, it differs in the fact that it has a vacuum distillation unit, a catalytic reformer, and hydrotreating units for treating the naphtha, distillate, and vacuum gas oil fractions. A simplified drawing of a hydroskimming refinery is provided in Figure 2.16. With the added catalytic reformer, the hydroskimming refinery has the capability to generate higher octane reformate; benzene, toluene, and xylene; and hydrogen for hydrotreating units. The added catalytic reforming and hydroprocessing assets enable oil refiners to operate

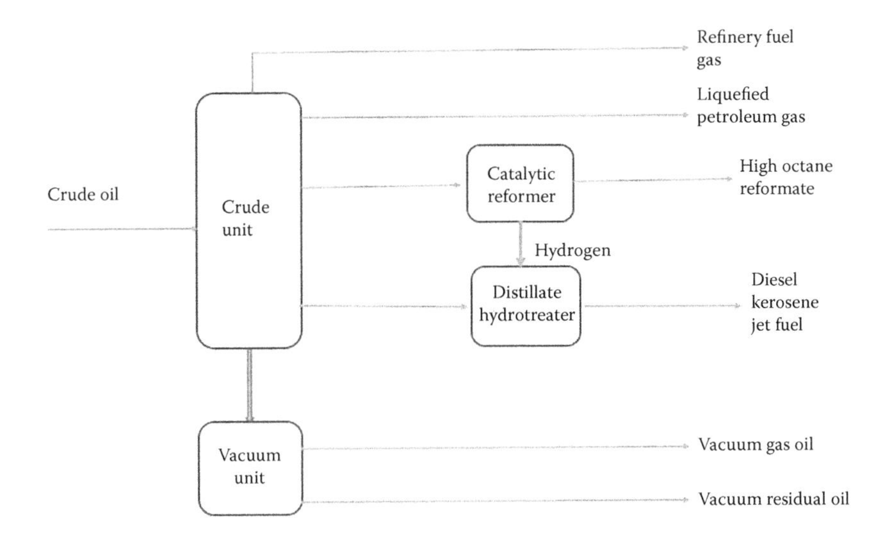

FIGURE 2.16 A simplified hydroskimming refinery.

their hydroskimming refineries for much higher profitability relative to topping refineries. Some of the hydroskimming refineries have light naphtha isomerization units, which can generate a premium high-octane component for gasoline blending. Light naphtha isomerization units can also be used to saturate benzene to aid in meeting gasoline benzene quality and for toll production of low-benzene gasoline components for an oil refining client.

As the degree of complexity of the refinery configuration increases, so will the need for detail and timely planning to ensure that hydrogen is available for the operation of the hydroskimming refinery. In addition, downtimes for the catalytic reformer should be minimized as much as feasible. Planning for the turnaround work on a catalytic reformer may have to be accomplished in tandem with planning for a turnaround for the hydroskimming refinery unless an external hydrogen supply is available. An alternative supply of hydrogen would be required for catalyst activation during the regeneration process and outages of the catalytic reformer.

Oil refiners that have hydroskimming refineries with two-train crude units, multiple catalytic reformers, and readily accessible hydrogen supply from a gas producer company are better positioned to maximize their refinery reliability and profitability. Some operators of hydroskimming oil refineries have added a steam methane reforming (SMR) hydrogen plant to provide hydrogen either to supplement or provide hydrogen during production outages of the catalytic reformer. The experience of the author is that it is generally cost effective and the hydrogen quality and supply are enhanced by relying on the expertise of a gas supply company to operate such a plant within the oil refinery complex or provide pipeline hydrogen to the hydroskimming refinery. Another factor that could impact the long-term viability and profitability of hydroskimming refineries is that despite the increased number of processing units, oil refiners are usually limited with respect to crude oil selections to the more expensive sweet crude oils. Vacuum gas oil and residual oils are not upgraded significantly in hydroskimming refineries, and this leads to much lower profitability for such refineries.[1,13] Hydroskimming refineries are usually less profitable relative to the cracking and coking refinery configurations, which are reviewed next in Sections 2.3.3 and 2.3.4, respectively.

2.3.3 Cracking Refineries

Cracking refinery configurations are often referred to as medium conversion refineries due to the fact that either fluid catalytic crackers and possibly vacuum gas oil hydrotreaters and alkylation units have been added to the hydroskimming refinery configuration through expansion projects over time. A simplified drawing of a cracking refinery configuration is provided in Figure 2.17.

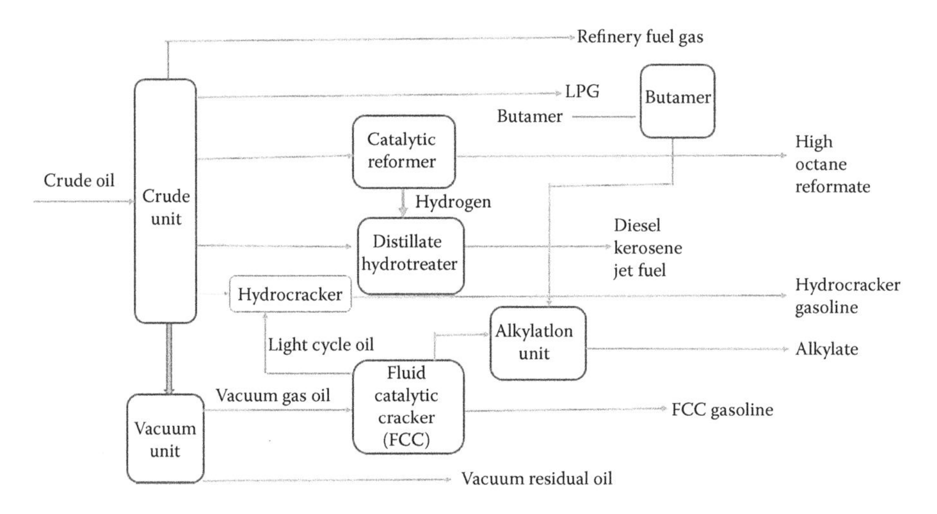

FIGURE 2.17 Cracking refinery configuration including FCC and alkylation unit.

Fluid catalytic cracking and alkylation units increase the operational flexibilities of the oil refiners to produce significantly greater percentages of gasoline and diesel products from crude oil relative to those from hydroskimming refineries. Basically, fluid catalytic cracker units permit the cracking of complex hydrocarbons with more than 20 carbon atoms to 70+% of hydrocarbons with fewer than 12 carbon atoms and lighter butylene and pentene compounds. The pentenes and butenes are usually fed to the alkylation units, where they react with isobutane to produce alkylate, a higher-octane gasoline blending component. Fluid catalytic cracking and alkylation units are covered in some detail in Sections 2.9 and 2.10.

Hydroskimming and cracking refineries are used predominantly in most of the regions of the world such as in Africa; the Commonwealth of Independent States (CIS); and Latin American, Middle Eastern, and Asian countries, as shown by the range of average complexity indices for their refineries in Table 2.9. The type of older refinery configurations in those regions are due to product mix demands, capital constraints, and the availability of sweet crude oil refineries in those regions that prevailed at the time the refineries were installed.

2.3.4 Coking Refineries

The most flexible, capital-intensive, and profitable refineries are those whose configurations include upgrading and hydrotreating units that are used in the conversion of vacuum residual oil to produce greater percentages of lighter transportation fuels from crude oils relative to cracking configuration refineries. Delayed coking, fluid coking, and hydrocracking units are typically added today in revamps of existing refineries and in grassroots refineries. Solvent deasphalting units such as residual oil solvent extraction (ROSE) units are sometimes used with coking units to enhance the refinery's crude oil bottom upgrading capabilities. A generalized schematic diagram of a coking configuration refinery is as given in Figure 2.18. In addition, coking refineries must have access to a reliable high-purity hydrogen supply because of the high hydrogen consumption of the hydrocracker and the hydrogen consumption required by the hydrotreaters for treating various products of the coking units. Oil fractions produced from coking units typically contain higher percentages of heteroatom contaminants and unsaturated compounds relative to similar straight-run fractions from crude distillation units. In addition, an adequate number of amine treating and sulfur recovery units should be provided to meet increased hydrotreating and sulfur management demands of coking configuration refineries. Since most or all of the vacuum residual oils from crude oils in the refinery are converted to lighter premium transportation fuels, the conversion of crude oil is maximized and these coking refinery configurations are usually referred to as maximum or deep conversion refineries.

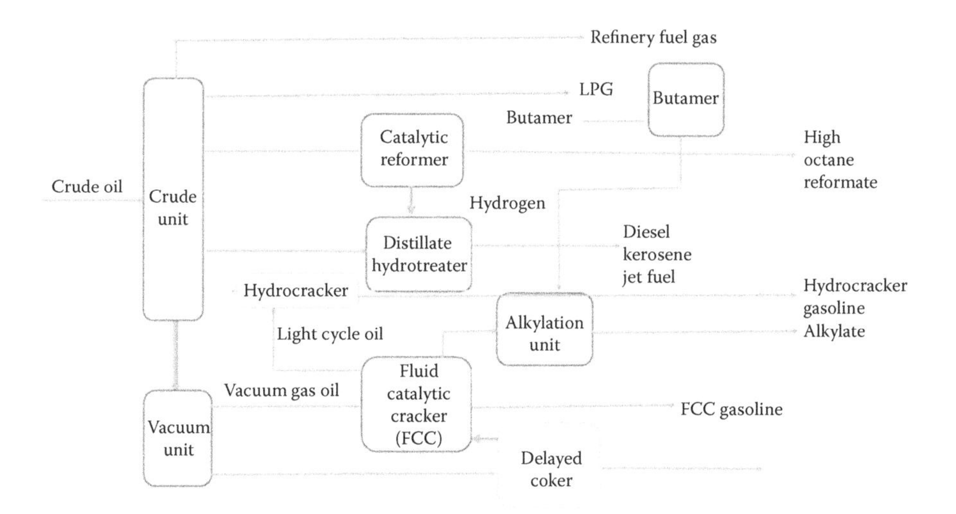

FIGURE 2.18 A coking refinery configuration.

The Nelson complexity index ranges are provided for topping, hydroskimming, cracking, and coking refinery configurations in Table 2.11. A coking refinery complexity index of greater than 12 is indicative of the fact that delayed and fluid coking units confer significant advantages to oil refiners who either possessed coking

TABLE 2.11
Complexity Indices for Refinery Configurations

Refinery Configuration	Oil Conversion	Complexity Index
Topping	Low	<2
Hydroskimming	Moderate	2–6
Cracking	Medium	6–12
Coking	Maximum/Deep	>12

Source: Johnston, D., *Oil Gas J.*, 94(12), March, 1996; Ramana, U. V., Presentation on Characteristics and Refining Products, *Workshop on Refining and Petrochemicals*, New Delhi, India, August 25–28, 2010.[19,24]

units in their original grassroots refineries or have retooled their refineries to process cost-attractive heavier sour crude oils. This point is further emphasized by the fact that average sulfur and residual oils for available global crude oils were increasing by 2010 due to availability of the cost-effective heavier sour and unconventional bitumen oils.[24] The average crude oil data for 1985–2010 are as summarized in Table 2.12.

There is a significant increase in the percentages of average residual oil and of sulfur and contaminant metals in residual oil over the 25 years reported in Table 2.12. Recent increases in the production of light tight oils in the United States and potential production increases in shale or tight oils in the United States, Russia, China, Canada, and Argentina are not likely to impact the need for coking units, as the tight oil fractions are slated to be coprocessed in downstream units after separate crude distillations of tight oils to eliminate concerns with crude oil asphaltene precipitation in crude distillation units. A more detailed schematic drawing of a cracking configuration refinery is as shown in Figure 2.19.

A coking refinery configuration usually contains several hydrotreating units for naphtha, kerosene, distillate, and vacuum gas oil and hydrotreaters for the production of low-sulfur gasoline and ultralow-sulfur diesel (ULSD). Additionally, it would have

TABLE 2.12
World Crude Oil Quality Trend

Properties of Crude Oil	1985	1990	1995	2001	2010
Sulfur, wt. %	1.14	1.12	1.31	1.41	1.51
Gravity, API	32.7	32.6	32.4	32.2	31.8
Metals in Residual Oil, wppm	275	286	297	309	320
Residual Oil In Crude Oil, vol. %	19.0	19.4	19.8	20.2	21.3
Sulfur in Residual Oil, wt. %	3.07	3.26	3.61	3.91	4.0

Source: Ramana, U. V., Presentation on Characteristics and Refining Products, *Workshop on Refining and Petrochemicals*, New Delhi, India, August 25–28, 2010.[24]

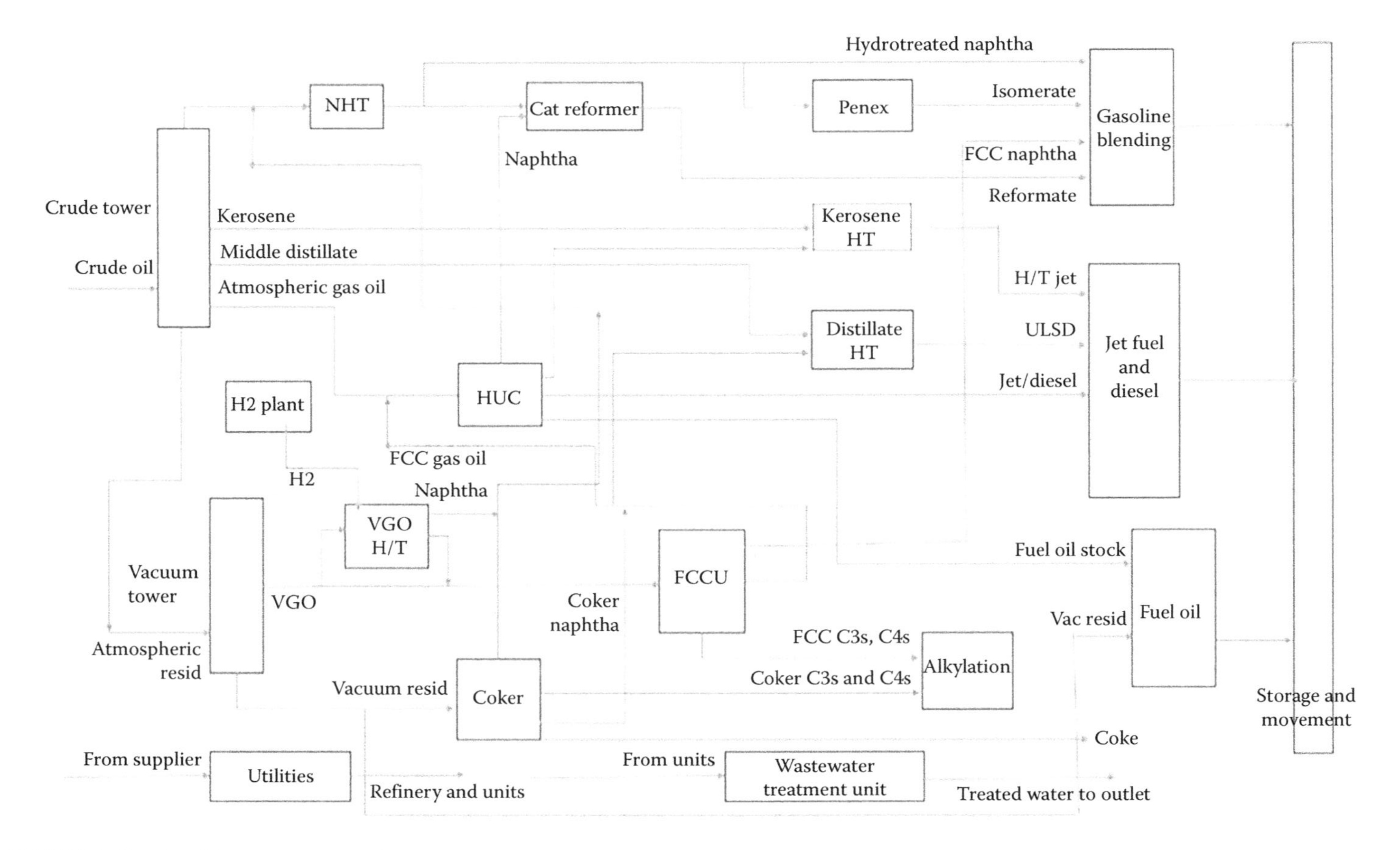

FIGURE 2.19 A coking refinery complex.

a light naphtha (Penex) isomerization unit, butane isomerization unit, alkylation, catalytic reformer, hydrocracker, fluid catalytic cracking unit, coking unit, and hydrogen production plant. Several amine treating and sulfur production units should be available for adequate sulfur recovery. Coking refineries provide oil refiners with huge profitability enhancement advantages over refiners with hydroskimming and topping refinery configurations. The average capacity of announced global projected grassroots oil refinery capacities are in excess of 200 MBPD per refinery. In order to be competitive, existing refineries are being revamped to provide increased oil processing flexibility. Oil refiners are incorporating more bottom-of-the-barrel upgrading technologies for heavy oil processing. Essentially, there has been significant rationalization in oil refineries to eliminate or expand hydroskimming and cracking configuration refineries to add crude distillation capacities and bottom upgrading technologies. Global capacity increases are actually matched by a reduction in the number of oil refineries, as shown in Figure 2.20. Despite the fact that there was a decrease in the number of oil refineries from about 725 to 665 in the period between 2002 and 2011, the total global oil refining capacity increased by about 9% during the same period.[25]

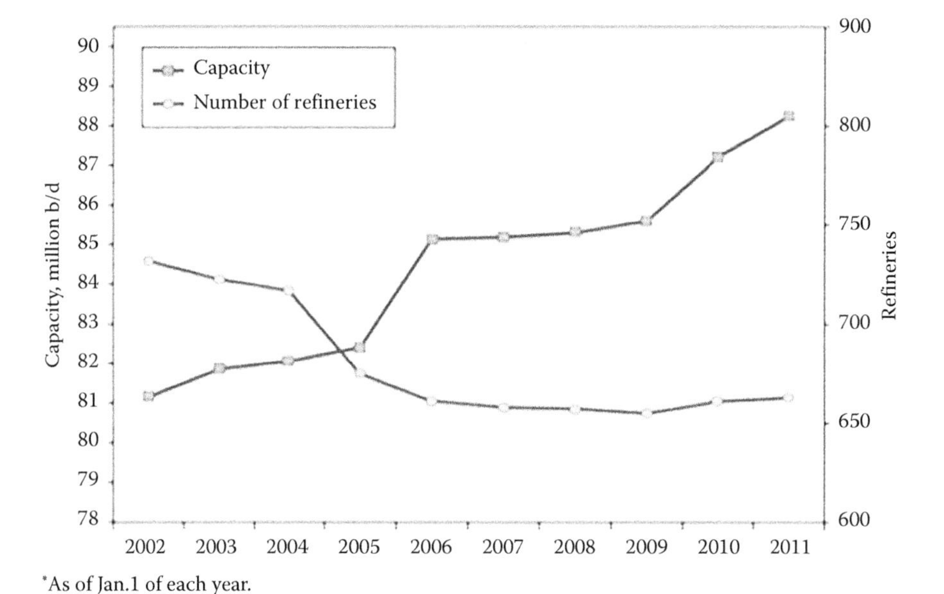

FIGURE 2.20 Global crude oil refining capacity and the number of refineries. From Brelsford, R., Kootungal, L., *Oil Gas J.*, 112(12), 34–45, December 1, 2014.[25]

Due to a number of oil refinery revamps in the United States after 2010 and the acquisition of BP's Texas City Refinery in Texas City, the ranking of the top largest 15 refineries (Table 2.13) changed and was modified to include two refineries of the Marathon Petroleum Corporation. The two Marathon Petroleum Corporation refineries are the Garyville Refinery in Garyville, Louisiana, and the Galveston Bay

TABLE 2.13
The Top 15 Global Oil Refineries

No.	Name of Refinery	Location	MBPD
1	Jamnagar Refinery (Reliance Industries, Ltd)	Gujarat, India	1240
2	Paraguana Refinery Complex (PDVSA)	Paraguana, Falcon, Venezuela	940
3	SK Energy Co., Ltd. Ulsan Refinery (SK Energy)	Ulsan, South Korea	850
4	Ruwais Refinery (Abu Dhabi Oil Refining Company)	Ruwais, United Arab Emirates	817
5	GS Caltex Yeosu Refinery (GS Caltex)	Yeosu, South Korea	730
6	S-Oil Onsan Refinery (S-Oil)	Ulsan, South Korea	670
7	ExxonMobil	Singapore	605
8	Port Arthur Refinery (Motiva Enterprises)	Port Arthur, Texas, USA	600.3
9	Baytown Refinery (ExxonMobil)	Baytown, Texas	560.5
10	Ras Tanura Refinery (Saudi Aramco)	Saudi Arabia	550
11	Garyville Refinery (Marathon Petroleum)	Garyville, Louisiana, USA	539
12	Baton Rouge Refinery (ExxonMobil)	Baton Rouge, Louisiana, USA	502.5
13	Galveston Bay Refinery (Marathon Petroleum)	Texas City, Texas, USA	459
14	Abadan Refinery (NIOC)	Abadan, Iran	450
15	Shell Pernis (Royal Dutch Shell)	Rotterdam, The Netherlands	416

Source: US EIA Report, Top 16 Oil Refineries.[27]
Note: Updated by Soni Oyekan in July 2016.

Refinery in Texas City, Texas. Marathon Petroleum Corporation's Garyville and Texas City refineries were listed as the 11th and 14th refineries globally in 2015. The highest-capacity refinery in the world is the Reliance Company's Jamnagar Refinery in Gujurat, India, with a capacity of 1240 MBPD as of July 2016, shown in Figure 2.21.[44] Marathon Petroleum Corporation's 539-MBPD Garyville Refinery and ExxonMobil's Singapore are as shown in Figures 2.22 and 2.23, respectively.[26]

FIGURE 2.21 Reliance Company's Jamnagar oil refining complex in Gujurat. Courtesy of Bechtel Corporation.

FIGURE 2.22 Marathon Petroleum Company Garyville refinery, Garyville, Louisiana. Photo Courtesy of Marathon Petroleum Corporation. (From Oyekan, S. O., Catalytic Applications for Enhanced Production of Transportation Fuels, 2009 NOBCChE Percy L. Julian Award Lecture, April 14, 2009, St. Louis, MO.[43])

FIGURE 2.23 ExxonMobil's Singapore refinery.
Courtesy of ExxonMobil Corporation.

It is also informative to be cognizant of the major integrated oil and solely oil refining companies that own significant numbers of the highly complex, multipurpose, high-capacity oil refineries. They include five top major corporate and national oil companies, ExxonMobil, Royal Dutch Shell, Sinopec, BP, and Saudi Aramco, based on oil refining capacities as listed in Table 2.14.

A majority of the refineries of the top oil refining companies are highly complex plants, and their refining operations produce low-cost petroleum products such as gasoline, diesel, asphalt, propylene, isobutane, fuel-grade coke, and sulfur relative to oil refiners with hydroskimming refineries. In addition, those that have petrochemical complexes produce lube oils and petrochemical products. Nelson complexity indices for a number of the coking configuration–type refineries of the top oil refining companies are usually greater than 10.

TABLE 2.14
Top Oil Refining Companies Based on Crude Oil Refining Capacities

No	Company	Crude Capacity, MBPD
1	ExxonMobil	5589
2	Royal Dutch Shell PLC	4109
3	Sinopec	3971
4	BP PLC	2859
5	Saudi Aramco	2852
6	Valero Energy Corporation	2777
7	PDVSA	2678
8	China National Petroleum Corp,	2675
9	Chevron Corp.	2540
10	Phillips 66	2515
11	Total SA	2304
12	Petroleos Brasileiro	1997
13	Marathon Petroleum Co. LP	1714
14	Petroleos Mexicanos, PEMEX	1703
15	National Iranian Oil Co.	1451

It is accepted that the key goal of oil refiners is to maximize production of high-value gasoline and diesel products from crude oil. Highly complex refineries are better equipped to refine a wide variety of crude oils and flexible to take advantage of changes in price differentials between gasoline and diesel. The flexibility of operating bottom-of-the-barrel upgrading assets and an excellent refining organization are major factors in maximizing the profitability of top oil refining companies. That operating flexibility enables the shifting of transportation fuel products between gasoline and diesel to take advantage of price differentials. Price differentials can be as high as 59 cents per gallon of diesel relative to gasoline, or about $24.8 per barrel, as was reported in 2013 for the United States.[28] The price differentials between diesel and gasoline are due to rising global demand for diesel, environmental regulations that require additional processing in order to substantially reduce distillate sulfur for the production of ultralow-sulfur diesel, differences in diesel and gasoline taxes, and increases in seasonal prices of gasoline and diesel.

Oil processing, distribution and marketing, and tax costs are much higher for diesel than for gasoline in the United States, as shown in Table 2.15. The listed factors lead to higher prices for diesel relative to gasoline prices. In addition to the need to move some of the higher-boiling compounds of heavy naphtha into the distillate range to maximize distillate volume and diesel production, diesel as a refined product also competes for distillate processing volumes with jet fuel, kerosene, and heating oil. Gasoline prices are also seasonal since gasolines are typically more expensive most of the year and especially in the summer months relative to winter months. Relaxation

TABLE 2.15
Average ULS Diesel and Gasoline US Cost Components, Jan 2013

Component Cost	UL Sulfur Diesel	Regular Gasoline
	$4.00/gallon	**$3.45/gallon**
Crude Oil	$2.32 (58%)	$2.31 (67%)
Oil Refining	$0.52 (13%)	$0.32 (9%)
Distribution & Marketing	$0.68 (17%)	$0.41 (12%)
Taxes	$0.48 (12%)	$0.41 (12%)

Source: Hansen, L. R., OLR Backgrounder: Diesel vs Gasoline Prices, Oil Research Report, 2013-R-007, January 28, 2013.[28]

of gasoline qualities in the winter months and especially lowering of the Reid vapor pressure specifications for gasoline enable the blending of reasonable amounts of cheaper butane into gasolines.

In addition to the above-discussed factors, gross increases in the average price of benchmark crude oils such as Brent and WTI also impact the prices of gasoline and diesels. An example of this is illustrated by Figure 2.23, where diesel and regular gasoline were sold at peak prices of $4.77 and $4.00 per gallon, respectively, when the average prices of domestic and imported crude oils in the United States were in the range of $90 and $120 per barrel. With the collapse of crude oil prices in 2015 and 2016, diesel and regular gasoline were sold on average in the United States at $2.41 and $2.17 per gallon, respectively, when the average prices of domestic and imported crude oils were in the $45–$48 range. Data shown in Figure 2.24 were taken from the average prices of diesel and regular gasoline for the month of July of each year.[29,30]

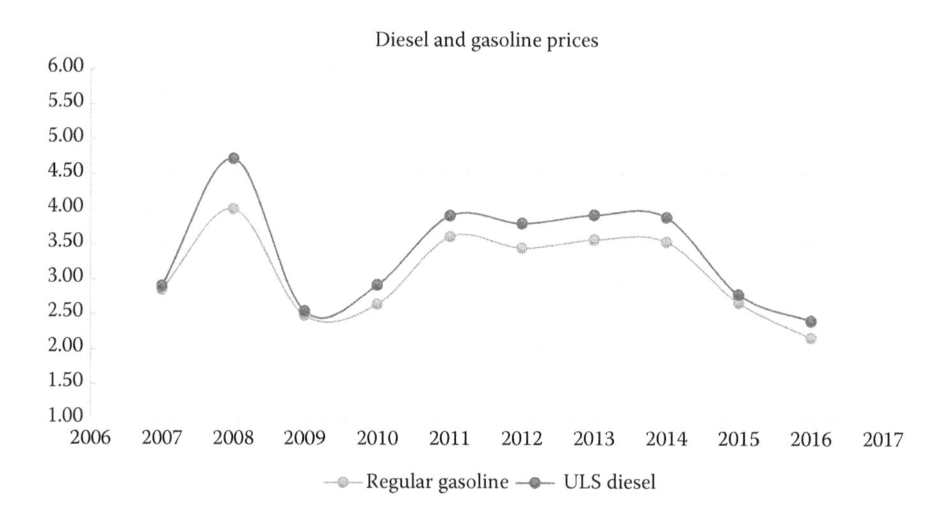

FIGURE 2.24 Average differences in dollars/gallon between diesel and gasoline.

A study was conducted to develop a better understanding of the price differential between diesel and gasoline in 161 countries. In the study, gasoline prices, unlike those in the United States transportation fuel marketing environment, were much higher than the prices of diesel for 84% of the countries.

It was determined that diesel was about 10% cheaper relative to the price of gasoline.[31] Gasoline prices were higher than diesel in the countries studied in 2016 despite the fact the countries had not established the use of regulated ultralow-sulfur diesel as a transportation fuel. The relative differences between gasoline and diesel prices for the countries were dependent on oil refining costs and relative taxes imposed on gasoline and diesel. In those countries, higher percentages of taxes were imposed on gasoline relative to the tax levied on diesel. In another similar study conducted in 2016, the projections for most countries showed that gasoline would also be cheaper than diesel in 2017.[31] For Asian countries that have high transportation fuel consumption such as India and China, and for select South American countries such as Argentina and Brazil, gasoline prices are also higher than the prices of diesel. The United States with its ultralow-sulfur diesel is one of the unique countries for which diesel prices exceed gasoline prices, and sometimes the spread could be substantial and as high as $32.00 per barrel. Price differentials between gasoline and diesel within countries and in the world transportation fuel markets drive how oil refiners configure and use their oil-processing units in order to maximize gasoline and diesel production.

2.4 OIL REFINERIES, FUELS, AND GLOBAL WARMING

The world is experiencing evidence of climate change due to an increase in the temperature of the Earth, which was determined to be about 1.5 degrees over the last century. The temperature increase for the Earth after the next 100 years is projected to be in the range of 0.5–8.6 degrees.[32] Since this projected temperature increase could lead to major and potentially catastrophic changes in the climate of the Earth, there are major concerns and heightened interest regarding substantially reducing the projected increase of the temperature of the Earth. It has been determined that the temperature increase is due primarily to increases in the concentrations of greenhouse gases (GHGs) such as carbon dioxide (CO_2), methane (CH_4), nitrous oxide (N_2O), ozone (O_3), and fluorinated gases such as hydrofluorocarbons (HFCs), perfluorocarbons (PFCs), and sulfur hexafluoride (SF_6) in the Earth's atmosphere. The aforementioned gases are identified as greenhouse gases because of the impacts they have on increasing the temperature of the Earth's atmosphere by basically operating the Earth's atmosphere in a manner similar to that desired in an appropriately functioning greenhouse used for plants. The accumulation of greenhouse gases in the Earth's atmosphere aids in concentrating short-wavelength radiation that arrives as heat from the sun (solar energy). Some of the radiation reaching the Earth is reflected by the Earth's surface; however, the vast majority of the radiation is absorbed, leading to warming of the Earth. As the Earth's surface heat increases, it starts to emit long-wavelength, infrared radiation back to into the atmosphere.[33,34] A small optimal amount of gases, less than 0.1% of greenhouse gases and primarily nitrogen and oxygen, are vital to life on Earth due to their blanketing effect in maintaining appropriate temperatures over the Earth. As the concentrations of greenhouse gases increase due to human or anthropogenic activities, the blanketing effect is exacerbated there by negatively impacting the natural balance between incoming and outflowing radiation from the Earth. The concentrations of greenhouse gases, if not controlled, would lead to a rapid rise in the Earth's temperature due to high rates of consumption of energy by people.

Global greenhouse gases emitted through human activities, shown in Figure 2.25, indicate that CO_2 generated from the combustion of fossil fuels and industrial processes accounts for 65% of the greenhouse gases. An additional 11% of the GHG, that is, CO_2, is generated from human activities in forestry and other land use such as deforestation, land clearing for agriculture deforestation, and degradation of soils. Methane and nitrous oxide are produced during agricultural activities, energy use, biomass burning, and waste management. The other GHGs are fluorinated gases such as hydrofluorocarbons and perfluorocarbons that are emissions from industrial processes, refrigeration, and various consumer product uses.

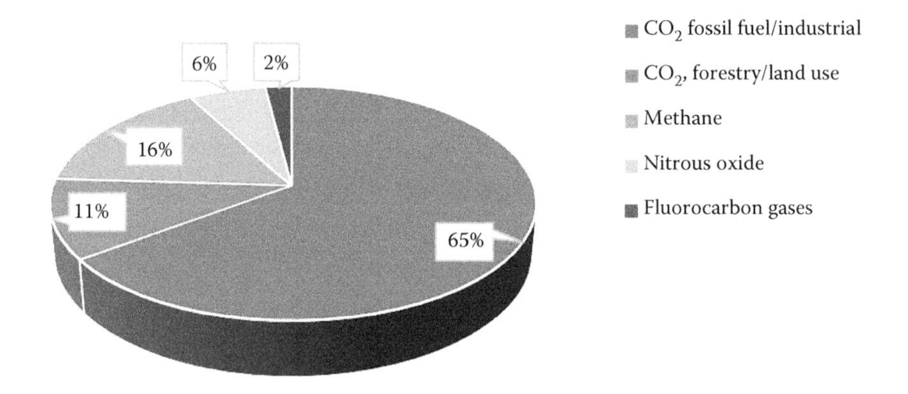

FIGURE 2.25 Global greenhouse gas emissions.
IPCC (2014) based on global emissions from 2010. Details about the sources can be found in the Contribution of Working Group III to the fifth assessment report of the Intergovernmental Panel on Climate Change.

A survey of greenhouse gas emissions reported for 2010 by global economic sectors showed that electricity and heat production accounted for 25% of the GHGs produced, followed by agriculture, forestry, and land use at 24%. The burning of coal, natural gas, and oil for electricity and power generation led to the production of the 25% of the global greenhouse gases reported for 2010; industry and transportation contributed 21% and 14%, respectively, of the GHGs generated. Survey data for global greenhouse gas emissions by economic sector are reported in Table 2.16.

The GHG emissions generated in the transportation fuel sector are due to the burning of fossil fuels for road, rail, air, and marine transportation. It was reported that 95% of the world's transportation energy comes from oil-based fuels, and that was largely for gasoline and diesel. It was further reported that the top CO_2 emitter was China, followed in order by the United States, the European Union, India, Russian Commonwealth or Commonwealth of Independent States, Japan, and

TABLE 2.16
Global Greenhouse Gas Emissions by Economic Sectors

Economic Sector	% of GHG
Electricity & Heat Production	25
Agriculture, Forestry & Land Use	24
Industry	21
Transportation	14
Other Energy	10
Buildings	6

TABLE 2.17
Greenhouse Gas Emissions by Country

Country	% of GHG
China	25
United States	16
European Union	10
India	6
Russian Federation	6
Japan	4
Canada	3
Other Countries	30

Source: Chevron, Global Marketing, Diesel Fuels Technical Review.[35]

Canada. The other countries of the world contributed 30% of the GHG emitted in 2010. Greenhouse emissions by countries and regions are reported in Table 2.17.[35]

Over the last 30 years, climate change and key factors that are possibly responsible for the change have become better understood. This development has led to a strong, broad global response to reduce the concentrations of greenhouse gases emitted that could lead to significant climate changes. At the first meeting in Rio de Janeiro in 1992, 180 countries signed the United Nations Framework Convention on Climate Change (UNFCCC), which outlined the need to reduce greenhouse gas emissions as a global response to climate change. Since there were no major obvious responses and actions from the countries that participated at the 1992 meeting, another meeting held in Kyoto produced an agreement, the Kyoto Protocol, on December 11, 1997, to force action by state parties of the UNFCCC. The Kyoto Protocol became enforceable after February 16, 2005. Two notable industrialized countries, Canada and the United States, did not ratify the July 2016 Doha Amendments. Despite some of the challenges with getting a consensus on a fully comprehensive agreement, the ultimate objective of the UNFCCC is the "stabilization of greenhouse gas concentrations in the atmosphere at a level that would stop dangerous anthropogenic interference" with the climate system.[36] For the state parties that ratified the Kyoto Protocol, several national programs have been suggested and are in use in their efforts to reduce GHG emissions. In the US economy, transportation is second only to electricity generation in terms of the percentage of GHG emissions associated with the sector. As discussed previously, CO_2 accounts for over 95% of the GHG emissions in the transportation sector.

Energy efficiency, system efficiency, reduction of transportation activity, and alternative fuels have been suggested, and such programs are being managed for reducing GHG and especially CO_2 emissions. One key lead for the reduction of CO_2 in transportation involves use of nonfossil components in transportation fuels. Specific programs that impact transportation fuel qualities and are of the utmost interest to oil refining and automotive manufacturing business sectors are those that would require changes in fuel quality. Greenhouse gas reduction programs in countries include the use of nonfossil components in transportation fuels. The programs are similar to that in the United States that requires the use of oxygenates such as ethanol in gasoline. The US EPA program also requires the use of biodiesel blended fuels for heavier trucks, buses, and rail transportation. Fuel regulations in the US stipulate the use of gasolines with a low benzene content limit of 0.62 volume percent and low sulfur concentration of 10 wppm. Biodiesel blended fuels are expected to have a sulfur concentration limit of 10 wppm and significant percentages of nonfossil biodiesel components. Biodiesels are clean-burning nonfossil diesel fuel components that are produced from vegetable oils and animal fats.[37] Biodiesels can be blended in regulated proportions into diesel to produce desired biodiesel blends.

2.4.1 Overview of Global Transportation Fuels

An overview of gasoline and diesel specifications is provided for countries in this section. The intent is to provide a good understanding of what is required from oil refiners to meet gasoline and diesel specifications in many countries. Sulfur concentrations for gasoline and diesel are some of the key properties established by countries so as to minimize negative impacts of sulfur oxides (SO_x) in tailpipe emissions from automotive vehicles. Sulfur reduction in gasoline is driven by two important factors. First, there is a major need to effect reduction of sulfur oxides generated during the combustion of high-sulfur-containing fuels. Second, reducing sulfur concentrations in gasoline minimizes the rate of sulfur poisoning of catalytic converter catalysts in automobiles. A catalytic converter is a catalytic technology that is installed in automotive exhaust systems with the goal of minimizing tailpipe emissions. Unreacted hydrocarbons or particulates, volatile organic compounds (VOCs), carbon monoxide (CO), nitrogen oxides, and sulfur oxides are emitted from automotive vehicles. Catalytic converters are installed to convert harmful gases and hydrocarbons in the exhaust gas. As is the case with catalytic processes, catalyst life in catalytic converters is shortened due to the deposition and poisoning effects of sulfur. As the concentrations of sulfur contaminants increase on the catalytic converter catalyst, the activity and efficacy of the catalyst are degraded, leading to activity losses. When the catalyst in the converter is spent, untreated tailpipe contaminants would then be continuously released into the environment. Since the rate of catalyst deactivation is a function of the concentration of sulfur oxides in the exhaust gases, minimizing its concentration via lowering the sulfur concentrations in gasoline greatly extends the effective life of catalysts in catalytic converters.

2.4.2 Diesel Fuels

Diesel fuels are used for on-road transportation; off-road uses such as in farming construction and logging; and in electric power generation, marine shipping, and military transportation.

In the United States, on-road transportation accounted for nearly 60% of the diesel fuel used.[35] Due to the high consumption of diesel in the transportation sector and high tailpipe emissions of harmful SO_x and NO_x gases, there is a need to minimize the harmful emissions. There have been progressive developments with respect to sulfur limits in countries, as shown in Table 2.18. European Union EU-27 countries started with a low 50 wppm limit for sulfur in diesel fuels in 2005, reduced the diesel sulfur to 10 wppm in 2010, and operate at that level beyond 2020. China began with making drastic sulfur reductions in diesel from a high of 2000 wppm in 2005 to 350 wppm in 2011 and 50 wppm by 2015, and plan to have a diesel sulfur limit of 10 wppm by 2018. Other countries with low sulfur diesel limits are Japan and the United States at 10 wppm and 15 wppm, respectively. As an example of diesel sulfur limits for some African countries, South Africa with diesel sulfur in the 50–500 wppm range in 2010 plans to remain in that sulfur range and then drastically reduce its national diesel sulfur concentration to 10 wppm by 2020. Diesel supply and demand and prices of diesels containing greater than 15 wppm sulfur would most likely influence accelerating plans to reduce diesel sulfur to less than 10 wppm by 2020 to facilitate imports of diesel fuel.

TABLE 2.18
Sulfur Limits for Diesel Fuels for Selected Countries

Country	2005	2010	2015	2020
Brazil	3500	500–1800	500	500
China	2000	500–1800	50	10
EU-27	50	10	10	10
India	500	350	350	350
Japan	50	10	10	10
Russia	500	500	50	10
Thailand	150	150	50	50
United States	500	15	15	15
South Africa	3000	50–500	50–500	10

Source: Global Comparisons of Gasoline Prices, May 21, 2018. www.globalpetrolprices.com/gasoline_prices/.[38]

TABLE 2.19
Parameters of Selected Global Diesels

Fuel Parameter		China V	Bharat IV (1)	Euro IV	Euro V	Conv	Ref(2)	WFC(3)
Poly Aromatics, max	vol. %	11.0	11.0	11.0	8.0	NS	1.4	2.0
Sulfur, max	wppm	10.0	50.0	50.0	10.0	15.0	15.0	10.0
Cetane Number, min		51.0	51.0	51.0	51.0	(4)	48.0	55.0
Density @ 15C	Kg/m^3	800–850	800–845	845.0	845.0	NS	NS	820.0
Flash Pt., min	Deg. C	5	NS	55.0	55.0	NS	54.0	55.0
Ash Content, max	m/m	0.01	NS	0.01	0.00	NS	NS	0.001

Source: Global Comparisons of Gasoline Prices, May 21, 2018. www.globalpetrolprices.com/gasoline_prices/.[38]

Notes:
1. Diesel standard for India via Bharat.
2. Reference diesel (California diesel).
3. WFC is World Fuel Charter category, and it is applicable to markets requiring Euro4. Euro 5 heavy.
4. Greater than 40 Cetane number or less than 35 volume percent aromatics.
5. The flashpoint specifications are specified for different pour points (PP) that are in degrees C as well. Flash Points are 55°C, 50°C, and 45°C, respectively.
6. The worldwide fuels charter category diesel suggests a very low poly aromatic concentration. Limit of 2.0 volume percent and a high cetane number of 55. www.transportPolicy.net, Global Comparisons of Fuels.[38]

Diesel fuel specifications are provided in Table 2.19 for select countries and diesel markets.

A number of countries are making plans to limit diesel sulfur to 10 ppm to reduce tailpipe SO_x emissions. Countries such as the United States also mandate the use of biodiesel from renewable resources as a blend component in diesel. Biodiesels are defined as mono-alkyl esters of long-chain fatty acids derived from vegetable oils or animal fats that meet the ASTM D6751 specification for use in diesel engines. In the United States, it is permissible to blend up to 20 volume percent biodiesels into petroleum diesel. Biodiesel is designated as B100 to signify that it is 100 percent biodiesel. If 15% of the biodiesel is blended into 85% petroleum diesel, the resultant fuel is designated as B15 diesel. Current surveys show significant use of biodiesels containing 5–10 volume percent biodiesel blends or B5 to B10 diesels in North America, South America, and some of the EU-27 countries. Concentrations of biodiesel in biodiesel blends are projected to increase to 20 volume percent in North America and to an average maximum of 5 volume percent for some South American countries by 2020.[39]

2.4.3 Gasoline Fuels

Sulfur limits for gasolines of selected countries are as listed in Table 2.20. They show the progressions in great initiatives by countries to reduce the concentrations of sulfur

TABLE 2.20
Gasoline Sulfur Limits for Selected Countries

Country	2005	2010	2015	2020
Brazil	1000	1000	50	50
China	500	150	50	10
EU-27	50	10	10	10
India	500	150	150	150
Japan	50	10	10	10
Russia	500	500	50	10
Thailand	150	150	50	50
United States	30/90/300 (1)	30/80 (2)	30	10 (3)
South Africa	1000	500	500	10

Source: U.S. Department of Transportation, Transportation and Climate Change.[37]

Notes:

1. Some of the states in the United States permitted sulfur limits that were higher than the 30 wppm sulfur for California, and those of the other states of the country were at the higher concentration levels in 2005.
2. The stipulation of the Tier 2 Gasoline Sulfur program is for average refinery gasoline sulfur of 30 wppm with an 80 wppm cap assured meeting the nitrogen oxides, NO_x, emission specification, and that was binding after 2009 in the United States.
3. United States Tier 3 Gasoline Sulfur program dropped average gasoline sulfur to 10 wppm in 2017.

in gasolines. Brazil, as an example in South America, reduced its gasoline sulfur limit from a high of 1000 wppm in 2005 to 50 wppm in 2010 and 50 wppm through 2020. The European Union EU-27 countries and the United States continue their leadership roles with respect to reducing tailpipe emissions and overall environmental stewardship. EU-27 countries established gasoline sulfur limits of 10 wppm as early as 2010. The United States established its Tier 2 Gasoline Sulfur program in 2007, and that led to 90% reduction in refinery average gasoline sulfur to the 30 wppm limit. Its Tier 3 Gasoline Sulfur program limiting refinery average gasoline sulfur to 10 wppm was enforced in 2017. South Africa, which was chosen to represent gasoline sulfur regulatory activities in a broad sense in Africa, is operating with a gasoline sulfur limit of 500 wppm and plans to move to a 10 wppm sulfur limit by 2020.

A selected list of parameters for various global grades of gasolines from representative countries is provided in Table 2.21. Current gasoline sulfurs as of 2016 were taken from projections for the countries for 2015 and 2020. It is assumed that the selected countries are implementing their gasoline sulfur limits as advertised. Research and motor octane numbers have a wide range as expected due to the fact that a number of countries have several grades of gasoline. In the United States, gasolines can be sold as regular, premium, supreme, and reformulated, with different octanes

TABLE 2.21
Parameters for Selected Gasoline for Global Markets

Fuel Parameter		China V	Bharat IV (1)	Euro IV	Euro V	EPA RFG (2)	EPA Conv (3)	WFC (4)
RON, min	(5)	81/92/95	91.0	91–95	91–95	NA	NA	91/95/98
MON, min	(6)	NS	81.0	81–85	81–85	NS	NS	82.5/85/88
Sulfur,	wppm	50		10	10	10		10
AKI	(7)	84/87/91	NS	NS	NS	87/87/91	87/87/91	NS
Aromatics (8)	vol.%	40.0	35.0	35.0	35.0	20.7/19.5	27.7/24.7	35.0
Olefin (9)	vol.%	25.0	21.0	18.0	18.0	11.9/11.2	12.0/11.6	10.0
Benzene (10)	vol.%	1.0	1.0	1.0	1.0	0.66/0.66	1.21/1.15	1.0
Lead	mg/l	5.0	5.0	5.0	5.0			NS
Density	Kg/m3	NS	720–775	NS	720–775	NS	NS	715–770
RVP (11)	KPa	(12)	60.00	60–70	60–70	47.6/82.0	57.2/83.6	(13)

Notes:

1. Bharat IV is the set of specifications for gasoline in India.
2. United States EPA reformulated gasoline.
3. United States conventional gasoline.
4. World Fuels Charter category.
5. RON is research octane number.
6. MON is motor octane number.
7. AKI is antiknocking index.
8. The aromatic concentrations vary seasonally for US EPA RFG and EPA Conv. The first number is for the summer months and the second number is for the winter season.

9–11. Olefin and benzene concentrations also vary seasonally and the RVP, Reid vapor pressure, limit varies seasonally in the US.

12. The limits of the Reid vapor pressures for the summer and winter month for the China V gasolines are 40–65 and 45–95 kilopascals, respectively.
13. For WFC, the RVP specifications vary with the temperature of the atmosphere. For temperatures that are greater than 15°C, RVP is 45–60.
    For temperatures that are less than 15°C and greater than 5°C, RVP is 65–80.
    For temperatures that are greater than –15°C and less than –5°C, RVP is 75–90.
    For temperatures less than –15°C, RVP is 85–105.

that are referred to as road octane numbers. The road octane number is defined as the average of the research and motor octane numbers. Hence, there are no specific standards that are comparable to those for the other global markets in Table 2.21.

The range of research octane number (RON) suggested by the World Fuels Charter (WFC) category for gasoline is 91–98. The range covers those global gasoline octanes with the exception of the minimum 81 RON of China V gasoline specification. Aromatic and olefin concentrations in global gasolines vary considerably from those of the United States. While the aromatics concentrations of the other countries are in the 35–40 volume percent range, those of the United States are in the 19.5–27.7 volume percent range, and that range varies for summer and winter months. Olefin compound concentration limits in United States gasoline grades are about 45%–62%

of those of the other global gasoline grades. Lead limits were reported as 5 mg per liter for global gasoline grades, and the US uses only unleaded gasoline. As shown in Table 2.21 and in note number 13, the WFC category gasoline Reid vapor pressure varies widely to accommodate a broad range of climactic zones and minimize fugitive volatile organic compounds during fueling and combustion of gasolines.

Most of the gasoline limits in Table 2.21 were established to reduce tailpipe emissions from automotive vehicles that are harmful to the atmosphere, humans, and animals and lead to drastic global climate change. One approach for reducing GHG emissions that has been alluded to earlier in this book is the need in gasoline to pursue a similar biodiesel fuel blend strategy by replacing higher percentages of the fossil fuel contribution in gasoline with blends from renewable resources. Oxygenate compounds from renewable resources are added to gasoline. The added oxygenates enhance the combustion of gasoline fuel and minimize tailpipe emission of carbon monoxide and particulate matters. In some countries, oxygenates are added solely for their octane enhancement contributions in gasoline. Alcohols such as methanol, ethanol, isopropyl alcohol, normal butanol, and tertiary butanol and ethers are used. Methyl tertiary butyl ether (MTBE), tertiary amyl butyl ether (TAME), and ethyl tertiary butyl ether (ETBE) are most of the ethers used in gasoline.[40] In the United States, ethanol is the mandated oxygenate, and it is used mostly as a 10% blend in gasoline. The 10% ethanol in gasoline blend is known as E10 and the 85% ethanol gasoline blend as E85.

Concentration limits for US EPA gasoline aromatics and olefins are low relative to those of the other gasoline aromatic and olefin specifications in Table 2.21. To meet future gasoline specifications with respect to aromatic and olefin concentrations limits, oil refiners may have to rely on greater use of isomerate and alkylate, which are aromatic-free blending components. For oil refiners that do not have appropriate processing units to produce low aromatic and olefin blend components, they would likely have to purchase appropriate gasoline blending components. The use of significant volumes of purchased gasoline blend components to meet gasoline specifications could negatively impact some of the oil refiners' profitability.

In addition, since most of the countries would have to meet diesel sulfur limits of 10 wppm or 15 wppm, as shown in Table 2.21, oil refiners would have to ensure that they have significant hydrotreating and hydrocracking capacities. MARPOL, the International Convention for the Prevention of Pollution from ships, is also committed to minimizing marine pollution from ships. In 1997, MARPOL adopted Annex VI, which established limits on nitrogen oxides, sulfur oxides, and particulate matter emission and stipulated the use of fuels with low sulfur content so as to protect the health of people and the environment by reducing ozone-producing pollution, which can cause smog and asthma.[41] Some of the changes in MARPOL Annex VI that came into effect in 2015 include the reduction of sulfur in marine fuels from 1.0 wt. % to the 0.1 wt. % limit for ocean-going ships that operate in designated emission control areas (ECAs).[42] The emission control areas are defined for ships that operate within 200 miles of the shores of North America and the Baltic and North Seas, and the sulfur limit of their fuels is set at a maximum of 0.1 wt. %.

In the next chapter, process and distillation units are briefly introduced to provide a common basis for understanding the oil refinery, principal process objectives, and products generated from the units. The overview should enhance an understanding of the pivotal role of the catalytic naphtha reforming unit in the production of high-octane reformate; benzene, toluene, and xylene; and hydrogen. Butane isomerization, light naphtha isomerization, alkylation, and hydrocracking units are reviewed in terms of their continued importance in meeting gasoline quality regulations that may require fewer aromatics in gasolines. It should also be noted that, as indicated earlier, within the biodiesel limits, biodiesel blends can be generated to produce low-aromatic, ultralow-sulfur diesel fuels. Similarly, oxygenates such as alcohols and ethers at appropriate concentrations based on regulatory limits could be blended into gasolines to meet lower sulfur, aromatic, and olefin limits.

REFERENCES

1. Marcogliese, R., Refining 101: Refinery Basic Operations. Refining 201: Coking Technologies and Applications. Presentation to Lehman Brothers, Jan 16, 2007 media. corporate-ir.net/media_files/NYS/VLO/presentations/Lehman_011007.pdf.
2. Speight, J. G., *The Chemistry and Technology of Petroleum*, 2nd edition, Marcel Dekker, New York, 1991.
3. Gary, J. H., Handwerk, G. E., *Petroleum Refining—Technology and Economics*, Marcel Dekker, New York, 1975.
4. Wilhelm, S. M., Liang, L., Cussen, D., Kirchgessner, D. A., Mercury in Crude Oil Processed in the United States (2004), Private notes.
5. Hase, B., Vickery, V., Radford, R., Mercury in Hydrocarbon Process Streams; Sampling/Analysis Methods, Exposure Monitoring, Equipment Decontamination, Waste Minimization, Paper Am-12-22, *AFPM 12 March 2012 Meeting*, San Diego, California.
6. Satterfield, C. N., *Heterogeneous Catalysis in Practice*, McGraw-Hill Book Company, New York, 1980.
7. Oyekan, S. O., Voss, K. E., Desulfurization Technologies for Control of Atmospheric Emission of Sulfur Compounds in Brazil, Oyekan's Private Notes, Engelhard Corporation, Edison, New Jersey, 1984.
8. Schuman S. C., Shalit, H., Hydrodesulfurization, *Catal. Rev. Sci. & Engr.*, 4, 245–318, 1971.
9. Mullins, O. C., et al., Editors, *Asphaltenes, Heavy Oils and Petroleomics*, Springer, New York, 2007.
10. Podgorski, D. C., et al., Heavy Petroleum Compounds. 5. Compositional and Structural Continuum of Petroleum Revealed, *Energy and Fuels*, 27(3), 1268–1276.
11. Islam, M. R., Potential of Ultrasonic Generators for Use in Oil Wells and Heavy Oil/Crude Oil/Bitumen in Transportation Facilities in *Asphaltenes: Fundamentals and Applications* by Subrana, M., Sheu, E. Y., Springer Science and Business Media, Berlin, Germany.
12. Asomaning, S., Heat exchanger fouling by petroleum asphaltenes, *Ph.D. Thesis*, University of British Columbia, 1997.
13. Jechura, J., Petroleum Refining Overview, Colorado School of Mines, January 2016.
14. Natural Resources Canada, Government of Canada, www.nrcan.gc.ca/energy/crude-petroleum/4561.
15. US Energy Information Administration, www.eia.gov/todayinenergy/index.php?tg=crack spread.
16. US Energy Information Administration, Low Crude Oil Prices, Increased Gasoline Demand Lead to High Refiner Margins, May 15, 2015, www.eia.gov/todayinenergy/detail.php?id=21312.
17. Jones, J. J., Design-Build Method Useful for Refinery Project Execution, NPRA Presentation, *Oil & Gas Journal*, 106(18), May 12, 2008.

18. Cross, P., Desrochers, P., Shimizu, H., Canadian Fuels, The Economics of Petroleum Refining, Understanding the Business of Processing Crude Oil into Fuels and Other Value Added Products, December, 2013, http://www.canadianfuels.ca/website/media/PDF/Publications/Economics-fundamentals-of-Refining-December-2013-Final-English.pdf
19. Johnston, D., Refinery Report: Complexity Index Indicates Refinery Complexity and Value, *Oil Gas J.*, 94(12), March, 1996.
20. Jechura, J., Colorado School of Mines, Petroleum Refining Overview, inside.mines.edu/~jjechura/Refining/01_Introduction.pdf
21. US EIA Petroleum Refineries Vary by Level of Complexity, October, 2012, www.eia.gov/todayinenergy/index.php?tg=refining
22. Fabusuyi, T., Downstream Sector: Crude Oil Refining, http://docplayer.net/37422266-Downstream-sector-crude-oil refining.html
23. US EIA, International Energy Outlook 2016, Chapter 2. Petroleum and Other Liquids, May 11, 2016, www.eia.gov/outlooks/aeo/pdf/0383(2016).pdf
24. Ramana, U. V., Presentation on Characteristics and Refining Products, *Workshop on Refining and Petrochemicals*, New Delhi, India, August 25–28, 2010.
25. Brelsford, R., Kootungal, L., Asia Pacific Refining Primed for Capacity Growth, *Oil Gas J.*, 112(12), 34–45, December 1, 2014.
26. Duddu, P., The 10 Biggest Oil Refining Countries. December 2013, www.hydrocarbons-technology.com/features/featurethe-10-biggest-oil refining-countries/
27. US EIA Report, Top 16 Oil Refineries, www.eia.gov/energyexplained/index.php?page=oil_refiningwww.eia.gov/beta/international/
28. Hansen, L. R., OLR Backgrounder: Diesel vs Gasoline Prices, Oil Research Report, 2013-R-007, January 28, 2013, www.cga.ct.gov/2013/rpt/2013-R-0007.htm.
29. US EIA, Diesel and Gasoline Update, August 2016, www.eia.gov/petroleum/gasdiesel/
30. GlobalPetrolPrices.com, Understanding the Price Gap between Gasoline and Diesel Fuels, www.globalpetrolprices.com/articles/4/
31. GlobalPetrolPrices.com, Gasoline and Diesel Price Outlook for Europe, August 2016, www.globalpetrolprices.com/articles/51/
32. U.S. EPA, Climate Change: Basic Information, https://19january2017snapshot.epa.gov/climatechange/climate-change-basic-information_.html
33. U.S. EPA, Global Greenhouse Emission Data, Updated August, 2016, https://19january2017snapshot.epa.gov/ghgemissions/global-greenhouse-gas-emissions-data_.html
34. Delio, I., Warner, K. D., Wood, P., *Care for Creation*, Franciscan Media, Cincinnati, Ohio.
35. Chevron, Global Marketing, Diesel Fuels Technical Review, www.chevron.com/-/media/chevron/operations/documents/diesel-fuel-tech-review.pdf
36. United Nations Framework Convention on Climate Change. UNFCCC, Retrieved July 23, 2016, https://unfccc.int/resource/docs/publications/handbook.pdf
37. U.S. Department of Transportation, Transportation and Climate Change, www.transportation.gov/climate-change-clearinghouse
38. Global Comparisons of Gasoline Prices, May 21, 2018, www.globalpetrolprices.com/gasoline_prices/
39. Dixson-Decleve, S., World Wide Automotive Fuel Quality Developments, IFQC moderated Panel Session, SAE Toulouse. June, 2004.
40. Leisenring, R. G., Jr., Technology' Low Sulfur Diesel Controversy continues, *Oil & Gas Journal*, 94(17), April 22, 1996
41. Smith, W., Bandemehr, A., Overview of US Commitments Under MARPOL Annex VI, US EPA, April 16, 2010, Veracruz, Mexico, www.epa.gov/sites/production/files/2014-05/documents/annexv.pdf

42. International Maritime Organization. Prevention of Air Pollution from Ships, http://www.imo.org/en/OurWork/Environment/PollutionPrevention/AirPollution/Pages/Air-Pollution.aspx
43. Oyekan, S. O., Catalytic Applications for Enhanced Production of Transportation Fuels, 2009 NOBCChE Percy L. Julian Award Lecture, April 14, 2009, St. Louis, MO.
44. Bechtel Corporation, Jamnagar Oil Refinery, http://www.bechtel.com/projects/jamnagar-oil-refinery/

3 Overview of Oil Refining Process Units

3.0 INTRODUCTION

Brief introductory overviews are provided for crude and vacuum distillation units and key oil refining process units. The introductory reviews are appropriate, as they aid in reinforcing the pivotal and critical position of catalytic reforming units in oil refineries. Naphtha feed type and quality supplied by crude and other downstream processing units have significant impacts on the catalytic performance and productivity of catalytic reformers. Catalytic reforming process units' net hydrogen is used as the coreactant in downstream hydroprocessing units. The quality of the net hydrogen and effectiveness of the chloride guard bed treatment could negatively impact the reliability of downstream piping, equipment, and hydroprocessing units. Additionally, due to the high sensitivity of the platinum-containing reforming catalysts to poisoning by a variety of heteroatoms, catalytic performances of catalytic reformers are typically highly negatively impacted by a short time of contact with catalytic poisons. Due to this timely response to naphtha feed contaminants, catalytic reformers play critical roles in providing operating and technical personnel with timely information for the proper management of their naphtha hydrotreater operations and crude slate compositions. In top-tier, high-performing oil refining companies, immediate effective timely corrective actions can be taken either at the crude distillation units or naphtha hydrotreaters based on naphtha feed quality alerts from catalytic reforming units. Timely on-stream data from the plant information management system (PIMS) and laboratory information management systems (LIMS) and continuous monitoring of crude distillation and the other refinery process unit operations should enable the oil refiner to determine if high heteroatom contaminants are being introduced by a crude oil within the crude slate and make necessary operational adjustments. In addition, if the naphtha quality and contaminants from a downstream unit such as either a delayed coker or hydrocracker have changed, catalytic reforming units can provide timely indications as well so as to effect necessary operational changes at the catalytic reformer, delayed coking, and hydrocracker. Based on the listed "sensing" nature and position of catalytic reforming units, we can conveniently label them as the "gate-keeping" process units of oil refineries.

3.1 CRUDE DISTILLATION UNIT

The total capacity of the crude distillation units within a refinery is used as the capacity of an oil refinery. As usual with processing units, the capacity of a crude unit is defined as the volume of crude oil that can be processed through it. Crude distillation units are also variously referred to as crude distillation towers, atmospheric distillation units, atmospheric distillation towers, and often simply as crude units.

Before delving into a brief overview of oil refining units, it may be a useful time to review some of the major factors that should be considered during the feasibility and preliminary planning stages for grassroots refineries. Based on the proposed crude oil distillation capacity of the refinery, it may be prudent in the feasibility stage to determine if the refinery should have one to three trains of processing unit type configuration. That is, at the grassroots or revamp stages of the oil refinery, should plans be formulated for two or three crude and vacuum distillation units and associated processing units? Decisions on multitrain processing units should be based on the nameplate capacity, location with respect to available engineering services, and expansion programs of the refinery in the future. Multitrain refineries provide flexibility in refining a wide variety of crude oils and selective planning for turnaround maintenance of crude and vacuum distillation and other process units. In the operation of multitrain refineries, oil refiners do not need to effect a complete shutdown of the refinery due to shutdown of one of the crude units. Turnaround maintenance of a crude distillation unit can be conducted without taking a shutdown of the entire refinery if the refinery configuration has a multitrain design concept of providing for at least two crude distillation units and downstream associated processing units. However, multitrain refining plants are feasible if the oil refiner has the necessary capital to incorporate multiple crude units during the detailed design and construction phases of the grassroots refinery. With multitrain configuration and operations, the oil refiner can continue to produce significant amounts of oil-refined products while one of the crude oil distillation towers or some of the downstream processing units are in shutdown mode for turnaround maintenance or routine repairs of equipment. To fully realize the benefits of several-train complex refineries, the oil refiner should have an adequately sized tank farm with a number of tanks to handle a variety of crude oils of varying qualities. An appropriately sized tank farm will provide flexibility for the oil refiner's crude oil and oil fraction acquisition group so as to enable that group to minimize crude oil and other feedstock acquisition costs and possibly contribute to the overall profitability of the refinery. In tank farms, the tanks should be inspected and cleaned out as often as required to ensure utilization of most of the crude oils and minimize loss of significant oil fractions in extensively contaminated sludge. The oil sludge could be treated, recycled, and used as needed based on the capability of the oil refiner to reprocess liquid from the sludge in a reliable manner.

The first major and essential unit of an oil refinery is the atmospheric distillation unit, and this is covered in the next section. A crude oil distillation unit is utilized in the fractionation of a crude oil or a mix of crude oils to oil fractions according to the boiling ranges of the hydrocarbon compounds in crude oils. There are typically thousands of hydrocarbon compounds within an oil fraction with boiling points that are within the boiling range of that crude oil fraction. Oil fractions and gas products are as depicted in the simplified crude unit diagram in Figure 3.1. The designated oil fractions are taken from a figure and data provided in a US Energy Information Administration report.(1)

Crude oil as delivered to a refinery usually contains appreciable amount of inorganic salts, basic sediments, and some organic chlorides. Most of the inorganic

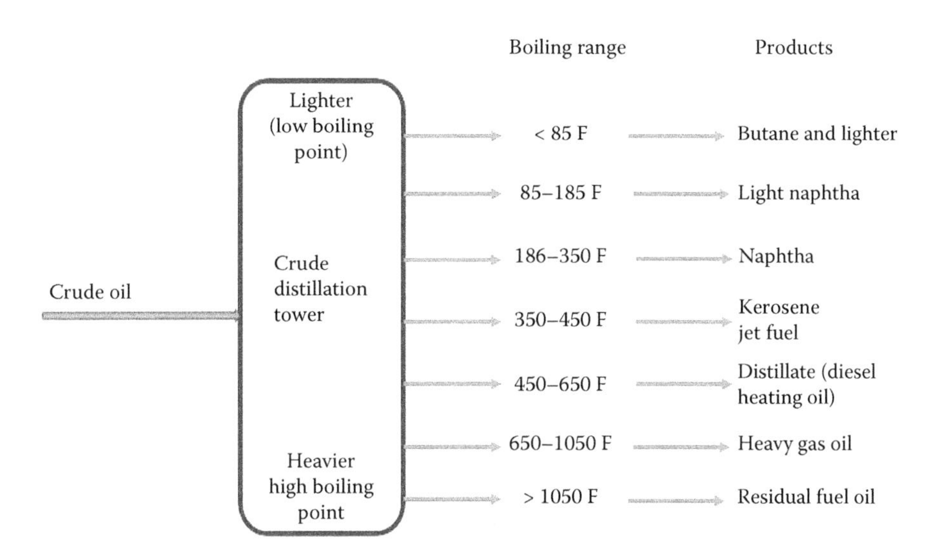

FIGURE 3.1 Products from a crude distillation tower adapted from US EIA report. From US EIA, Today in Energy, Crude Oil Distillation and the Defining of Refinery Capacity, July 2012, Accessed Sept 13. 2016. www.eia.gov/todayinenergy/detail.php?id=6970.[1]

salts are chlorides of sodium, calcium, and magnesium. It is absolutely necessary to effect removal of most of the inorganic salts and sediments from the crude oil mix in a desalter before heating the desalted oil to elevated temperatures in the range of 720–760 F and feeding the oil to the atmospheric crude tower. Removal of salts and sediments from the crude oils ensures that the crude towers as well as downstream piping and equipment are protected from acidic gases and acids that could corrode them.

In order to maximize removal of inorganic salts and basic sediments, the desalter should be operated in a temperature range of about 300–325 F, water should be added in the range of 6–8 volume percent of the crude oil, and emulsion breaking chemicals used to facilitate inorganic salt and sediment removal.[5] A mix of oil, water, and chemicals is heated to the desired desalter temperatures by exchanging heat with some of the hot distilled fractions and other streams from the crude tower. After the desalter, the crude oil mix is heated further via exchange of heat with some of the distilled fractions and other streams, as shown in Figure 3.2.[78] It is then heated in a fired heater to a temperature between 720 and 760 F and the heated crude oil blend sent to the bottom of the atmospheric tower. Cooling and condensing of the crude distillation column overhead is provided partially by exchanging heat with the incoming crude oil and partially by an air-cooled or water-cooled condenser. A pumparound system also aids in removing additional heat from the distillation tower.

Crude distillation tower products taken as sidecuts between the top and bottom of the tower include naphtha, distillate, light and heavy atmospheric gas oils, and residual oil. Gas is also produced and exits from the top of the crude oil overhead condenser.[2–4] The oil fractions can be fed directly to the appropriate downstream processing units or sent to intermediate tankage before processing as desired by

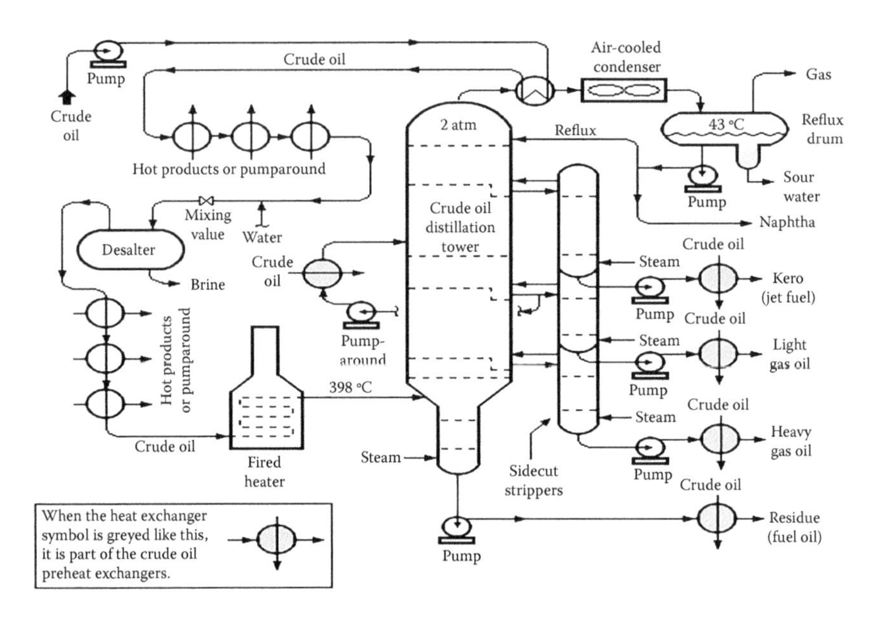

FIGURE 3.2 A schematic diagram of an atmospheric unit.
From Maverick Engineering, Inc., Industrial: Refining: Atmospheric Distillation, Accessed August 12, 2016. http://www.maveng.com/index.php/business-streams/industrial/refining/atmospheric-distillation.[78]

the oil refiner. The oil fractions from the crude distillation tower are processed in appropriate processing units that are downstream of the atmospheric distillation unit in the refinery.[10]

One of the key challenges in crude oil distillation is maintenance of the reliability of the operations in the crude tower and associated crude tower equipment. A good chemical treatment program is maintained to minimize corrosion of relevant equipment due to small residual salts and basic sediments and water (BSW) that usually remain, as the salt and BSW removal efficiency in the desalter are about 95% and 85%, respectively.[5] Another major challenge encountered in the distillation of blends of crude oils in crude distillation units is the precipitation of asphaltenes in desalters, tower bottoms, and heat exchangers.[6,7] Mason reported that while the concentration of asphaltenes in very low density, sweet crude oil is about a fraction of 1 wt. %, it is about 20 wt. % in dense bitumen oils.[6] The deposition of precipitated asphaltenes usually leads to fouling of heater tubes, a lower rate of productivity, and extensive maintenance of affected equipment. A most recent development in the last 5 years has been the degree of asphaltene precipitation and negative productivity impact that are experienced in oil refineries due to the distilling of blends of shale oil together with some of the conventional crude oils in crude distillation towers. It is fairly well known that shale oils are light, highly paraffinic unconventional oils and, though cost effective for oil refiners, may pose significant challenges in the coprocessing of shale and other crude oils due to extensive precipitation of asphaltenes that is induced by the more paraffinic shale oil in the crude oil blends.[8,9]

FIGURE 3.3 Atmospheric crude towers.
From Discover Petroleum, Vacuum Distillation, resources.schoolscience.co.uk/SPE/knowl/4/2index.htm?vacuum.html.[12]

The crude distillation unit without a vacuum distillation and any downstream processing units is essentially a topping refinery configuration, as discussed in Section 2.3.1. An atmospheric crude unit is shown in Figure 3.3.[12]

3.2 VACUUM DISTILLATION UNIT

In the crude oil distillation tower, the fractionations have to be conducted at about one atmosphere pressure and at a temperature of about 750 F, as discussed in the preceding section. It is known that further attempts to conduct distillations in the atmospheric distillation tower at temperatures greater than 750 F would lead to significant thermal cracking and decomposition of heavy hydrocarbons to produce coke and gas products. Such inefficient separation of the atmospheric residual would severely devalue the atmospheric residual oil and lead to coke deposition and plugging of heater tubes in the furnace and in the piping between the fired heater and vacuum distillation tower. To effect further distillation in order to yield higher-value products relative to atmospheric residual oil, it is necessary to conduct the distillation at drastically reduced pressures in the range of 25–40 millimeters of mercury in the flash zone and at temperatures between 700 F and 720 F. By decreasing the pressure in the separation column, the boiling points of compounds are substantially reduced and permit separation of more useful oil fractions without degrading the atmospheric residual oil. The distillation of the atmospheric residual oil at low pressure is referred to as the vacuum distillation process. The lower the pressures that can be achieved, the more efficient the distillation of the atmospheric residual oil to produce vacuum light gas oil, vacuum heavy gas oil, and vacuum residual oil. The vacuum pressure can be

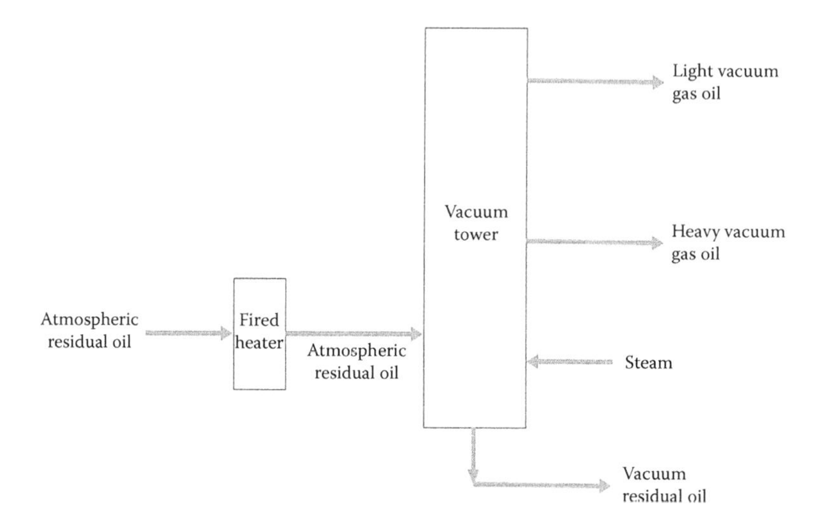

FIGURE 3.4 Simplified drawing of the vacuum distillation unit.

FIGURE 3.5 A vacuum distillation column in a petroleum refinery.
From Discover Petroleum, Vacuum Distillation, resources.schoolscience.co.uk/SPE/knowl/4/2index.htm?vacuum.html.[12]

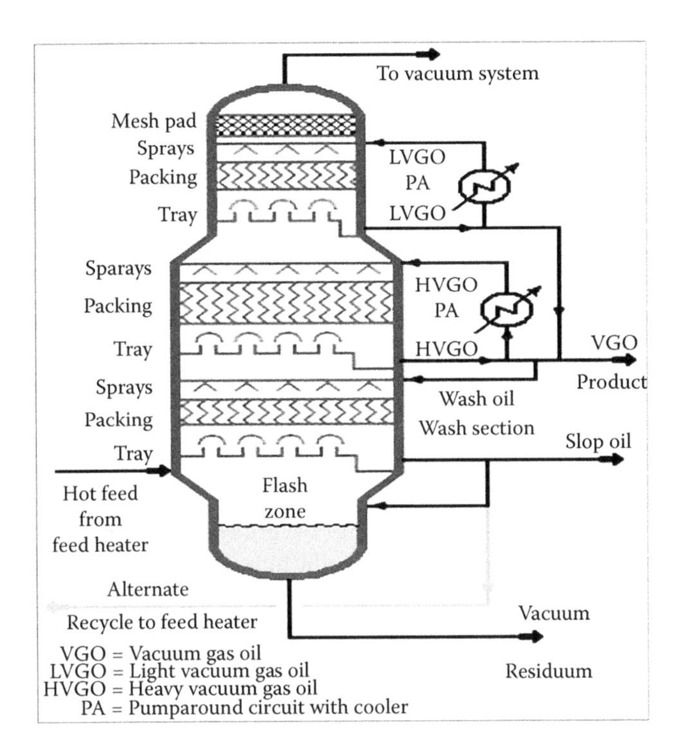

FIGURE 3.6 Internals of a vacuum distillation tower and heat exchanger system.
From Discover Petroleum, Vacuum Distillation, resources.schoolscience.co.uk/SPE/knowl/4/2index.htm?vacuum.html.[12]

further lowered to about 10 mm of mercury by adding steam to the furnace inlet and at the bottom of the vacuum tower. This leads to increased vaporization of the feed to the vacuum distillation unit and more effective separation.[11]

A simplified drawing of a vacuum distillation unit is given in Figure 3.4, showing the vacuum-fired heater and the main fractionation cuts, light vacuum gas oil (LVGO), heavy vacuum gas oil (HVGO), and vacuum residual oil. Usually, vacuum gas oils are used as feed to the fluid catalytic cracker and hydrocracker (HC) units in order to produce higher volumes of gasoline and diesel from a barrel of crude oil and especially for crude slates processed through the refinery that contain a higher percentage of heavy crude oil.[14] A picture of a vacuum distillation unit is provided as Figure 3.5.

The internals of the vacuum distillation tower showing the trays, sprays, and packings as well as the associated heat exchanger system are shown in Figure 3.6. In order to operate reliably at the low vacuum pressures of 10 mm Hg for maximizing yields of vacuum gas oils, key items associated with successful operation of the ejector system require that a number of items be monitored. The list includes the atmospheric tower overflash, atmospheric stripper performance, vacuum tower top temperature, heat balance and ejector internal erosion, and product buildup.

Global oil refining process unit capacity data assembled by Joules Burn in 2010 by region are listed in Table 3.1.[79] The 2010 data showed that vacuum distillation capacity was 54% relative to that of the crude distillation capacity for United States

TABLE 3.1
Global Oil Refining Process Unit Data in Millions of Barrels per Day

	USA	Europe	China	Japan	India	Korea
Crude Distillation	17.87	13.80	9.46	4.73	4.00	2.72
Vacuum Distillation	9.68	5.65	0.50	1.76	0.81	0.48
Catalytic Cracking	5.71	2.25	1.56	0.98	0.50	0.21
Catalytic Reforming	3.54	1.16	0.59	0.83	0.05	0.27
Coking	2.47	0.36	1.14	0.12	0.17	0.02
Hydrocracking	1.67	1.21	1.32	0.18	0.17	0.31

Source: Burns, J., Global Refining Capacity, Supply and Productions. *The Oil Drum*. Accessed September 13, 2016.[79]

oil refineries. On the same basis, relative to the crude oil distillation capacities, the vacuum distillation capacities for Europe and China were 40% and 5%, respectively. The relative vacuum distillation capacities for Japan, Korea, and India also showed lower vacuum distillation unit utilization that was low and similar to that of China. The difference in utilization is possibly due to the type of crude oils used and to the demand for atmospheric residual oil produced in the refineries in China, Japan, Korea, and India.

Over 80% of the refineries in the United States have vacuum distillation units, and in the past 30 years, reliable operations and increased profitability of vacuum distillation units have been achieved due to better understanding of how the vacuum distillation unit works. As indicated previously, vacuum distillation units can operate at pressures as low as 10 mm Hg. In a 1994 paper, Martin, Lines, and Golden showed that a benefit of $2 million per year could be realized for a 0.75 volume percent gain in vacuum gas oil for a 150 MBPD oil plant refining WTI crude oil and operating reliably at vacuum pressures below 10 mmHg.[13]

3.3 BUTANE ISOMERIZATION UNIT

Butane is produced with other gases in an oil refinery from atmospheric distillation units and several downstream catalytic cracking, hydrotreating, hydrocracking, coking, paraffin isomerization, and catalytic reforming units. The listed refinery process units that produce butane are covered in subsequent sections. Butane can be disposed of by selling the gas either as butane gas or in combination with propane as in liquefied petroleum gas. Butane is useful as an excellent gasoline blend component in winter months in most of the Asian, North American, and European countries when the mandated vapor pressures of gasolines are higher than those for gasolines in the summer months. With the current drive to lower the concentrations of aromatic compounds in gasoline, alkylates produced via the reactions of butenes and pentenes with isobutane in alkylation units are premium high-octane gasoline blend components. Isobutane can be sold to petrochemical manufacturers and is more often used as feed in the production of isobutylene.

Isobutylene can then be reacted with methanol to produce methyl tertiary butyl ether. MTBE is used as an oxygenate and octane blend component for gasoline and was a recommended oxygenate additive for gasoline before being replaced with ethanol in the United States due to some concerns with possible contamination of groundwater.

Chloride-promoted alumina catalysts that have hydrogenation/dehydrogenation and moderate acidic isomerization functionalities are used in the conversion of butane to isobutane. For most low-temperature paraffin isomerization catalysts, platinum concentrations are in the range of 0.12–0.25 wt. %, chloride is 4.5–5.5 wt. %, and the balance is alumina. Chloride addition to the feed is required in order to maintain catalyst acidity, performance, and stability. Key catalyst contaminants are water, sulfur, olefins, fluorides, oxygenates, carbon oxides, and nitrogen, and the contaminants usually exhibit different modes of selective poisoning of the hydrogenation and acidic functionalities of the catalyst. Effective dryer operations and adsorbent maintenance programs in addition to consistent, continuous monitoring of process unit key performance indicators are required for optimum performance of Butamers.TM[18] A simplified drawing of a butane isomerization unit is as shown Figure 3.7. In the process, butane gas and makeup hydrogen are fed into the reactor section after removal of water and some of the feed sulfur in the dryers. Operating temperatures in the reactors are typically between 280 F and 360 F, pressures between 300 and 450 psig, liquid hourly space velocity (LHSV) of about 2 and hydrogen to hydrocarbon molar ratio of less than 0.5. Reactor outlet temperatures are usually maintained at less than 400 F to minimize excessive hydrocracking and temperature runaways.

Isobutane is recovered at the bottom of the stabilizer, and the overhead gas is routed to the caustic scrubber where acidic gases are neutralized before sending

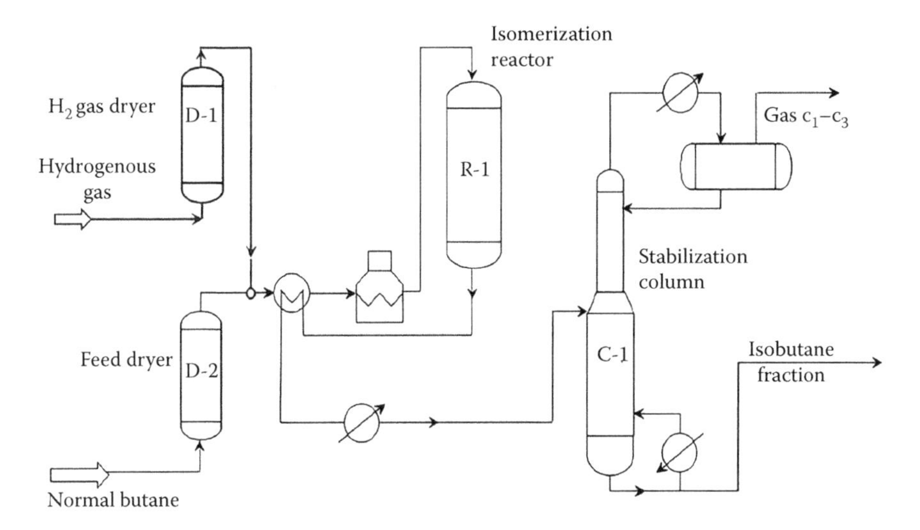

FIGURE 3.7 A schematic diagram of a butane isomerization process unit.
From Robles, R. Y., Hydroisomerization of Butane to Isobutane. Submitted to Dr. Arno de Klerk, University of Alberta, December 8, 2010; UOP LLC, ButamerTM Process, http://www.uop.com/objects/IsobutaneProcess.pdf.[15,17] (Courtesy of Honeywell UOP.)

the treated gas to the gas concentration plant of the oil refinery. Two reactors are usually preferred and used in series for maximizing butane isomerization, and the dual reactor operation provides much-needed flexibility for catalyst replacement and maintenance programs for the paraffin isomerization units.

Special metallurgy is used in areas around the stabilizer overhead and piping to minimize fouling and corrosion due to acidic compound deposition. Isobutane products as determined by achievable butane isomerization are assessed via isobutane molar ratios defined as iC4/nC4. Higher isobutane molar ratios are usually enhanced by operating at lower reactor temperatures. The simplified unit drawing in Figure 3.7 represents a hydrogen-once-through (HOT) operation that does not include the recycling of unreacted hydrocarbons and hydrogen to the reactors. While a hydrogen-once-through unit provides initial capital benefits for the addition of a Butamer, catalyst performance stability challenges may require frequent catalyst replacements and associated increases in production costs relative to those of butane isomerization units operating with a recycle gas system. Overall, butane isomerization units are cost effective and reliable for generating the required isobutane within the refinery for the alkylation unit.

3.4 LIGHT NAPHTHA ISOMERIZATION UNIT

The next oil fraction of hydrocarbons from the atmospheric crude distillation unit is light naphtha, which could be as high as 10–15 volume percent of a crude oil, as light naphtha compounds are also produced in downstream process units. Light naphtha typically consists of mostly pentanes and hexanes as well as five and six carbon naphthenes and aromatic compounds. Natural gasolines that are recovered during the production of oil and gas contain significant quantities of pentane and hexane and can be used directly as gasoline blend components as long as the ultimate gasoline blends meet sulfur, octane, and benzene specifications. When this is not feasible, the light naphtha is subjected to additional processing in order to substantially reduce the concentrations of benzene and sulfur so that the treated natural gasoline can be blended conveniently into gasoline to meet octane, sulfur, and benzene quality specifications. The research octane number of light naphtha is usually in the range of 62–73 octane. By subjecting the light naphtha to further processing in a paraffin isomerization unit, the naphtha can be upgraded to produce an isomerate with a range of 78–84 octanes. Companies that license paraffin isomerization technologies also provide variations of the standard light naphtha process unit with additional equipment that can enable oil refiners to achieve greater than 15 octane number upgrades for their light naphtha feeds. Standard light naphtha isomerization units are quite versatile and useful, as their process conditions can be adjusted to permit units operated to enable oil refiners to meet gasoline octane, benzene, and Reid vapor pressure specifications.

There are two light naphtha isomerization technologies available for use by oil refiners. The technology that is in greater use is the one based on the active platinum alumina-chlorided catalyst and operations at lower operating temperatures. Penex.TM[32] is the Universal Oil Products (UOP) trademark label used for the light naphtha isomerization unit that operates with a platinum alumina-chlorided catalyst at low operating temperatures in the range of 260–400 F. The alumina-chlorided catalysts used in pentane/hexane isomerization units are similar to those used for the butane isomerization process and are reviewed in Section 3.3. A simplified drawing of a low-temperature, chlorinated alumina catalyst, light naphtha isomerization process unit is essentially the same as that for butane in butane isomerization units, as shown in Figure 3.7. The other technology that is used for light naphtha paraffin isomerization is the total isomerization process (TIP). TIP technology operates with a zeolitic catalyst and at reactor temperatures that are greater than 500 F. A key feature is the recycling of unreacted feed after separating out unreacted paraffins from the reactor products via an adsorbent system. Regardless of whether it is a TIP or Penex™ process, the primary objective of paraffin isomerization is essentially to produce more branched isomers of the feed pentane and hexane and upgrade the octane of the light naphtha. There is usually an appreciable amount of benzene and up to 5 mol percent in the feed to a light naphtha isomerization unit. The multifunctional chloride alumina catalyst is capable of hydrogenating the benzene to produce nonaromatic compounds. Light naphtha isomerization process units provide an excellent option for reducing the concentrations of benzene in refinery gasoline. Research octane numbers, motor octane numbers (MONs), and Reid vapor pressures for compounds present in light naphtha are listed in Table 3.2. The light naphtha isomerization process is similar to that of butane, and it is essentially an extension of the technology that was used initially for butane conversion. It is, in the main, as depicted in Figure 3.7.

Hydrotreated naphtha from the light naphtha hydrotreater is dried and fed with dried hydrogen into the reactors, as shown in Figure 3.7. Isomerization and hydrogenation

TABLE 3.2
Octane and Reid Vapor Pressures for Light Naphtha Hydrocarbons

Compound	RON	MON	RVP, psi
nC4	95.00	89.60	51.50
Iso C4	102.00	97.60	59.00
nC5	61.70	61.30	14.40
Iso C5	93.50	89.50	18.90
nC6	24.80	26.00	4.60
22 DMB	94.00	93.40	9.10
23 DMB	105.00	94.30	6.90
2 MP	74.40	74.90	6.30
3 MP	75.50	76.00	5.60
MCP	96.00	85.00	4.20
Benz.	120.00	114.80	3.00
CH	84.00	77.20	6.00
C7+	82.00	71.00	2.10
MTBE	115.20	97.20	8.00
Ethanol	108.00	92.80	18.00

Source: Oyekan, S. O., Private Notes; Ghosh, P., Hickey, K. J., Jaffe, S. B., Development of Detailed Gasoline Composition-Based Octane Model, *Ind. Eng. Chem.* Res., 45, 337–445, 2006.[16,18]

reactions of light naphtha compounds take place over chloride platinum alumina catalysts, as indicted in Section 3.3.

The recommended process conditions are quite broad, and they are dependent on whether the process unit is a grassroots or revamped unit from another process such as a fixed bed catalytic reformer. Pressures are in the range of 400–900 psig, temperatures 260–400 F, liquid hourly space velocity of 1–2, and the hydrogen to hydrocarbon molar ratio of the recycle gas between 1 and 2. Isomerate product is recovered via a product stabilizer, and the light ends are treated in the caustic scrubber and sent to a gas concentration plant. Since isomerization reactions are reversible, low-temperature operations in the reactors promote the formation of higher concentrations of branched isomers and higher octanes. Achievable individual iso product ratio is limited by the thermodynamic equilibrium for the specific isomerization of that normal paraffin to its isomer product. The performance of the light naphtha isomerization unit is based on monitoring isomerate octanes and isomeric product ratios such as iso pentane, 22 dimethyl butane, and 23 dimethyl butane ratios as listed in Figure 3.8. In addition, the paraffin isomerization number (PIN), which is defined as the sum of the iso pentane (iC5), 22 DMB, and 23 DMB ratios, is also monitored for the performance of the process unit. Paraffin isomerization number definition and relevant information are provided in Figure 3.9. In addition, a number of key performance indicators should be monitored as per monitoring programs recommended by the paraffin isomerization processes technology and catalyst provider companies such as Honeywell UOP, Axens, Shell CRI, Sud Chemie, Albemarle, Olcat, and others.[17,19,20] Key performance indicators that are recommended for monitoring include feed x factor, reactor inlet,

iC4 ratio = iC4/(iC4 + nC4) × 100
iC5 ratio = iC5 = iC5/(iC5 + nC5) × 100
2,2-DMB ratio = 2,2-DMB/(2,2 DMB + 2,3 DMB + 2MP + 3MP + nC6) × 100
2,3-DMB ratio = 2,3-DMB/(2,2 DMB + 2,3 DMB + 2MP + 3MP + nC6) × 100

FIGURE 3.8 Key isomerized product ratios.

PIN = Paraffin isomerization number
PIN = iC5/C5P + 22-DMB/C6P + 23-DMB/C6P
Suggested SOR PIN is estimated as the sum of 77% + 32% + 10.5% or 119.5%

Where: iC5/total C5 paraffins and estimated to be 77%
22 DMB/C6 paraffins is estimated to be 32% and
23 DMB/C6 paraffins is estimated to be 10.5%
and
Feed X factor = % MCP + % BZ + %CH + %C7

FIGURE 3.9 Performance indicator number & feed X factor.
From UOP LLC, Gasoline-Industry Leading Solutions for Efficient Upgrading and for benzene Removal, www.uop.com/processingsolutions/refining/gasoline/naphtha-isomerization. Accessed September 28, 2016; Graeme, S., Ross, J., Advanced Solutions for Paraffin Isomerization, *Paper AM-04-49, NPRA Annual Meeting*, March 21–23, 2004.[19,20]

reactor outlet temperatures, reactor delta temperatures, isomerized product ratios, and PINs.

Most light naphtha isomerization units operate with hydrogen recycle with the exception of those that operate solely with makeup hydrogen and do not have recycle gas compressors. The units are referred to as hydrogen-once-through or HOT units. Hydrogen-once-through process units were introduced by Honeywell UOP when catalyst stability was observed to be much better than expected due to much slower rates of coke deposition on platinum alumina-chlorided catalysts. Achieving good process unit operations and reliability requires that effective management programs for feed and hydrogen driers, timely monitoring of piping and equipment that could foul and corrode, effective management of caustic scrubber operations, and good unit startup and shutdown procedures are being utilized.[16,17,19,20]

Total isomerization process technology for paraffin isomerization uses platinum/zeolite/alumina catalysts and operates at higher reactor temperatures of 500–600 F, pressures in the range of 400–900 psig, and hydrogen to hydrocarbon molar ratios of 1–2.[21] There are some distinct advantages for the TIP technology units, as they permit good operations on light naphtha feeds with higher sulfur and water concentrations of about 30 wppm, each relative to the feed for the low-temperature platinum chlorinated-alumina catalyst technology. They also do not generate the significant challenges associated with chloride and acidic compound management to maintain unit reliability and productivity.

In summary, low-temperature paraffin isomerization units based on platinum chlorinated-alumina catalysts are more prevalent in oil refineries relative to high-temperature total isomerization process units. Pentane/hexane isomerization units provide flexibility in managing isomerate octane, benzene, and Reid vapor pressure, and as such, their isomerate products can be used in managing octane, benzene, and Reid vapor pressures of gasoline blends for oil refiners who have light naphtha isomerization units.

3.5 HYDROTREATING UNITS

An introduction to hydrotreating processing units in oil refineries is provided because of the need to establish a good understanding of key oil refinery operations. Hydrotreating units are operated primarily to reduce the concentrations of various heteroatoms in oil fractions. Heteroatoms include hydrocarbon-containing elements such as sulfur, nitrogen, oxygen, silicon, chlorides, and mercury. Hydrotreaters also aid in removing metal contaminants such as iron, chromium, and molybdenum that are corrosion products in the feeds to the units. Hydrotreating process units are also referred to as guard beds, as impurities such as iron scales and inorganic sediments accumulate on the catalyst surfaces. Olefins and di-olefins are saturated and hydrogenation of aromatic rings occurs to produce naphthenes over appropriate catalysts and process conditions. In addition, for higher-boiling oil fractions such as gas oils and residual oils, the hydrotreating processes also reduce concentrations of metal contaminants such as nickel, vanadium, and copper.

3.5.1 Hydrotreating Reactions

The major reactions that occur in hydrotreaters are hydrogenation, which involves saturation of unsaturated compounds; hydrodesulfurization (HDS); hydrodenitrogenation (HDN); hydrodearomatization (HDA); and hydrodeoxygenation (HDO). During the processing of vacuum gas oil and residual oils, asphaltenes are also converted via saturation of some of the aromatic rings in the asphaltenes. Examples of the generalized, representative hydrotreating reactions are given in Figure 3.10.[16]

1. Hydrodesulfurization

   Organo-Sulfur + H_2 → Organic Compound + H_2S

Organo sulfur compound could be mercaptans, organic sulfide and disulfide compounds, thiophenes, benzothiophenes, dibenzothiophenes, benzonaphthothiophenes and other sulfur compounds.

2. Hydrodenitrogenation

   Organo-Nitrogen + H_2 → Organo Compound + NH_3

Organo nitrogen compounds could be amines, pyrroles, anilines, pyridines, carbazoles, benzocarbazoles, dibenzocarbazoles, acridines, indoles, quinolines and indolines and other nitrogen compounds as discussed.

3. Hydrogenation

   Unsaturated Compound + H_2 → Saturated Compound

Unsaturated compounds include olefins, diolefins, aromatics, etc.

4. Hydrocracking

   $RCH_2CH_2R' + H_2 \rightarrow RCH_3 + R'CH_3$

The hydrocracking reactions lead to the production of light end products such as methane Ethane, propane and butane.

5. Hydrodemetalization (most in residual oils)

   Organo-Ni + H_2 → Organic Compound + Ni

   $2Ni + 3H_2S \rightarrow Ni_2S_3 + 3H_2$

FIGURE 3.10 Generalized hydrotreating reactions.
From Oyekan, S. O., Private Notes.[16]

3.5.2 Hydrotreating Process Feedstocks and Catalysts

Key features of various hydrotreating processing units are covered in this section. Processing objectives, hydrocarbon feeds, catalysts, process conditions, catalyst contaminants, and reliability concerns are highlighted in brief introductions of hydrotreating units used in the processing of oil fractions. Hydrotreating processes in oil refineries can be broadly classified into two types according to their processing objectives. The first set of hydrotreating units are those that are required to generate hydrotreated products from feeds and achieve substantial reduction of concentrations of heteroatom and metal contaminants. Hydrotreater products that meet feed contaminants' specifications can then be fed to the downstream catalytic conversion unit. Sulfur, nitrogen, metals, and other contaminants in an inefficiently treated feed would lead to negative performance and reliability impacts in downstream catalytic conversion units. An optimally operated hydrotreating unit upstream of a catalytic conversion unit produces feed that can typically enhance the productivity and reliability of the downstream catalytic conversion unit. Another type of hydrotreating units are those that are operated in order to generate finished hydrocarbon products. That is, hydrotreated products are not fed to downstream conversion process units. Hydrotreated product streams are used directly in the blending process to produce desired oil-refined products.

Hydrocarbon feeds for hydrotreating units cover different oil fractions with respect to boiling ranges, concentrations of heteroatoms and metal contents, asphaltenes, and relative hydrogen deficiencies as indicated by the hydrogen to hydrocarbon molar ratios for crude oil, light naphtha, heavy naphtha, kerosene, diesel, vacuum gas oil, and vacuum residual oils. The qualities of the oil fractions are as shown for Hibernia crude oil from Canada in Table 3.3, and for Canadian Cold Lake crude oil in Table 3.4.[24,25]

As shown in Table 3.4 for Cold Lake crude oil, the concentrations of sulfur and nitrogen increase progressively from heavy naphtha oil through to vacuum residual oil. The complexities of the compounds containing heteroatoms and contaminant metals increase substantially when processing vacuum gas oil and vacuum residual oil fractions. Organo-sulfur and organo-nitrogen compounds were covered in some detail in Chapter 2. Low average hydrogen contents of the hydrocarbons, as shown by the hydrogen to hydrocarbon molar ratios for the higher-boiling oil fractions and high concentrations of heteroatom compound contaminants, lead to increased hydrogen requirements and consumption in the hydrotreating of such crude oil fractions. It should also be emphasized that the qualities of ultimate crude oil fractions to be processed in the hydrotreaters are highly dependent on the processing strategies that an oil refiner has chosen for maximizing profitability and reliability in each of its refineries and for its overall company. For most oil refiners, the slate of crude oils that could be processed on a certain day or week is based on linear programming studies that incorporate relevant assay data for the crude oils in the studies. A number of oil refiners process two to eight and more crude oils with widely varying properties, oil miscibility properties, and oil fraction yields in their atmospheric and vacuum crude distillation units. The output of the studies generally provides a projection of expected qualities and yields of the blend oil fractions. The set of oil fraction data generated for each refinery is typically supported by necessary online oil flowrates and laboratory analytical data to provide determination of oil fractions yields and qualities.

Hydrotreating catalysts are generally similar in basic major constituents, as they are usually composed of gamma alumina as the support, substrate, or "carrier" and a combination of base metal oxides deposited on the alumina support. Silica, aluminosilicate, and other inorganic oxides have been used as hydrotreating catalyst supports.[24] The base metal oxide combinations on catalysts could be cobalt oxide (CoO) and molybdenum oxide (MoO_3), nickel oxide (NiO) and molybdenum oxide, nickel oxide, and tungsten oxide ($NiOWO_3$). In some cases, nickel oxide, cobalt oxide, and molybdenum oxide have been deposited on the same catalyst alumina support. Base metal oxides such as phosphorus oxide, silicon oxide, titanium oxide, and sodium oxide have also been added by some catalyst manufacturing companies, and they have been ascribed the role of modifiers for the alumina matrix.[27] Simple

TABLE 3.3
Properties of Hibernia Crude Oil and Its Oil Fractions

	Whole Crude	Butane&C4-	Light Naph	Heavy Naph	Kerosene	Diesel	Vac Gas Oil	Vac Resid Oil
Boiling Range, F	−200–1000+ F	−200–60	60–185	165–330	330–430	430–650	650–1000	1000+ F
Cut Volume. %	100.0	1.5	5.7	14.8	14.7	17.0	29.9	17.3
API Gravity, degrees	33.9	121.4	81.0	54.9	43.1	34.0	24.7	12.7
Carbon, wt. %		82.4	84.0	85.9	86.2	86.5	90.7	
Hydrogen. wt. %		17.6	16.1	14.1	13.8	13.2	12.8	
H/C, ratio		2.6	2.3	2.0	1.9	1.8	1.7	
Sulfur, wt. %	0.5	0.0	0.0	0.0	0.0	0.2	0.7	1.4
Nitrogen, wppm	1350	0.0	0.0	0.0	0.2	88.5	1196.1	4868
Ni, ppm	1.3							6.5
V, ppm	0.7							3.5
Ca, wppm	0.5							
Paraffins		100.0	84.3	57.6	47.1	41.8	26.4	
Naphthenes		0.0	14.1	31.3	12.7	34.1	37.1	
Aromatics		0.0	0.0	16.4	16.9			

Source: Adapted from the Chevron's crude oils properties and data set of Hibernia crude oil by Soni Oyekan. Colorado School of Mines, Refinery Feedstocks and Products – Properties and Specifications inside.mines.edu/~jjechura/Refining/02_Feedstocks_&_Products.pdf; Chevron, Chevron Crude Oil Marketing, http://crudemarketing.chevron.com/crude/north_american/hibernia.aspx.[22,76]

TABLE 3.4
Properties of Cold Lake Crude Oil from Canada

	Crude Oil	Butane & Lighter	Light Naph	Heavy Naph	Kerosene	Diesel	Vac Gas Oil	Vac Resid Oil
Boiling Range, F	−200 – 1000+ F	−200 – 60	60–185	165–330	330–430	430–650	650–1000	1000+ F
Cut Volume. %	100.0	0.8	11.8	7.7	4.8	10.3	29.2	35.4
API Gravity	19.7	114.5	88.7	60.1	35.3	26.0	15.1	0.7
Carbon, %	84.1	82.7	83.6	85.8	85.7	85.8	84.9	83.5
Hydrogen, %	12.1	17.3	16.4	14.2	13.3	12.3	11.5	11.0
H/C, ratio	1.7	2.5	2.4	2.0	1.9	1.7	1.6	1.6
Sulfur, wt.%	3.8	0.0	0.0	0.1	0.9	2.0	3.4	6.2
Nitrogen, ppm	3862.0	0.0	0.0	0.5	20.4	106.3	1581.3	8290.6
Ni, ppm	63.4	0.0	0.0	0.0	0.0	0.0	0.0	155.9
V, ppm	163.0	0.0	0.0	0.0	0.0	0.0	0.0	400.8
Ca, ppm	1.2	0.0	0.0	0.0	0.0	0.0	0.0	0.0
Paraffins	18.3	99.9	93.4	47.0	16.1	9.4	7.0	1.5
Naphthenes	20.9	0.0	6.6	39.1	62.0	51.6	28.9	6.6
Aromatics	54.5	0.0	0.0	14.0	20.5	36.9	63.2	91.9

Source: Table of data adapted from the ExxonMobil's crude oils properties and specifications data set for Cold Lake Oil Blend by Soni Oyekan. ExxonMobil, Worldwide Operations, Crude Trading – Assays Available for Download www.corporate.exxonmobil.com/en/company/worldwide-operations/crude oils.[23]

designations in oil refinery and catalyst supply business are CoMo, NiMo, NiCoMo, and so on when discussing hydrotreating catalysts. Combinations of NiMo and CoMo catalysts have also been loaded either in the same reactors or in separate reactors in series with a NiNo catalyst in the lead reactor and a CoMo catalyst in the lag reactor in hydrotreating units. Research and development studies on base metal catalysis have led to inventions for optimizing the porosity and average pore sizes of the alumina support and the relative concentrations of base metal oxide compositions.[26,28,29] These novel catalysts have led to significant improvements in the processing of hydrotreater feeds with higher concentrations of more refractory compounds that have heteroatoms and especially for atmospheric and vacuum residual oils.

Most of the hydrotreating catalysts perform adequately in the hydrotreating of two or more oil fractions. This fact presents opportunities for oil refiners to reuse "fairly clean" spent hydrotreating catalysts that are not irreversibly deactivated with metal contaminants. Regenerated hydrotreating catalysts can be reused or "cascaded" from the hydrotreating of one oil fraction to another as long as it has been adequately determined that the regenerated catalysts meet the standard qualities for good regenerated hydrotreating catalysts.

3.5.3 A Generalized Hydrotreating Process Unit

Hydrotreating processes are fairly similar, with some variations in the number of reactors, as significant percentages of reactor volumes of up to 40% could be required to provide for hydrodemetalization and metal contaminant removal. To some extent, the hydrotreater reactor operates as a catalytic guard bed for the removal of nickel, vanadium, and silicon.

A simplified flow diagram of a typical hydrotreating process unit is as shown in Figure 3.11. Feed and hydrogen are preheated via the heat exchanger and heated in the fired heater to reactor temperatures typically in the range of 570–750 F. The start-of-run temperature is dependent on the specific oil fraction that is being hydrotreated and process operating conditions. The feed and hydrogen flow downward through various sections of the reactor bed that are packed with inert materials and catalysts according to prescribed catalysts and inert material loading diagrams. Reactor pressures can range from 225 to 2000 psig and hydrogen treat gas rates of up to 3000 SCF/B. Two-phase flow of liquid and gas is the usual mode of operation, with the exception of the vapor phase in naphtha hydrotreaters. In the reactors, desired hydrodesulfurization, hydrodenitrogenation, hydrodemetalization, hydrodeoxygenation, and saturation of olefins, di-olefins, and aromatics take place over the catalysts. Additionally, feed contaminants such as iron scales, arsenic, and silicon are deposited on the catalysts.

Hydrotreating processes do not typically lead to any significant production of gas, as minimal hydrocracking reactions occur in the reactors and, as a consequence, the yields of liquid products are high. When feedstocks from catalytic crackers, hydrocrackers, and cokers, which are essentially cracked stocks, are present in the feed, the associated reactions of the unsaturated compounds lead to increased temperatures in the reactors, as the reactions are exothermic.

Reactor effluent is cooled and scrubbed with an amine such as monoethanolamine (MEA) in a high-pressure separator as shown in Figure 3.12. A liquid product is then

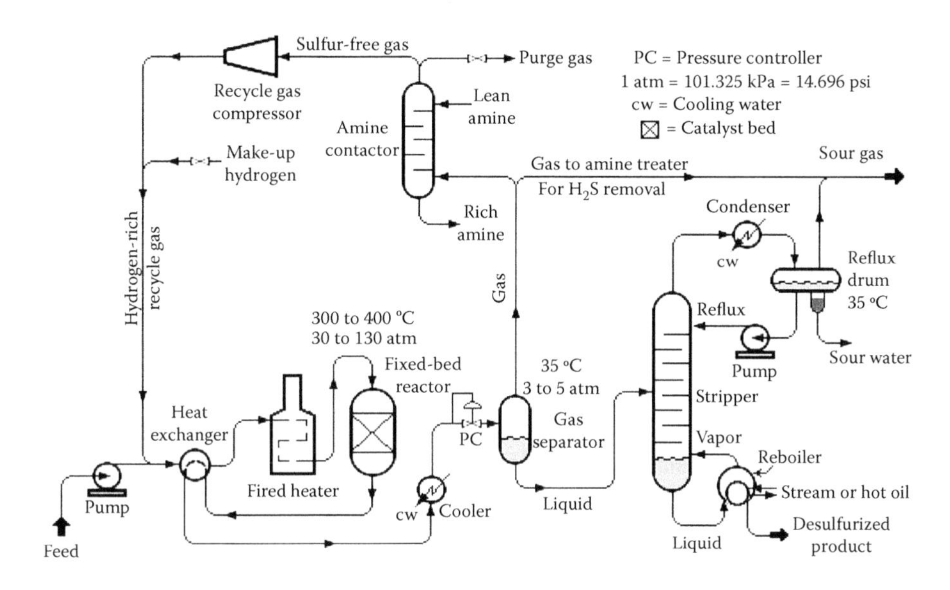

FIGURE 3.11 A simplified flow diagram of a hydrotreating unit.
From Discover Petroleum, Vacuum Distillation, resources.schoolscience.co.uk/SPE/knowl/4/2index.htm?vacuum.html.[12] (Courtesy of Axens.)

separated from the hydrogen, hydrogen sulfide, and ammonia gas. The gas product is sent to an amine treater for hydrogen sulfide removal. After the removal of purged gases, hydrogen is recycled with make-up hydrogen and fed to a reactor. Liquid product from the high-pressure separator is charged to the stripper tower to remove hydrogen sulfide, ammonia, and light end gases. The bottom-stripped liquid from the stripper tower is the desired liquid product from a hydrotreater.

3.5.4 Light Naphtha Hydrotreating

Light naphtha, as described in Chapter 2 for the light naphtha isomerization process, is mostly pentane and hexane with some five and six carbon naphthenes and other hydrocarbons containing seven carbon atoms. The boiling range of the light naphtha feed is typically between 65 F and 165 F, and this range may vary depending on the cut point that is selected between light and heavy naphtha. In some oil refineries, light naphtha is hydrotreated in the same reactor system with heavy naphtha and the hydrotreated naphtha product is then separated via a naphtha splitter to yield the light and heavy fractions. The hydrotreated light naphtha product can then be used in gasoline blending.

Most light naphtha hydrotreaters have as their main objective the preparation of the light naphtha for the light naphtha isomerization process. The feed is hydrotreated to produce a liquid hydrocarbon that meets feed quality sulfur and other quality specifications of the downstream paraffin isomerization process. In recent years, in another variation of light naphtha hydrotreating, some process units are used primarily to produce light naphtha with low sulfur and benzene concentrations for gasoline blending.

Catalysts used in light naphtha hydrotreaters are usually the CoMo type. The process conditions are liquid hourly space velocities in the range of 1–5, 150–500 psig pressures, reactor temperatures of 500–750 F, and hydrogen treat gas rates of 500–1500 standard cubic feet per barrel (SCF/B).

3.5.5 Heavy Naphtha Hydrotreating

A naphtha hydrotreater functions primarily as the process unit for preparing the feed for the catalytic reformer by operating essentially as a catalytic guard bed. The main function is to substantially reduce the concentrations of sulfur, nitrogen, hydrogen sulfide, ammonia, water, and oxygenates contained in the hydrocarbon feed. In addition, the naphtha hydrotreater is referred to as a guard bed, as it is capable of removing iron scales, silicon and arsenic, and other metal contaminants in the naphtha feed via their depositions on the catalyst. Catalytic reformers are usually operated downstream of a naphtha hydrotreater. When a catalytic reformer has to be operated without the benefit of a naphtha hydrotreater, the catalytic reformer is usually limited to operating with a feed that contains high sulfur, nitrogen, iron scales, silicon, and arsenic contaminants. As a result, catalytic reforming operations for those types of units are usually conducted at low severities of 70–85 octane, which usually leads to rapid catalyst deactivation and poor reformate and hydrogen yields. The resultant effects of operating catalytic reformers with untreated naphtha containing high sulfur, nitrogen, oxygenates, and metal contaminants are low productivities, frequent and expensive catalyst replacements, and turnaround maintenance programs. There are additional significant refinery productivity losses if the refinery is dependent on hydrogen production from such inefficiently operated catalytic reformers.

Naphtha hydrotreater feeds vary widely for different refineries due to the fact that the naphtha feed can include straight-run naphtha from the crude distillation unit and naphtha from downstream processing units such as the hydrocracker, coker, visbreaker, and fluid catalytic cracker units. In addition, the naphtha feed to the hydrotreater could also contain purchased "opportunity" naphtha. Great care should be exercised in the acquisition and coprocessing of purchased naphthas in a blend with other naphthas. A highly contaminated naphtha feed can cause reliability problems for the naphtha hydrotreater and shutdown of the catalytic reformer. The properties of a variety of naphthas are listed in Table 3.5.

As shown in Table 3.5, sulfur concentrations vary from 0.5 wppm for the hydrocracker naphtha to 7000 wppm for a naphtha that was produced from solvent refining of coal and labeled as solvent refined coal (SRC-1). Coal to liquid processes were studied extensively in the early 1980s. Nitrogen varies from 0.5 wppm for hydrocracker naphtha to 2000 wppm for SRC-1 naphtha. Coker and catalytic cracker naphthas have high olefins, and significant concentrations of diolefins are usually present in coker naphthas.

Mercaptans, disulfides, sulfides, and thiophenes are the sulfur compounds in naphthas, and thiophenes are the more refractory of the sulfur compounds. The concentrations of the different types of sulfur compounds in naphthas vary based on the source of the naphtha. Comparative sulfur data for two straight-run and two coker naphthas are provided in Table 3.6.[28]

TABLE 3.5
Properties of a Variety of Naphthas

	WTI SR	Coker	FCC	H/C	Solvent Refined Coal
API Gravity, Degree	54.0	57.0	54.0	41.0	38.5
ASTM Dist. F	200–400	150–350	150–270	170–350	160–375
Paraffins	49.0	34.0	27.0	45.0	38.0
Naphthenes	36.0	14.0	21.0	43.0	36.0
Aromatics	15.0	7.0	17.0	12.0	13.0
Olefins	0.0	45.0	35.0	0.0	13.0
N + 2A	66.0	28.0	55.0	67.0	62.0
Sulfur, ppm	1500.0	3500.0	1000.0	0.5	7000.0
Nitrogen, ppm	1.5	100.0	40.0	0.5	2000.0

The assigned reactivities of sulfur compounds in hydrodesulfurization reactions are 30 for mercaptans, 35 for disulfides, 15 for sulfides, and 1 for thiophenes. Coker naphthas usually contain high concentrations of thiophenes. Based on the reactivity order of sulfur compounds in naphtha feeds, it is beneficial to achieve significant reductions in the concentrations of thiophenes in order to meet required sulfur specifications for hydrotreated naphtha. Two-stage naphtha hydrotreating units may also be required due to the need to saturate di-olefins in a pretreater reactor. Naphtha hydrotreater process conditions for straight-run or virgin naphtha, catalytic cracker, and coker naphtha are compared in Table 3.7.[28]

Process conditions listed in Table 3.7 for each of the naphthas are for producing hydrotreated naphtha products for further processing over bimetallic platinum containing catalysts in catalytic reformers. Bimetallic platinum-containing catalysts are highly sensitive to sulfur and other feed contaminants. The naphtha hydrotreater process conditions listed in Table 3.7 are selected to achieve less than 1 wppm each of sulfur and nitrogen as well as to saturate olefins and achieve a bromine number target of less than 1 centigram per gram for hydrotreated naphthas.

TABLE 3.6
Sulfur Compound Types in Naphthas

Naphtha Source	Straight Run (1)	Straight Run (2)	Coker (1)	Coker (2)
Mercaptans, ppm	30	44	20	20
Disulfides, ppm	24	32	30	110
Sulfides, ppm	150	82	0	20
Thiophenes, ppm	0	49	150	900
Total, ppm	213	207	200	1100

Source: Oyekan, S. O., Townsend, G. J., Mathur, K. N., Sopko, J. S., SRC-1 Naphtha Reforming Study. Final DOE Report, DOE/OR/03054-70. 1983.[28]

TABLE 3.7
Process Conditions for Hydrotreating of Naphthas

	Straight Run	Cat Cracker	Coker
	Feed Properties		
Sulfur, wppm	<300.0	400–700	10,000.0
Nitrogen, wppm	<1.0	10–30	20–50
Bromine No. cg/g[c]	<1.0	30–50	120.0
Diolefins, vol. %	0.0	0.0	10.0
	H/T Conditions[a]		
Temperature, F	500–650	500–650	500–650[b]
LHSV, V/Hr/V	4.0	2.0	>1.0
H2, Partial Pressure, psia	60–80	150–300	>250
Treat Gas Rate, SCF/B	300–500	1300–1800	3000.0
H2 Consumption, SCF/B	10.0	300.0	900.0

Source: Oyekan, S. O., Townsend, G. J., Mathur, K. N., Sopko, J. S., SRC-1 Naphtha Reforming Study. Final DOE Report, DOE/OR/03054-70. 1983.[28]

[a] Operating conditions are selected to produce feed that can be processed efficiently over the bimetallic catalysts used in catalytic reformers.

[b] An additional reactor and appropriate catalyst to saturate diolefins is typically used as a lead reactor followed by the main hydrotreater reactor when processing significant percentages of coker naphtha.

[c] Bromine number is a measure of the degree of unsaturation of the feed based on concentrations of olefins and diolefins, where concentration is defined as centigram per gram.

Wide ranges of possible process conditions are used, and they are dependent on the quality of the naphtha to be hydrotreated. Reactor pressures, liquid hourly space velocities, and hydrogen treat gas rates vary significantly due to the higher sulfur, nitrogen, and unsaturated compounds concentrations in catalytic cracker and coker naphtha relative to those for straight-run naphtha feeds. The higher hydrogen consumption of 900 SCF/B for coker naphtha is due to the processing of much higher sulfur, nitrogen, olefin, and diolefin concentrations relative to straight-run and fluid catalytic cracker naphthas. Comparative data are provided in Table 3.8 for two naphtha hydrotreaters that processed a mix of oil naphthas.[24]

Heavy naphtha hydrotreater process conditions are typically in the following ranges. Reactor temperatures are between 500 F and 700 F, pressures between 200 and 800 psig, liquid hourly specie velocities of 0.5 to 10, and hydrogen treat gas rates of 200 to 3000 standard cubic feet per barrel. Though this section is provided as an introduction, it is necessary to note that reactor temperatures could be limited, as the naphtha hydrotreater catalyst deactivates with time on stream. Toward the end of catalyst life, raising the reactor temperature may not result in increased hydrodesulfurization. What is sometimes observed is that increasing reactor temperature toward the end of catalyst life actually leads to higher sulfur concentrations in the hydrotreated naphtha.

TABLE 3.8
A Comparison of Two Naphtha Hydrotreaters

	Case 1	Case 2
Naphtha Feed Properties		
API Gravity	64.0	57.5
Sulfur, wppm	100.0	300.0
Nitrogen, wppm	0.3	9.0
Bromine No. cg/g	8.0	25.0
ASTM D86 Distillation, F		
10%	137	134
50%	220	242
90%	298	316
EP	332	381
Process Conditions		
Catalyst: $CoMo/Al_2O_3$		
Pressure, psig	300	300
LHSV	5	3
Temperature, F	610	650
H2 Treat Gas Rate, SCF/B	300	1000
H2 Consumption, SCF/B	40	120
Product Quality		
Sulfur, wppm	0.1	0.2
Nitrogen, wppm	0.1	0.3
Bromine No. cg/g	<0.1	2.0

Source: Chuang, K. C. et al., Catalytic Processes in Petroleum Refining. AIChE Today Series, 1992.[24]

It is well known that recombination reactions of olefins and hydrogen sulfide can occur at higher process temperatures in naphtha hydrotreaters and produce higher concentrations of mercaptans in hydrotreated naphthas. These mercaptan-forming reactions are sometimes referred to as mercaptan reversion or mercaptan recombination reactions. A naphtha hydrotreater flow diagram is as shown in Figure 3.11. Cobalt oxide/molybdenum oxide (CoMo) and Nickel oxide/molybdenum oxide (NiMo) catalysts are used depending on whether hydrodesulfurization or hydrodenitrogenation is the major reaction of interest. Generally, CoMo catalysts are more active for hydrodesulfurization reactions and are less active relative to NiMo catalysts for hydrodenitrogenation and hydrogenation reactions. NiMo catalysts are usually used for hydrodenitrogenation and in hydrotreaters processing a high percentage of coker naphtha.

Process conditions can vary considerably and are dependent on the concentrations of sulfur, nitrogen, and other contaminants in the naphtha feed. In order to reduce feed sulfur to less than 0.5 wppm and nitrogen to less than 0.5 wppm, the typical ranges for the process conditions are 0.5–10 LHSV, reactor temperature of 500–700 F, 200–800 psig pressure, and treat gas rates of 200–3000 SCF/B.[77]

If there are significant percentages of coker and catalytic cracker naphtha in the naphtha feeds, a pretreater reactor is usually located ahead of the naphtha hydrotreater reactor to saturate olefinic and di-olefinic compounds before hydrotreating the resultant naphtha in the main reactor.

There are significant concerns with silicon contaminants in hydrotreated naphtha, and coverage of this special metal contaminant is deferred to the sections on catalytic reforming processes.

3.5.6 Gasoline Desulfurization

In Chapter 2, current and future proposed gasoline sulfur regulations for countries were reviewed and it was noted that oil refiners continue to meet the production of low-sulfur gasolines and achieve low sulfur concentrations of 10 ppm as per the Worldwide Fuels Council and the United States Environmental Protection Agency. Member countries of the European Community have been operating with 10 wppm sulfur in gasolines since 2010. Other countries per the reviews in Chapter 2 are also planning to mandate the usage of lower-sulfur gasoline by 2020. Lower sulfur concentrations in gasolines are required to ensure good, stable performance of the catalytic converters in automobiles used for the reduction of emission of pollutants from automobile exhausts.

Past studies and reviews of oil refining operations and gasoline blend stocks indicated that fluid catalytic cracker naphthas or FCC gasolines, the naphtha boiling-range fractions produced from catalytic crackers, are 25%–40% in blended gasoline. FCC gasolines or naphthas are responsible for contributing 85%–99% of the sulfur in blended gasolines. These high FCC naphtha contributions to blended gasoline sulfurs are because the average sulfur of fluid catalytic cracker naphthas or FCC gasolines is in the 200–3000 wppm range.[29] Furthermore, studies also revealed that light catalytic cracker naphthas contained higher concentrations of olefins and mercaptans and low percentages of total sulfur compounds relative to the heavy catalytic cracker naphtha fractions. Additionally, heavier naphtha fractions of FCC gasolines contain higher percentages of sulfur and more refractory sulfur compound types such as thiophenes. It is also well known that high olefin concentrations in the light fraction of FCC gasolines are responsible for the higher octane numbers of the FCC gasoline blend stock.

In order to achieve the required deep desulfurization of the total catalytic naphtha and minimize octane losses at minimal capital and operational costs, a staged selective desulfurization with interstage separation process was developed for achieving the necessary desulfurization of the catalytic cracker naphthas and minimal octane losses. Other processing schemes involve separating the lighter fraction and processing the light and heavy catalytic naphthas separately. For successful, efficient FCC gasoline desulfurization and preservation octane, oil refiners are utilizing FCC naphtha desulfurization technologies that are being offered by process technology licensors. Gasoline desulfurization processes that are in use are based on exploiting key differences between the properties of light and heavy naphthas. The process

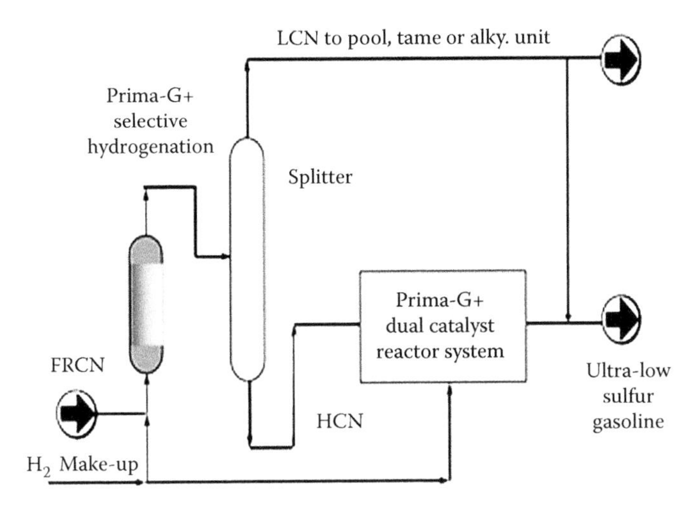

FIGURE 3.12 A schematic diagram of Axens prime-G+ FCC naphtha HDS unit.

technologies optimize staged selective hydrotreating of fractions of the catalytic cracker naphthas. Some of the gasoline desulfurization technologies licensors also use a naphtha splitter after a first stage of mild hydrotreating followed by higher-severity desulfurization of the heavier fractions. Process licensors also highlight utilization of the advantageous hydroisomerization of some of the heavy catalytic naphtha compounds that lead to gasoline octane increase for their licensed process. Most of the process technologies offered can process FCC naphtha, light coker naphtha, light straight-run naphtha, and steam cracker naphtha. For more details on each of the relevant technologies for oil refiners, readers are directed to the following process licensors. Axens offers its Prime-G+ processes, catalysts, and other variant gasoline desulfurization technologies.[30,31] Axens also offers the Olefins Alkylation of Thiophenic Sulfur (OATS) process technology that was developed by BP and licensed by Axens. Axens, in a 2003 presentation, stated that it had licensed numerous Prime-G+ units with capacities of 2.5 to 115 MBPD, feed sulfur range of 30–4100 ppm and olefin content of 15–55 volume percent. Additionally, all the licensed Prime-G+ units produced desulfurized naphtha meeting the less than 10 ppm of sulfur target and minimal octane losses.[30] A schematic diagram of the Axens Prime-G+ is given as Figure 3.12.[31] Honeywell UOP's FCC Gasoline Selectfining™ and catalysts are available for oil refiners.[32]

Other prominent processes offered for FCC naphtha desulfurization include ExxonMobil's SCANfining and OCTgain technologies and CD Tech's CD Hydro technology.[33,29]

3.5.7 Distillate Hydrotreating

The hydrotreating of the distillate fraction from the crude oil unit is required for the production of higher qualities of distillate hydrocarbons that are within the boiling range of 320–700 F. Improving the qualities of the distillate involves achieving substantial reductions of sulfur, nitrogen, and other contaminants and some aromatic saturation for the hydrotreated products. Based on these factors and the specific boiling ranges selected for the desired ultimate products, kerosene, jet fuel, diesel, and heating oils are possible oil refined products from the processing of distillate fractions. Appropriate hydrotreating catalysts, process conditions, and blending strategies are usually utilized to produce one of the products from distillate hydrotreating. For the goals of this book and in order to retain the focus on ultralow-sulfur transportation fuels, this section is devoted to distillate hydrotreating to meet ultralow-sulfur diesel sulfur specifications of less than 10 ppm. The key objectives of deep desulfurization of distillates are to produce diesel with less than 10 ppm sulfur and reduce aromatic concentrations to meet cetane number quality specifications.

Straight run distillate, coker distillate, light cycle oil, and blends of the distillates are feedstocks for a distillate hydrotreater unit. Key properties for three different sample feeds are provided in Table 3.9. As shown in Table 3.9, sulfur is in the range from 0.5 wt. % for the straight-run distillate to 0.9 wt. % for the FCC light cycle oil, and nitrogen is between 0.01 and 0.08 wt. %. Aromatics vary broadly between 35 wt. % for straight-run distillate to 71 wt. % for cycle oil. Studies on the relative reactivity of sulfur compounds in distillates showed that the bulkier dibenzothiophenes are less reactive relative to benzothiophenes, which in turn are less reactive relative to thiophenes, sulfides, and mercaptans. Furthermore, during the catalyst and process development studies conducted to improve the distillate desulfurization process and catalysts in order to reduce diesel sulfur from 500 to 30 wppm and to 10 wppm sulfur by 2016, it was observed that sterically hindered sulfur compounds such as 4, 6 dimethyl dibenzothiophenes were highly refractory and their concentrations would have to be significantly reduced in order to meet the ultralow-sulfur diesel of 10 ppm.

The relative reactivity of sulfur compounds for hydrodesulfurization for key sulfur compounds in distillates has been studied and is shown in Figure 3.13.[34–36]

TABLE 3.9
Distillate Boiling Range Feeds

	Straight Run Distillate	Blend of SR & Coker	FCC Cycle Oil
Gravity, API	33.0	31.0	21.3
Sulfur, wt. %	0.5	0.5	0.9
Nitrogen, wt. %	0.01	0.03	0.08
Bromine No, cg/g	0	5	7
Aromatics, vol.%	35	40	71
	Distillation, F		
IBP	420	400	320
10%	485	480	442
50%	535	530	531
90%	625	630	655
FBP	680	680	700
Cetane Number	44	40	20's

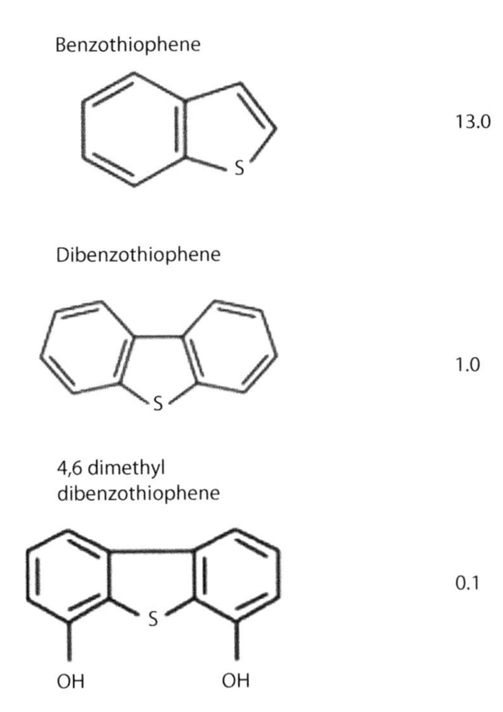

FIGURE 3.13 Benzothiophene and dibenzothiophenes, where "me" in 4, 6-dimethyl dibenzothiophene is the methyl group, CH_3.

As shown in Figure 3.13, the least reactive sulfur compound is the 4, 6 dimethyl dibenzothiophene. It is also known that a catalyst with good hydrodenitrogenation activity is required to convert nitrogen compounds and minimize their inhibiting effects on the hydrodesulfurization activity of the catalyst.[37–39]

Some of the strong inhibiting basic organo-nitrogen compounds are as shown in Figure 3.14. Improvements led to excellent stable catalysts that desulfurize sterically hindered 4. 6 dimethyl dibenzothiophenes (4, 6 dimethyl DBT) and sterically hindered DBT compounds. Additionally, improvements in reactor internal technologies that enhanced optimal flow of feed through the catalyst beds led to improved chemisorption of reactants on more accessible catalytic sites. Axens, Chevron, Haldor Topsoe, Honeywell UOP, and Shell Global Solutions license such efficient reactor internals.[39–43]

The ranges of process conditions used to produce ultralow-sulfur diesel are as provided in Table 3.10. Additionally, increasing the residence time of the reactants in the reactors would enhance both the hydrogenation of some of the aromatic rings of the sterically hindered compounds and enable maximizing hydrodesulfurization of the distillate feed. Process conditions and good hydrotreating catalysts have made it possible to meet the less than 10 wppm sulfur specifications by oil refiners.

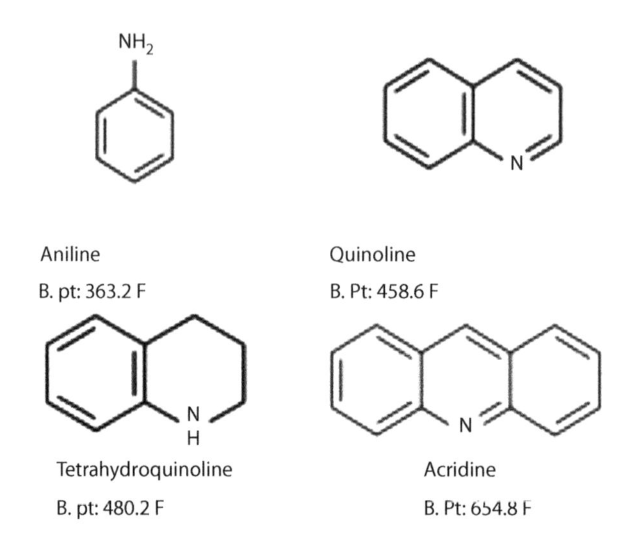

FIGURE 3.14 Basic nitrogen compounds that inhibit HDS reactions.

TABLE 3.10
Process Conditions for Distillate Hydrodesulfurization

Temperature	Deg. F	600–700
LHSV,	hr-1	0.5–5
Pressure	Psig	300–1000
Treat Gas Rate	SCF/B	1000–4000

Excellent hydrodesulfurization of the distillate also results in achieving good aromatic saturations and diesel products that meet mandated diesel cetane number specifications.

There are also diesel blend options available for meeting diesel fuel sulfur and cetane number specifications in the United States and other countries other than those based on fossil fuel operations. Fuel blending options permit using blends of suitable nonfossil, zero sulfur, and zero aromatics referred to as biodiesels to meet sulfur and cetane number specifications as required and within the guidelines of the percentages of biodiesel components permitted in the diesel blends.

3.5.8 Vacuum Gas Oil Hydrotreating

Most vacuum gas oil hydrotreaters have as a primary goal the pretreating of feed streams to produce hydrotreated feeds that are suitable for processing in fluid catalytic crackers and, as such, they are also referred to as catalytic cracker feed hydrotreaters and often simply as "cat feed hydrotreaters".[44] Vacuum gas oils are also used as feeds in steam cracking units even though naphthas are the more favored feeds for that application.

Gas oils are sometimes used as blend stocks for marine fuels. In this book, the focus is on the use of vacuum gas oils as feeds for catalytic cracker and hydrocracking units.

The process of hydrotreating vacuum gas oil before feeding the product to a fluid catalytic cracker provides numerous benefits for operations in the catalytic cracker as well as for the refinery's catalytic cracker naphtha and downstream gasoline desulfurization unit. Some of the benefits for the ultimate gasoline sulfur were reviewed in Chapter 2. Processing of a hydrotreated vacuum gas oil feed in a fluid catalytic cracker leads to higher yields of catalytic cracker naphtha, reduced light cycle oil yields, and lower catalytic naphtha sulfur relative to processing of the same unhydrotreated vacuum gas oil. Additionally, processing hydrotreated vacuum gas oil feeds in fluid catalytic crackers enables the achievement of substantial reductions of sulfur oxides emission (SO_x) in the regenerator effluent gases. Reduced sulfur in vacuum gas oil feeds to the fluid catalytic crackers also leads to significant reductions in the costs for sulfur getter additive management for fluid catalytic crackers. The products of vacuum gas oil hydrotreating have much lower sulfur, nitrogen, metal contaminants, and polyaromatics. They are excellent low-contaminant feeds for fluid catalytic crackers.

A wide variety of feedstocks with different properties and similar boiling ranges as crude oil vacuum gas oils are processed in catalytic cracker feed hydrotreaters. An example of three of such catalytic cracker hydrotreater feeds are provided courtesy of Stuart Shih in Table 3.11.[24]

With the increased production and availability of bitumen from oil sands, hydrotreating of vacuum gas oils has also been shown to be similarly applied in order to produce suitable feeds for further processing in other oil refining process units. Feed properties and data from a pilot plant test program for an extra-heavy bitumen and coker vacuum gas oil and a number of hydrotreating catalysts are summarized in Table 3.12.[45]

TABLE 3.11
Feeds for Catalytic Cracker Feed Hydrotreater

Properties	Arabian Light VGO	Flexicoker Heavy Gas Oil	Heavy Lube Extract
API Gravity, Degree API	23.5	12.6	10.4
Sulfur, wt. %	2.42	3.73	4.3
Nitrogen, wppm	650	2000	1900
Pour Point, degrees F	85	110	85
Conradson Carbon, wt. %	0.14	2.5	1.6
	Distillation, D-1160, F		
5	732	646	812
50	827	845	897
90	940	963	954
95	958	1007	973

Source: Chuang, K. C. et al., Catalytic Processes in Petroleum Refining. AIChE Today Series, 1992.[24]

TABLE 3.12
Pilot Plant Testing with Extra-Heavy VGO

	50% Bitumen VGO + 50% Coker VGO
Feed Properties	
API Gravity	12.89
Sulfur, wt. %	3.44
Nitrogen, wppm	4000
PNA, wt. % (1)	48
Operating Conditions	
Temperature, F	700–734
Pressure, psig	1450
H2 Treat Gas, SCF/B	6000
Product Properties	
Sulfur, wppm	30–500
Nitrogen, wppm	3–1500
PNA, wt. %	5–17

Source: Adkins, B., *Advances in Upgrading Heavy Crudes and Coker Gas Oil through Hydroprocessing*, Albemarle Catalysts Technical Seminar, Sonoma, California, July 2007.[45]

Since vacuum gas oil hydrotreating could include a variety of feeds with widely varying amounts of sulfur, nitrogen, and poly aromatics, process conditions for hydrotreating vacuum gas oil boiling range feeds are quite broad and highly dependent on the final disposition of the products. In the pretreating of feeds for catalytic cracking units, process conditions and especially reactor temperatures should be controlled and monitored so as not to overly saturate too many aromatics and thereby substantially reduce ultimate catalytic cracker naphtha aromatics and octanes.[47] The range of process conditions for vacuum gas oil hydrotreaters is provided in Table 3.13.

Catalysts that are used in vacuum gas oil hydrotreating units are essentially similar to the CoMo, NiMo, CoNiMo, and combination CoMo and NiMo catalysts that were discussed in Chapter 2. Some of the catalysts for hydrotreating vacuum gas oil and heavier oils may have some added beneficial features due to modifications of the alumina or catalyst substrate. These alumina physical and catalytic changes could include modifications that positively impact catalyst porosity and average pore sizes. The goal of such changes is to enhance the accessibility of active sites for hydrogenation, hydrodesulfurization, hydrodenitrogenation, and hydrodemetalization reactions for the more sterically hindered hydrocarbon molecules such as those of polynuclear aromatics and other complex organo-sulfur and organo-nitrogen compounds. A simplified process diagram of a vacuum gas oil hydrotreater is provided in Figure 3.15.[50]

TABLE 3.13
Generalized Process Conditions for VGO Hydrotreaters

Feed Boiling Range, F	600–1200
Process Conditions	
LHSV. Hr-1	0.5–3
Reactor temperature, F	600–800
Pressure, psig	500–2000
H2 Treat Gas Rate, SCF/B	1000–4000
Catalyst Life, Bbl/lb	50–150

Source: Process conditions per Soni Oyekan from references. Chuang, K. C. et al., Catalytic Processes in Petroleum Refining. AIChE Today Series, 1992; Adkins, B., *Advances in Upgrading Heavy Crudes and Coker Gas Oil through Hydroprocessing*, Albemarle Catalysts Technical Seminar, Sonoma, California, July 2007; Speight, J. G., Ozum, B., *Petroleum Refining Processes*, Marcel Dekker, Inc., 2002.(24,45,46)

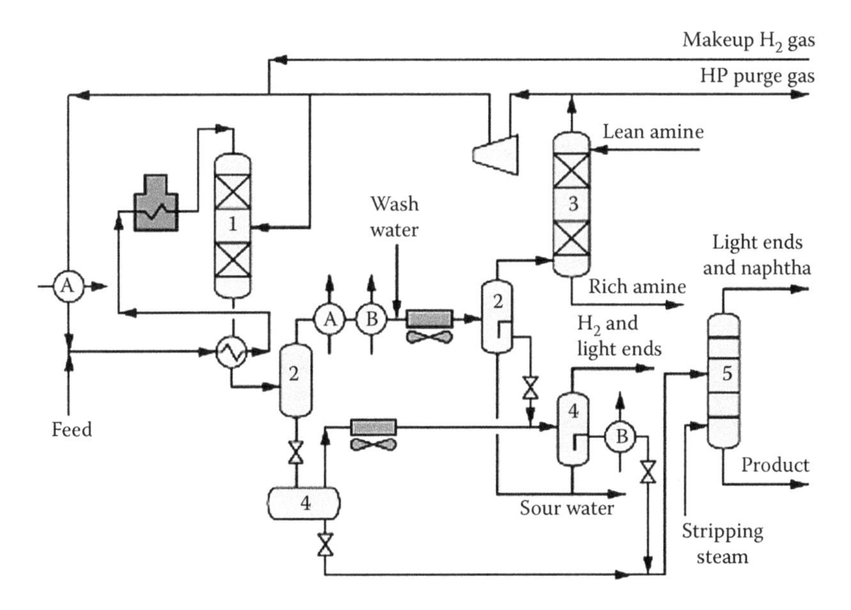

FIGURE 3.15 A simplified flow diagram of a vacuum gas oil hydrotreater.
From Jechura, J., Colorado School of Mines, Hydroprocessing: Hydrotreating & Hydrocracking, Chapters 7 and 9, August, 2016.(50)

3.5.9 Residual Oil Hydrotreating

Heavy sour crude oils that are being produced from recent heavy oil and nonconventional oil discoveries are available at moderate prices relative to the premium priced light sweet oils. Lighter low-sulfur shale oil production has also been ramped up, especially in the United States. The increase of heavy sour and nonconventional oils supplies has also resulted in the production of substantial amounts of vacuum residual oils. Properties of selected whole crude oils and those of their corresponding vacuum residual oils are given in Table 3.14. The heavier sour crude oils have high percentages of vacuum residual oils that are in the 30+ volume percent volume range, as shown for Maya, Boscan, WCS, and Athabasca bitumen oils.[48,49,52]

In the current environment that requires the use of ultralow-sulfur clean fuels and oil refiners experiencing gradual reduction in the supply of light sweet crude oils, it has been deemed necessary to upgrade residual oils to significantly increase oil refining profitability. Oil refiners are using residual oil upgrading processes that are based on either hydrogen addition or carbon rejection mode to produce high

TABLE 3.14
Vacuum Residual Properties for Some Crude Oils

	Bakken Shale Oil	West Texas Inter (WTI)	Mexican Maya	Western Canadian Select[a]	Athabasca Bitumen[b]	Venezuelan Boscan
Whole Crude						
Gravity, API	41.0	39.0	21.8	20.9	8.7	10.2
Sulfur, wt. %	0.2	0.3	3.3	3.5	5.3	3.0
Vacuum Residual Oil						
Vol % of Crude Oil	5.2	9.4	36.0	36.0	32.0	61.0
Gravity, API	14.0	11.4	0.5	2.5	4.0	1.7
Sulfur, wt. %	0.8	1.1	5.8	5.7	5.8	6.0
Nickel + Vanadium, wppm	9.0	128.0	920.0	NR[c]	448.0	2240.0
CCR, wt. %	11.3	18.2	26.0	NR	23.5	32.0

Source: Data in Table 3.14 were compiled and adapted by Soni Oyekan from references. Shih, S. S., Summary of Catalytic Resid Upgrading Technology provided as part of the AIChE Course on Catalytic Processes in Petroleum Refining; ExxonMobil, Guide to World Crudes, ExxonMobil Published Assay for Kearl Oil Sands Diluted Bitumen, *Oil and Gas Journal*, May, 2014; Jechura, J., Colorado School of Mines, Hydroprocessing: Hydrotreating & Hydrocracking, Chapters 7 and 9, August, 2016.[48–50]

[a] Western Canadian Select is a heavy sour blend of 20 heavy conventional oils produced in Canada, Athabasca bitumen, upgraded bitumen, sometimes referred to as light synthetic crude oil (SCO) and condensate added to meet pipeline viscosity requirements.

[b] Athabasca bitumen, which also has added condensate to meet pipeline viscosity requirements.

[c] NR means not reported.

value gasoline and diesel. By 1989, over 40 residual oil upgrading processes were reported by Court and Smith.[51] Coking, residual oil hydrotreating, residual oil fluid catalytic cracking (RFCC), visbreaking, and solvent deasphalting are some of the major upgrading technologies used by oil refiners.

Specific catalysts are used in residual oil hydrotreating for demetallation, hydrodenitrogenation, and hydrodesulfurization due to the amounts of nickel, vanadium, iron, and heteroatom compounds in residual oil. A variety of special catalysts with average pore sizes of 50–300 angstrom, surface areas of 100–150 meters squared per gram and various combinations of cobalt oxide, molybdenum oxide, and nickel oxides on mainly alumina bases are used.[24,26]

3.6 FLUID CATALYTIC CRACKING UNITS

Fluid catalytic cracking units are the major process units that enable cracking configuration refineries to produce high-value gasoline and fuels from a barrel of crude oil. Catalytic cracking process units significantly enhance the profitability of oil refineries. The Euro Community and the United States have substantial catalytic cracking capacities as a sizeable percentage of their atmospheric crude oil capacities. Atmospheric crude oil distillation, catalytic cracking, and other key oil refining conversion process units are compared for the United States in Table 3.15. Over the past 20 years, catalytic cracking capacities have been about 35% of the atmospheric crude distillation capacities.[52] FCC capacity as a percentage of the atmospheric crude capacity percentage has remained flat relative to the percentage increases for coking and hydrocracking process units relative to atmospheric crude distillation capacity. Catalytic cracking technologies and processes have evolved over the past 80 years, and two major broad classifications of catalytic cracking technologies are now in use.[53–55] Side-by-side reactor and regenerator and stacked reactor/regenerator process technologies are predominantly used.

Fluid catalytic cracking is a process through which high-boiling compounds in gas oils are converted to lower-boiling, more valuable products via catalytic and

TABLE 3.15
US Capacities for Atmospheric Crude Distillation and Conversion Units

	Atm. Crude Capacity. MMBPD	Catalytic Cracking, MMBPD	Coking, MMBPD	Hydrocracking, MMBPD	Catalytic Reforming, MMBPD
1995	15.1	5.6	1.8	1.2	3.9
2000	16.3	6.0	2.1	1.4	3.8
2005	17.0	6.2	2.5	1.5	3.8
2010	16.9	6.1	2.6	1.7	3.7
2015	17.8	6.0	3.0	2.1	3.7

Source: Adapted from US EIA Report. US EIA Report, Downstream Charge Capacity of Operable Refineries, June 22. 2016. Accessed by Soni Oyekan on Oct. 22, 2016.[52]

thermal cracking reactions on acidic zeolite-containing catalysts. In the process, the catalyst and oil mix is aerated and fluidized as if it were a liquid. The catalysts mix reacts with and cracks hydrocarbons in gas oil and residual oils in the reactor at low pressures of 30–50 psig and at reactor temperatures of 900–1400 F. Fluid catalytic cracker feed could be composed of heavy atmospheric gas oil (HAGO), vacuum gas oil, coker gas oil, and residual oil. Typically, FCC units convert 90% of their feeds to a variety of products that are blended directly into distillate and a naphtha (FCC gasoline), which is blended into gasoline after hydrotreating to meet ultralow gasoline sulfur specifications. Fuel oil is also a major product of the FCC unit. Catalytic naphtha produced by the FCC unit is high in octane due to the high olefin concentrations in its light catalytic naphtha fraction. Additionally, based on the specific catalyst mix used, higher concentrations of butenes, pentenes, and other olefins are produced. Where applicable and in oil refineries with alkylation units, light olefin FCC products are used as feeds for alkylation units to produce premium high-octane alkylates for gasoline blending. Liquefied petroleum gas and refinery fuel gas are also produced. During the catalytic cracking process, coke is deposited as a byproduct on the catalyst, which is burned off in the regenerator, producing heat that is carried by the regenerated catalyst and supplied for heating of the fresh feed to the FCC unit. As indicated in Chapter 2 on cracking configuration refineries, the catalytic fluid cracker is a major catalytic conversion unit, as it can provide 105–110 volume percent yield of products relative to the reactor feed.[53] A simplified schematic flow diagram of a fluid catalytic cracking unit is given in Figure 3.16.[81] An FCC unit is shown in Figure 3.17.[56]

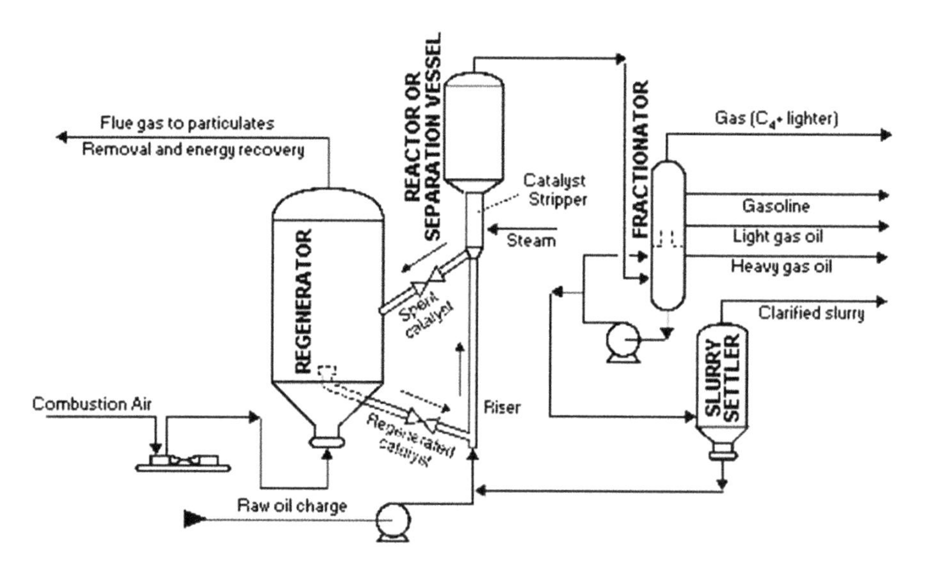

FIGURE 3.16 A schematic flow diagram of fluid catalytic cracker.
From US EIA, Fluid Catalytic Cracking is an Important Step in Producing Gasoline, Produced with Permission from Valero Corporation, www.eia.gov/todayinenergy/detail.php?id=9150.[56]

FIGURE 3.17 A picture of an FCC unit; the reactor, regenerator, and main column are highlighted.
From US EIA, Fluid Catalytic Cracking is an Important Step in Producing Gasoline, Produced with Permission from Valero Corporation, www.eia.gov/todayinenergy/detail.php?id=9150.[56]

Key sections and equipment of a fluid catalytic cracker such as the distillation column or main fractionator, slurry settler, electrostatic precipitator, CO boiler, cyclones, steam, and catalyst fine management systems are shown in Figure 3.16.

In the discussion, catalyst mix was intentionally used to describe the catalytic systems in the reactor and regenerator sections. The main catalyst in the catalyst mix is the cracking catalyst, which is typically a catalyst composed of a silica/alumina matrix and active zeolite component. Silica/alumina matrix and zeolites are combined by a binder material. Another FCC catalyst type is made via in-situ crystallization processes in which the active zeolites are created within a microsphere in a process referred to as in-situ crystallization. Due to demands for high-octane catalytic naphtha, octane additives such as ZSM-5 are used as additives to produce high-octane hydrocarbons. In order to minimize high concentrations of carbon monoxide in the dilute phase of the catalyst in the regenerator, CO promoters are used. Carbon monoxide promoters are essentially platinum and palladium promoter additives. Environmental regulations limit flue gas SO_x and NO_x emissions to 25 and 20 ppm, respectively, and this leads to the use of additives for reducing gaseous emissions.[57,58] Environmental regulations for particulates in flue gas also lead to the use of particulate reduction additives.[59] As a consequence, the catalytic performance and effectiveness of the individual additives in the catalyst mix of the FCC unit at any time are highly dependent on the variety of catalyst and additive interactions, the degree of catalyst deactivation, and management of catalyst fines. To ensure that the activity and effectiveness of the catalyst mix are maintained, makeup catalysts and additives are added as required.

3.7 COKING PROCESS UNITS

A coking process unit is the key unit in the coking refinery configuration that enables the oil refiner to significantly improve the profitability of its oil-processing business relative to those of cracking refineries, as indicated in Chapter 2. In the United States, key environmental regulations for cleaner fuels are driving the decline in residual fuel demand for fuel blending for use in the power generation sector and in asphalt production.[60] Oil refiners are upgrading heavy oil such as atmospheric and vacuum residual oil to cleaner high-value gasoline and diesel products. Crude distillation and process conversion unit capacities for the United States listed in Table 3.15 show that coking capacity increased by 67% from 1.8 to 3.0 MMBPD between 1995 and 2015.[52]

Other heavy oil-upgrading technologies that are in use include visbreaking, residual oil supercritical extraction (ROSE), solvent deasphalting, residual fluid catalytic cracking, and hydrotreating followed by hydrocracking. In a number of refinery revamps in the United States, coking units have been added to enable the oil refiner to process cost-effective heavy sour crude oils and bitumen.

There are three distinct coking technologies: fluid coking, delayed coking, and Flexicoking processes. Fluid coking is a continuous process in which the coke product is combusted to provide the heat in the process for the thermal cracking of hydrocarbons. A Flexicoker is a continuous fluid bed conversion process with integrated coke gasification to produce a clean fuel gas with a small amount of coke. Most of the coke produced is used in heat generation within the process. The delayed coking process is covered primarily as a general introduction to the coking process in this book.

Coking is a refinery unit operation that converts the bottom products from atmospheric and vacuum distillation columns into higher-value oil fractions via thermal cracking and condensation reactions. The products from coking units are typically used as feedstocks to other refinery processing units in order to produce ultralow-sulfur gasoline, diesel fuels, and coke. Feedstocks for coking units could include atmospheric and vacuum residual oils, solvent deasphalted oil, ROSE pitch and ebullated bed hydrocracker bottoms, and bitumen.[61] Characteristics for a crude oil, its atmospheric and vacuum residual oils, and a generalized range for coker feeds are given in Table 3.16. A schematic flow diagram of a delayed coking process unit is provided as Figure 3.18.

As shown in Figure 3.18, residual oil is heated at a temperature between 900 F and 960 F in a furnace. The residence time of the oil in the furnace is kept as low as feasible so as to minimize coke deposits due to condensation reactions of the hydrocarbons. Heated oil is then fed into the bottom of a large vessel called a coke drum wherein extensive controlled cracking and condensation reactions occur. A coke drum is usually operated at a pressure between 15 and 40 psig. Cracked lighter products rise to the top of the coke drum and are drawn off and sent to the main fractionator for

TABLE 3.16
Characteristics of Feedstocks for a Coker Unit

	Crude	Atmospheric Residue	Vacuum Residue	Coker Feed Range
Boiling Range, F	Whole	680+	1000+	800+
Yield, vol. %	100	40.6	16.3	45.0
API Gravity	34.0	15.7	6.4	2–10
Sulfur, wt. %	1.8	3.2	4.2	1–5
Pour Point, F		50	101	275–325
Nickel, wppm		10	25	10–100
Vanadium, wppm		37	89	50–700
CCR, wt. %[a]		8.9	20.1	10–30

Source: Data compiled and adapted by Soni Oyekan from reference data. Reid, T. A., *Coker Naphtha Hydroprocessing: Solutions for Trouble Free Operations*, AN-1997, Akzo Nobel Catalysts, Houston.[61]

[a] CCR is Conradson carbon residue.

separation into gas and liquid products such as naphtha, distillate, and gas oils. Due to the poor qualities of delayed coker unit products relative to straight-run oil fractions, its naphtha, distillate, and gas oils have to be further processed in the refinery to generate suitable blend components for gasoline and diesel fuels. Petroleum coke is produced and the quality of the coke is highly dependent on feedstock quality and

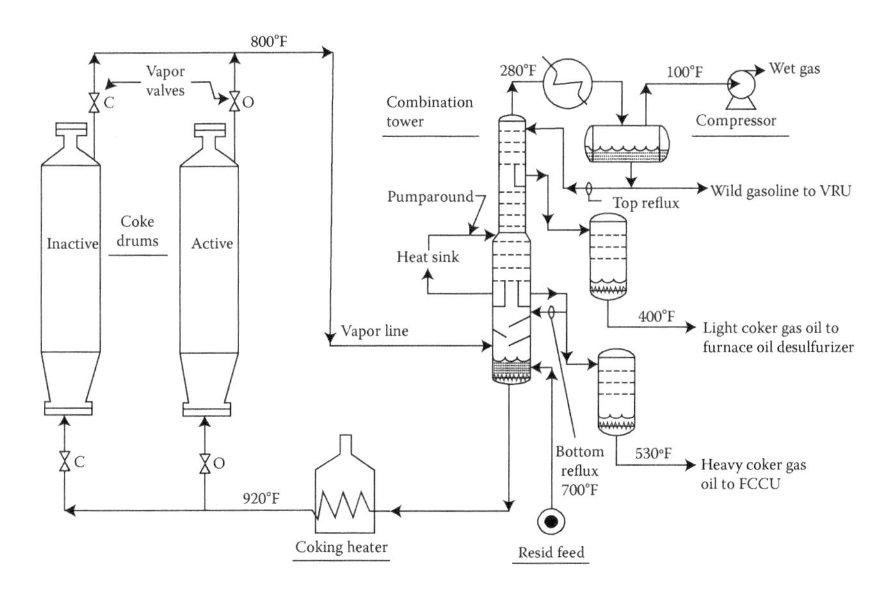

FIGURE 3.18 A schematic flow diagram of a delayed coker.
From US EIA, Coking is a Refinery Process that Produces 19% of Finished Petroleum Products Exports, Jan 28, 2013, www.eia.gov/todayinenergy/.[62]

FIGURE 3.19 A picture of a delayed coker process unit.
From US EIA, Coking is a Refinery Process that Produces 19% of Finished Petroleum Products Exports, Jan 28, 2013, www.eia.gov/todayinenergy/.[62]

process conditions. Pairs of coke drums are typically used in delayed coking units, as one coke drum holds the hot oil and serves as the "coking/reactor" drum while the other coke drum operation is progressed through the steps of cooling, decoking, and preparing to receive heated oil from the furnace. During the period of hydrocarbon cracking in the coke drum and possibly due to feed pretreatment, feed qualities, and unit operational factors, foaming may occur, and silicon-containing antifoam agents are often used to control the foaming. The silicon-containing antifoaming polymeric agents usually thermally decompose to produce silicon compounds, which are then distributed in the coker product streams and mainly in the naphtha fraction.[61] A picture of a delayed coking unit is provided as Figure 3.19.[62]

3.8 HYDROCRACKING PROCESSES

Hydrocracking is usually defined as a chemical reaction that involves primarily catalytic cracking and hydrogenation of larger hydrocarbons to produce smaller hydrocarbon compounds. In oil refining, the hydrocracking process refers more to the hydroprocessing process that combines hydrotreating and cracking of heavier gas oil–type hydrocarbons with hydrogen as the coreactant over a variety of catalysts to produce light high-value naphtha and distillate hydrocarbons. Hydrocrackers and coking units are becoming the favored process units in oil refinery revamps, as shown by the operating data in Table 3.15. For the United States, hydrocracker capacity increased by 75% from 1.2 MMBPD in 1995 to 2.1 MMBPD by 2015. Hydrocrackers

TABLE 3.17
Properties of Vacuum Gas Oil of Crude Oils

	Taching China	California USA	Maya Mexico	FCC Heavy Cycle Oil	Minas Indonesia	Arabian Heavy	Alaska North Slope
Yield, vol. %	37	24	23.0	100.0	37.0	24.0	31.0
API Gravity	31.0	14.0	21.0	8.5	34.0	22.0	20.0
			HC Type, vol. %				
Paraffins	37.0	1.0	11.0	11.0	44.0	17.0	8.0
Naphthenes	47.0	42.0	29.0	9.0	34.0	25.0	36.0
Aromatics	16.0	57.0	58.0	59.0	21.0	58.0	55.00
Sulfur, wt. %	0.1	1.0	3.1	15.0	0.1	3.0	1.20
Nitrogen, ppm	690.0	6470.0	1880.0	NR	610.0	150.0	1510.00

Source: Jechura, J., Colorado School of Mines, Hydroprocessing: Hydrotreating and Hydrocracking, Chapters 7 & 9, August 2016.[64]

Data were compiled by Soni Oyekan. Jechura, J., Colorado School of Mines Course Note; Hydroprocessing: Hydrotreating and Hydrocracking, Chapter 7 & 9.[64]

are versatile, as they can process a wide variety of feeds such as atmospheric and vacuum gas oils, coker gas oils, deasphalted oils, thermally cracked oils, visbreaker gas oils, shale oils, and blends of the oils listed and produce high yields of diesel.[66] Vacuum gas oils and other feedstocks processed in hydrocracking units vary widely, and the extent of variation is dependent on crude oils, as shown in Table 3.17.[64]

Two main sets of reactions occur in the hydrocracking process and they involve hydrocarbon cracking and hydrogen addition. Fundamental reaction mechanism studies have shown that hydrogenation of some of the aromatic rings occurs before carbon-to-carbon scission of aromatic bonds. Several fixed-bed hydrocracking, ebullated bed, and isotherming process technologies are available for use by oil refiners.[63,65,68] A schematic flow diagram of a hydrocracking unit is provided in Figure 3.20.[80]

Hydrocrackers are designed and operated to meet a variety of refinery objectives with respect to the processing of gas oils. However, they are also used in some applications in the hydrocracking of naphtha to produce liquefied petroleum gas or LPG. Usually, hydrocrackers are used primarily in the processing of gas oils. Sour refinery fuel gas, LPG, light naphtha gasoline, heavy naphtha, kerosene, jet fuel, diesel, and unconverted oil are the typical products from the hydrocracking of gas oil feeds. Catalysts used in hydrocrackers include combinations of specific catalysts for hydrodemetellation, hydrodesulfurization, and hydrodenitrogenation; a special key catalyst for cracking and saturation; and a "finishing" hydrotreating catalyst. Process conditions used in hydrocrackers are as listed in Table 3.18.[24,65] Hydrogen consumption is high and it is usually in the range of 1000–2000 standard cubic feet per barrel, which makes the hydrocracker the greatest consumer of hydrogen per barrel of feed processed.[67]

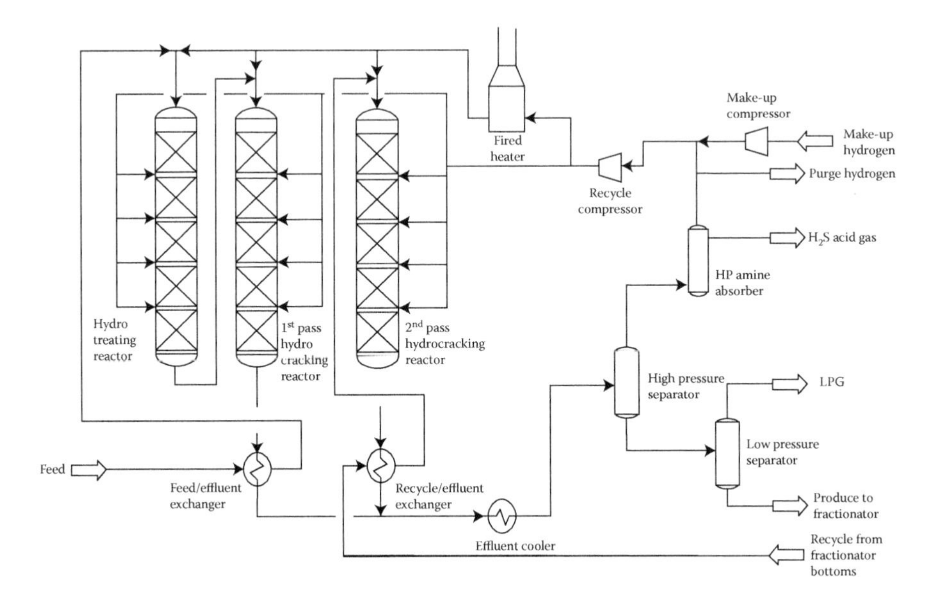

FIGURE 3.20 A schematic flow diagram of a hydrocracking unit.
From Colwell, R. F., Oil Refinery Processes – A Brief Review, Process Engineering Associates, Accessed August 10, 2016. www.processrngr.com.[80] (Courtesy of Process Engineering Associates.)

TABLE 3.18
Generalized Process Conditions for a Hydrocracker

Temperature, F	650–800
Pressure. Psig	1000–3000
LHSV, 1/hr	0.5–3.0
H2 Treat Gas, SCF/B	1000–4000
H2 Consumption, SCF/B	1000–2000

3.9 ALKYLATION PROCESSES

Alkylation process units are becoming more important and useful in plans and strategies of oil refiners to meet increasingly demanding qualities of gasoline products, as indicated in Chapter 2 with respect to reductions in the permissible concentrations of aromatics and olefins. Due to significant reductions in aromatic and olefin concentrations, higher percentages of alkylates and biofuels such as ethanol are expected to be used as blendstock for gasoline. Key reactions in the alkylation process are the reactions of C3–C5 olefins with isobutane. Secondary undesirable reactions that occur involve polymerization of some of the olefins and the other reaction-intermediate compounds. Reactions take place over either sulfuric or hydrofluoric acid via carbonium ion–type catalysis and produce isoparaffin hydrocarbons in the gasoline boiling range. The isoparaffin products, also referred to as alkylates, are premium gasoline blend stocks, as their research octane numbers are between 90 and 94. Though propylenes, butylenes, and amylenes are the olefin reactants, butylenes are preferred, as their reactions with isobutane produce higher-octane products relative to those of the other olefin compounds. In addition to alkylate, butane and LPG are the other products of alkylation units.

Sulfuric and hydrofluoric acid alkylation processes require efficient mixing systems for the reactants and catalysts, as alkylate and isobutane are sparingly soluble in the acid phase wherein the reactions occur. The solubilities of isobutane in sulfuric and hydrofluoric acids are 0.1 and 3.0 weight percent respectively. One of the key requirements of alkylation processes is the need for a high isobutane to olefin ratio feed mix so as to minimize undesirable polymerization reactions. A schematic flow diagram of a hydrofluoric acid alkylation unit is provided in Figure 3.21. As shown in the simplified flow diagram, olefin and isobutane feedstock are fed into a reactor that operates at 75–100 F for the hydrofluoric acid alkylation unit, and in the case of the sulfuric acid units, the reactor temperature is maintained at 30–65 F. For the hydrofluoric acid alkylation unit shown in Figure 3.21, the reactor effluent is sent to a settler where acid is separated from the product, and the product then goes through further separation via the deisodebutanizer before undergoing caustic treating to significantly eliminate acidic components in the alkylate product.

The safety of personnel, neighbors, and the environment are major concerns in the operation of all process units in an oil refinery, and safety concerns are elevated for these acid-catalyzed processes. Elaborate and multitier safety systems and procedures are used in oil refineries for the alkylation units. Oil refining industry and technology providers are also working on solid acid catalyst technologies to replace the current liquid acid-catalyzed alkylation processes.[68,69]

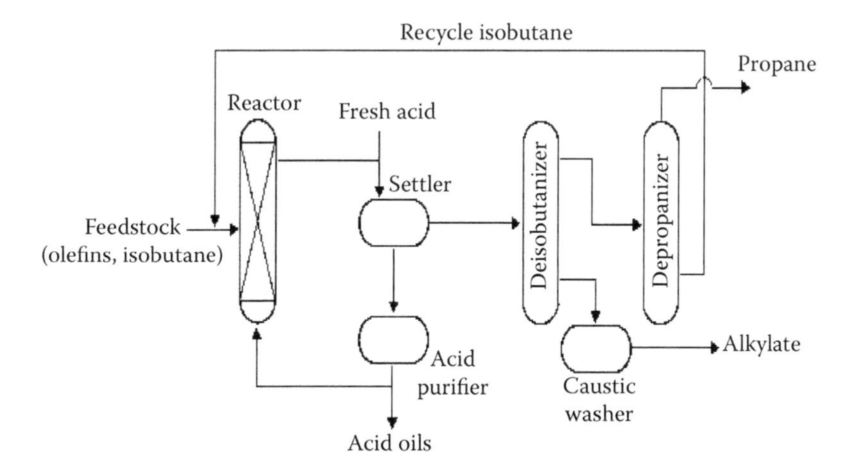

FIGURE 3.21 A hydrofluoric acid alkylation unit.

3.10 OTHER PROCESSES AND AUXILIARY SYSTEMS

In addition to crude and vacuum distillation, and thermal and hydroprocessing conversion units that have been reviewed briefly in this book, there are additional facilities such as tank farms that are used for crude oils, intermediate feed, and product blending and storage. These additional necessary facilities required in oil refineries are not covered in this book. In addition, utilities, wastewater treatment systems, separation methods, furnaces, pressure vessels, and vacuum systems are not covered. There are several good descriptions and management guidelines for acquiring, operating, and maintaining these facilities and equipment in order to provide the necessary capabilities and support for an oil refinery.[70,71]

There are increasing demands for hydrogen and sulfur recovery systems to meet the requirements of oil refiners as they process heavy, sour crude oils and unconventional bitumen-derived oils. Hydrogen supply and sulfur recovery process units are highlighted in this section, as these are required for increased hydroprocessing and coking capacities in oil refineries. Since there are pending regulations for cleaner transportation fuels, hydrogen supply and sulfur recovery units are highlighted and reviewed to complete the brief review of key oil refining process units.

3.10.1 Increased Hydrogen Demand

Hydrogen is used as a co-reactant in hydrotreating, hydrocracking, paraffin isomerization and catalytic reforming processes in oil refineries. The environmentally driven regulations for ultralow-sulfur clean fuels and fuel oils for heating have significantly increased the demand for hydrogen for oil refining processes by over 50% between 2008 and 2014, as shown in Figure 3.22. While hydrogen produced within the refineries by catalytic reformers has remained relatively constant at about 1.4 billion cubic feet per day, the volume of hydrogen supplied by gas production companies has increased from 1.3 to 2.8 billion cubic feet per day.[72] As reviewed

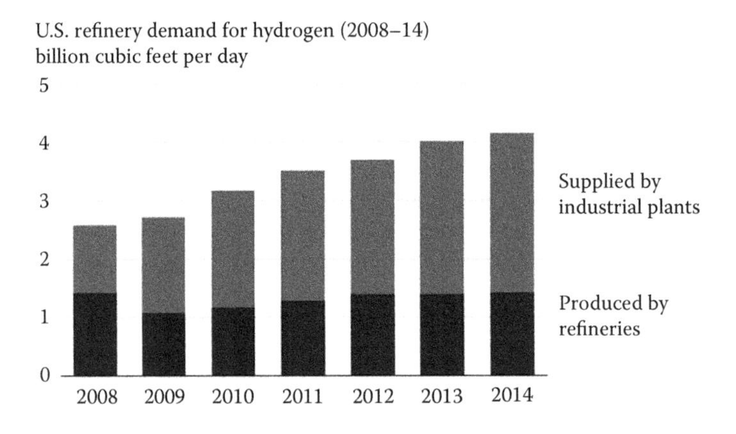

FIGURE 3.22 US hydrogen demand for 2008–2014.
From Srinivas, B. et al., Hydrogen Generation for Refineries, DOE Phase II SBIR, 2014 *Advanced manufacturing Office Peer Review*, May 5–6, 2014.[73]

previously, oil refiners are processing heavier sour crude oils and bitumen that have higher sulfur, nitrogen, metal contaminant, and asphaltene concentrations. Hydrotreating processes that are required for removal of heteroatom contaminants and hydrocracking processes drastically increase hydrogen demand. Furthermore, ultralow sulfur contents of about 10 wppm mandated for gasoline and diesel fuels increase hydrogen consumption. Low-sulfur requirements for marine fuels also lead to an increase in the demand for hydrogen for hydrotreating atmospheric and vacuum residual oils.

Hydrogen consumption for hydrotreating a selected number of atmospheric and vacuum residual oils is as provided in Table 3.19.[73] Hydrogen consumed in the hydrotreating of the West Texas Intermediate atmospheric and residual oils are 520–670 and 675–1200 standard cubic feet per barrel of feed, respectively. The hydrogen consumption in the hydrotreating of Bachaquero vacuum residual oil is in the range of 1080–1260 SCF/B due to the high concentrations of sulfur, nitrogen, and Conradson carbon residue in the oil.

Catalytic reformers supplied most of the required hydrogen for process units in most hydroskimming and cracking refineries. Hydrogen is used as a coreactant in most oil refining processes. Oil refiners that require substantial amounts of hydrogen for hydrocracking and hydrotreating process units to produce high-quality ultralow-sulfur fuels and heating oils have augmented their catalytic reformer hydrogen production with pipeline hydrogen, and steam methane reforming plant-produced hydrogen within or close to the refineries. Most SMR hydrogen production plants are operated by gas producers that sell the hydrogen gas within the refinery fence in many cases to oil refiners. Hydrogen can also be produced via the autothermal reforming process. Synthetic gases, carbon monoxide and hydrogen are produced via partial oxidation of hydrocarbons with oxygen and steam, followed by catalytic reforming.[82] As shown in Figure 3.22, gas companies produced more than 65% of the hydrogen used in oil refineries in 2014 relative to less than 50% in 2008.

3.10.2 Sulfur Recovery Processes

Sulfur recovery process units are becoming highly essential for reliable and optimal operations of oil refineries due to the increased need for extensive hydroprocessing of heavier sour crude oils and meeting stringent environmental regulatory quality standards for transportation and heating fuels. As reviewed in the previous section on hydrotreating process units, hot reaction products from the reactors are partially cooled through heat exchange with the fresh feed and sent to the gas separator. Most of the hydrogen-rich gas is treated in an amine contactor to remove acidic gases, primarily hydrogen sulfide. The hydrotreater liquid product is stripped with an amine to remove acidic gases and hydrogen sulfide.

The overhead sour gas is sent to the refinery's gas processing plant for removal of hydrogen sulfide in the amine gas treating system. After the amine gas treating system, residual products are sent through distillation towers to recover methane through pentane and heavier hydrocarbons. Hydrogen sulfide removed and recovered by the amine gas treating system is converted to elemental sulfur via a Claus process unit or to sulfuric acid via a wet sulfuric acid process unit.

TABLE 3.19
Hydrogen Consumption in the Hydrotreating of Crude Oil Fractions

	Venezuela		West Texas Inter.		Kuwait		Bachaquero
	Atmos	Vacuum	Atmos	Vacuum	Atmos	Vacuum	Vacuum
Gravity. API	15.3–17.2	4.5–7.5	17.7–17.9	10.0–13.8	15.7–17.2	5.5–8.0	5.8
Sulfur, wt. %	2.1–2.2	2.9–3.2	2.2–2.5	2.3–3.2	3.7–4.0	5.1–5.5	3.7
CCR, wt. %	9.9–10.4	20.5–21.4	8.4	12.2–14.8	8.6–9.5	16.0	23.1
Nitrogen, wt.%	NR	NR	NR	NR	0.20–0.23	NR	0.56
H2, SCF/B	425–730	825–950	520–670	675–1200	470–815	490–1200	1080–1260

Source: Srinivas, B. et al., Hydrogen Generation for Refineries, DOE Phase II SBIR, 2014 *Advanced manufacturing Office Peer Review*, May 5–6, 2014.(73)

Claus technology consists of thermal and catalytic processes, and yields of sulfur are highly dependent on the concentrations of hydrogen sulfide or its "richness." Feeds containing over 50% hydrogen sulfide are considered ideal for the process.(74) In the thermal process, hydrogen sulfide reacts with oxygen at temperatures greater than 1650 Fahrenheit.(75) The reaction is as shown below.

$$2H_2S + 3O_2 \rightarrow 2SO_2 + 2H_2O$$

In the catalytic step, residual hydrogen sulfide reacts with sulfur dioxide, one of the intermediate products of the thermal step, over either an activated aluminum (III) or titanium (IV) oxide catalyst to produce more sulfur, as shown below.

$$2H_2S + SO_2 \rightarrow 3S + 2H_20$$

Elemental sulfur produced in oil refineries is used in the production of a variety of chemicals in the chemical industry. Most of the sulfur is used in the production of sulfuric acid. Sulfuric acid is used in the production of sulfates and phosphate fertilizers. Sulfur is also used in the pharmaceutical industry in the production of vitamins. It is used in the production of chemicals and petroleum products and for a variety of industrial products. Consumers use sulfur products when they use matches, insecticides, and fungicides and when they drive automobiles, as the tires also contain sulfur used in the production of rubber.

REFERENCES

1. US EIA, Today in Energy, Crude Oil Distillation and the Defining of Refinery Capacity, July 2012, Accessed Sept 13. 2016. www.eia.gov/todayinenergy/detail.php?id=6970.
2. Gary, J. H., Handwerk, G. E., *Petroleum Refining Technology and Economics*, 2nd edition, Marcel Dekker, New York, 1984.
3. Leffler, W. L., *Petroleum Refining for Non-Technical Persons*, 2nd edition, Pennwell Books, Oklahoma, 1985.
4. Kister, H. Z., *Distillation Design*, 1st edition, McGraw Hill, New York, 1992.
5. NPRA Q&A 1996 Report on Corrosion, Optimizing Desalter Operations can Help Refiners Cope with Heavy Oils, *Oil and Gas Journal*, October 1997.
6. Mason, D., Part 7: Decontamination of the Crude Unit – Asphaltenes, *Organic Chemistry*, Feb 2016.
7. Wiehe, I., Crude Oil Compatibility, *COQG Meeting*, Houston, TX, October 2003.
8. Falkler, T., Sandhu, C., Fine-Tune Processing Heavy Crudes in your Facility, *Hydrocarbon Processing*, Sept 2010.
9. Baker Hughes, Overcoming Shale Oil Processing Challenges Assets. www.bakerhughes.com/system/00/ed8e0094a211e6809c4fad20864324/Shale-Oil-Processing-White-Paper.pdf.
10. Marcogliese, R., Refining 101: Refinery Basic Operations, Refining 201: Coking Technologies & Applications, January 16, 2007, Presentation to Lehman Brothers. http://media.corporate-ir.net/media_files/NYS/VLO/presentations/Lehman_011007.pdf
11. Emerson Process Management, Overview of Crude and Vacuum Distillation, Process Solution Section, www2.emersonprocess.com
12. Discover Petroleum, Vacuum Distillation, resources.schoolscience.co.uk/SPE/knowl/4/2index.htm?vacuum.html.

13. Martin, G. R., Lines, J. R., Golden, S. W., Understand Vacuum-System Fundamentals, *Hydrocarbon Processing*, October 1994.
14. Beychok, M., *Vacuum Distillation in Petroleum Refinery*, ChemEngineering, https://chemengineering.wikispaces.com/Vacuum+distillation+in+petroleum+refinery
15. Robles, R. Y., Hydroisomerization of Butane to Isobutane, Submitted to Dr. Arno de Klerk, University of Alberta, December 8, 2010.
16. Oyekan, S. O., Private Notes.
17. UOP LLC, Butamer™ Process, http://www.uop.com/objects/IsobutaneProcess.pdf
18. Ghosh, P., Hickey, K. J., Jaffe, S. B., Development of Detailed Gasoline Composition-Based Octane Model, *Ind. Eng. Chem. Res.*, 45, 337–445, 2006.
19. UOP LLC, Gasoline-Industry Leading Solutions for Efficient Upgrading and for Benzene Removal, www.uop.com/processingsolutions/refining/gasoline/naphtha-isomerization, Accessed September 28, 2016.
20. Graeme, S., Ross, J., Advanced Solutions for Paraffin Isomerization, *Paper AM-04-49, NPRA Annual Meeting*, March 21–23, 2004.
21. Mirimanyan, A., Vikhmon, A., Rudin, M., FSU Refiners to Build More Isom Capacity, *Oil and Gas Journal*, February 2007.
22. Colorado School of Mines, Refinery Feedstocks and Products – Properties and Specifications, inside.mines.edu/~jjechura/Refining/02_Feedstocks_&_Products.pdf
23. ExxonMobil, Worldwide Operations, Crude Trading – Assays Available for Download www.corporate.exxonmobil.com/en/company/worldwide-operations/crude oils
24. Chuang, K. C., Kokayeff, P., Oyekan, S. O., Shih, S. S., Catalytic Processes in Petroleum Refining. AIChE Today Series, 1992.
25. Parrott, S. L., Gardner, L. E., US Patent 4762814A, Hydrotreating Catalyst and Process for Its Preparation, Phillips Petroleum Company, August 9, 1988.
26. Riley, K. L., Pine, L. A., US Patent 4113656A, Hydrotreating Catalyst and Process Utilizing the Same. Exxon Research and Engineering Company, Sept 12, 1978.
27. Aldridge, C. L., Bearden, R., Riley, K. L., US Patent 4988434A, Removal of Metallic Contaminants from a Hydrocarbonaceous Liquid, Exxon Research and Engineering Company, January 19, 1991.
28. Oyekan, S. O., Townsend, G. J., Mathur, K. N., Sopko, J. S., SRC-1 Naphtha Reforming Study. Final DOE Report, DOE/OR/03054-70. 1983.
29. Loescher, M. E., *FCC Gasoline Treating Using Catalytic Distillation*, Texas Technology Showcase.Houston, Texas, March 2003.
30. Debuisschert, Q., *FCC Gasoline Desulfurization Update*, Axens European Refining Seminar, London, November 2003.
31. Debuisschert, Q., Nocca, J. L., *Prime-G+™: Commercial Performance of FCC Naphtha Desulfurization Technology*, Axens/IFP Group Technologies, Texas. www.axens.net/our-offer/by-market/oil-refining/bottom-of-the-barrel/46/gasoline-selective-desulfurization.html.
32. Honeywell UOP, FCC Gasoline Desulfurization. www.uop.com/processing-solutions/refining/gasoline/desulfurization/
33. Riley, K. L., Kaufman, J. L., Zaczepinski, S., ExxonMobil; Desai, P. H., Akzo Nobel Chemicals; Gentry, A., R., Kellogg Brown & Root (now KBR Technologies), Selective FCC Naphtha Desulphurization.
34. Brody, S. S., Chaney, J. E., Flame Photometric Detector: The Application of a Specific Detector for Phosphorus and Sulfur Compounds – Sensitive Sub Nanogram Quantities, *Gas Chromatogr.*, 4(2), 42, 1967.
35. Frye, C. G., Mosby, J. F., Kinetics of Hydrodesulfurization, *Chem. Eng. Prog.*, 63(9), 66–70, 1967.
36. Gates, B. C., Katzer, J. R., Schuit, G. C. A., *Chemistry of Catalytic Processes*, McGraw Hill, 1979, New York, p. 407.

37. Turaga, U. T., Wang, G., Ma, X., Song, C., Inhibitive Effects of Basic Nitrogen in Deep Hydrodesulfurization of Diesel, Preprint Paper – *Am. Chem. Soc, Div. Fuel*, 48(2), 550, 2003.
38. Satterfield, C. N., Modell, M., Wilkens, J. A., Simultaneous Catalytic Hydrodenitrogenation of Pyridine and Hydrodesulfurization of Thiophene, *Ind. Eng. Chem. Process. Des. Dev.*, 19, 154, 1980.
39. Cseri, T., Wanbergue, S., New ACETM Hydrotreating Catalysts, Axens European Refinery Seminar, November 2003, London.
40. Shell Global Solutions, Inside Refinery Technology Licensing. http://www.shell.com/business-customers/global-solutions/refinery-technology-licensing
41. Otwinowski, D., Streff, M., Optimal Hydroprocessing Reactor Performance, Reactor and Catalysts, Honeywell UOP, Hart Energy, September 2015.
42. Zahiravic, E., *Reactor Internals Overview*, Haldor Topsoe, Denmark, www.topsoe.com/reactor-internals-overview
43. Chevron Technology, Better Reactor Internals, www.chevrontechnologymarketing.com/CLGtech
44. Haldor Topsoe, VGO Hydrotreating/FCC Pretreatment, www.topsoe.com/processes/hydrotreating
45. Adkins, B., *Advances in Upgrading Heavy Crudes and Coker Gas Oil through Hydroprocessing*, Albemarle Catalysts Technical Seminar, Sonoma, California, July 2007.
46. Speight, J. G., Ozum, B., *Petroleum Refining Processes*, Marcel Dekker, New York, Inc., 2002.
47. Chevron Lummus Global LLC Process Flow, *2011 Refining Processing Handbook*, Hydrocarbon Processing, 2011. www.scribd.com/document/292489966/2011-Refining-Processes-Handbook-pdf
48. Shih, S. S., Summary of Catalytic Resid Upgrading Technology, provided as part of the AIChE Course on Catalytic Processes in Petroleum Refining.
49. ExxonMobil, Guide to World Crudes, ExxonMobil Published Assay for Kearl Oil Sands Diluted Bitumen, *Oil and Gas Journal*, May, 2014.
50. Jechura, J., Colorado School of Mines, Hydroprocessing: Hydrotreating & Hydrocracking, Chapters 7 and 9, August, 2016. http://inside.mines.edu/~jjechura/Refining/08_Hydroprocessing.pdf
51. Court, A. L., Smith, W. H., Canada's First Heavy Oil Upgrader Complex on Stream, *Oil and Gas Journal*, pp 46–58, June 26, 1989.
52. US EIA Report, Downstream Charge Capacity of Operable Refineries, June 22. 2016. Accessed by Soni Oyekan on Oct. 22, 2016.
53. Hemler, C. L., Smith, L. F., Chapter 3.3, UOP Fluid Catalytic Cracking Process in Handbook of Petroleum Refining Processes by Robert A Meyer, McGraw-Hill, 2004.
54. Venuto, P. B., Habib, E. T., *Fluid Catalytic Cracking with Zeolite Catalysts*, Marcel Dekker, New York, 1979.
55. Schnaith, M. W., Gilbert, A. T., Lomas, D. A., Myers, D. N., Advances in FCC Reactor Technology, Paper AM 95–36, *NPRA Annual Meeting*, San Francisco, March 19–21, 1995.
56. US EIA, Fluid Catalytic Cracking Is an Important Step in Producing Gasoline. Reproduced with Permission from Valero Corporation, www.eia.gov/todayinenergy/detail.php?id=9150
57. Nee, J. R. D., Harding, R. H., Yaluris, G., Zhao, X., Dougan, T. J., Riley, J. R., Fluid Catalytic Cracking (FCC), Catalysts and Additives, Wiley Online Library, October 2004. onlinelibrary.wiley.com/doi/10.1002/0471238961.fluidnee.a01/otherversions.
58. Oberlin, J., *A New Catalyst to Reduce Particulate Emissions*, Albemarle Catalyst Technical Seminar, Sonoma, California, July 2007.

59. McGreevy, T. D., Residual Oil Deposition: Path Forward for US & Latin American Refiners, *10th Annual Bunker and Residual Fuel oil Conference*, June 11, 2013.
60. Gianzon, G. M., *Introduction to Coking Process*, Private Communication for Soni Oyekan.
61. Reid, T. A., *Coker Naphtha Hydroprocessing: Solutions for Trouble Free Operations*, AN-1997, Akzo Nobel Catalysts, Houston.
62. US EIA, Coking Is a Refinery Process That Produces 19% of Finished Petroleum Products Exports, Jan 28, 2013, www.eia.gov/todayinenergy/
63. Scherzer, J., Gruia, A. J., *Hydrocracking Science and Technology*, Marcel Dekker, New York, 1996.
64. Jechura, J., Colorado School of Mines, Hydroprocessing: Hydrotreating and Hydrocracking, Chapters 7 & 9, August 2016. http://inside.mines.edu/~jjechura/Refining/08_Hydroprocessing.pdf.
65. Koldachenko, N., Yoon, N., Maesen, T., Torchia, D., Brossard, D., Hydroprocessing to Maximize Refinery Profitability. Paper Am-12-41, *AFPM Annual Meeting*, San Diego, March 11–13, 2012.
66. Sinopec Tech, Oil Refining: Hydrocracking Technology, www.sinopectech.com
67. Oyekan, S. O., Catalytic Applications for Enhanced Production of Transportation Fuels, 2009 NOBCChE Percy L, Julian Lecture, April 14, 2009, St. Louis, MO.
68. Brelsford, R., Chevron's Salt Lake City Refinery Plans Alkylation Unit Recovery, *Oil and Gas Journal*, October 10, 2016.
69. Albemarle's AlkyStar™ Catalyst Successfully Employed in the World's First Solid Acid Alkylation Unit in Shandong, China, Albemarle Corp, December 16, 2015. Accessed September 10, 2016.
70. Lieberman, N. P., Lieberman, E. T., *A Working Guide to Process Equipment*, McGraw Hill, New York, 1997.
71. Lieberman, N. P., *Troubleshooting Refinery Processes*, Pennwell Books, Oklahoma, 1981.
72. US EIA, Hydrogen for Refineries Is Increasingly Being Supplied by Gas Suppliers, January 20, 2016. www.eia.gov/todayinenergy/detail.php?id=24612.
73. Srinivas, B., Gebhard, S., Copeland, R., Martin, J., Hydrogen Generation for Refineries, DOE Phase II SBIR, 2014, *Advanced Manufacturing Office Peer Review*, May 5–6, 2014.
74. McIntyre, G., Yddon, L. L., *Claus Sulphur Recovery Options*, Petroleum Technology Quarterly Spring, Texas, 1997.
75. Schreiner, B., Der Claus-Prozess, Reich an Jahren und Beduetender Denn je. *Chemie in Unserer Zeit*, 42(6), 378–392, 2008.
76. Chevron, Chevron Crude Oil Marketing, Accessed September 15, 2016. http://crudemarketing.chevron.com/crude/north_american/hibernia.aspx
77. Oyekan, S. O., Voss, K. E., Desulfurization Technologies for Control of Atmospheric Emission of Sulfur Compounds in Brazil, 1983.
78. Maverick Engineering, Inc., Industrial: Refining: Atmospheric Distillation, Accessed August 12, 2016. http://www.maveng.com/index.php/business-streams/industrial/refining/atmospheric-distillation
79. Burns, J., Global Refining Capacity, Supply and Productions, *The Oil Drum*. Accessed September 13, 2016.
80. Colwell, R. F., Oil Refinery Processes – A Brief Review, Process Engineering Associates. Accessed August 10, 2016. www.processrngr.com
81. U.S. Energy Information Administration, modified from OSHA Technical Manual, Accessed September 9, 2016. http://www.eia.gov/todayinenergy/detail.php?id=9150
82. Air Liquide, Reforming (ATR) – Syngas Generation. Accessed September 18, 2017. www.engineering-airliquide.com/autothermal-reforming-atr-syngas-generation

4 Basic Principles of Catalytic Reforming Processes

4.0 INTRODUCTION TO BASIC PRINCIPLES

A catalytic reformer is one of the major process units in oil refining, as this process unit is used in almost all of the oil refineries in the world. There are typically no catalytic reformers in topping refineries. Topping refineries, as reviewed in detail in the previous chapter, are mainly crude distillation towers that generate crude oil fractions. Catalytic reforming units are usually operated to convert low-octane heavy naphtha to a higher-octane liquid product referred to as reformate. The aromatic fraction of reformates contain benzene, toluene, and xylenes, which are separated out as desired and used as feedstocks in chemical plants. In addition to reformate, catalytic reformers produce most of the hydrogen used in many oil refineries. Reformate and raffinate, the residual product after separation of benzene, toluene, and xylenes, are used as gasoline blend components. As indicated in Chapter 2, hydrogen supply from gas-producing companies to oil refineries is increasing to supplement the hydrogen produced by catalytic reformers to meet the hydrogen requirements of oil refineries. Hydrogen requirements are substantially higher than in the past 10 years due to increased processing of higher percentages of heavier, sour crude oils to produce cleaner ultralow-sulfur transportation fuels and heating oils. External hydrogen is supplied by gas producer companies such as Air Liquide, Air Products, Linde, Praxair, and Technip. In the past 5 years, catalytic reforming capacity has averaged 3.75 MMBPD relative to the 17.25 MMBPD for the atmospheric crude oil distillation capacity for the United States.[1] Catalytic reforming capacity is about 22% of the US crude oil distillation capacity. In a 2006 survey and report, about 80% of the world's refineries had catalytic naphtha reformers and the total catalytic reforming capacity was 13% of the global crude oil distillation capacity.[2] It was reported in 2010 that the total worldwide catalytic reformer capacity was 13.9 MMBPD and that it was increasing at about a 4.5% rate annually. Significant capacity growth is occurring in countries such as China and India and in Middle Eastern countries.[3] Despite changes that are occurring in oil refining processing objectives and oil product quality requirements, catalytic reformers still remain major oil conversion units for the production of gasoline; petrochemical feedstocks such as benzene, toluene, and xylenes; and hydrogen for hydroprocessing units.

4.1 A BRIEF HISTORY OF PRE-1950 CATALYTIC REFORMING PROCESSES

Most of the current catalytic hydroprocessing processes had thermal cracking as the mode of operations for generating petroleum products. The "oil refined" products were typically highly contaminated with heteroatom compounds such as sulfur and nitrogen, high percentages of unsaturated hydrocarbons, and high asphaltene concentrations in the case of thermal cracking of heavier fractions of crude oil. Thermal cracking units had to be operated in a fashion that is similar to the current delayed coking process with at least two reactors. One of the reactors was in thermal reaction service and the other was subjected to coke removal and preparation for use in the thermal cracking of incoming heated feed. By alternating between the reaction drum and coke free drum, processing was inefficiently conducted with very low yields of desired products. Most of the crude oil feed ended up as coke, asphalt, and gas products.

In the era before 1939, thermal cracking of the broad-range naphtha portion of the crude oil was conducted at temperatures of 1000 F and 500–1000 psig reactor pressures. As expected, gasoline yields were quite low; the quality of gasoline was poor in terms of current reformate qualities such as octane, sulfur, and nitrogen concentrations. In addition, since the process operated with no catalyst, excessive coke was produced, extremely short cycles were achieved, and frequent alternate cleaning of coke deposits from reactors was necessary in order to achieve modest production of gasoline.[4] By 1939 and for some years, a molybdenum oxide/alumina catalyst was used and significant increases were realized in reformate yields. The rate of coke deposition was drastically reduced, and this led to less frequent decoking of catalyst and reactors. Hydroforming was the name of the catalytic reforming process that was based on the use of molybdenum oxide catalysts. The process, as indicated, included alternate processing and decoking of reactors.

The next major historic development in the catalytic naphtha process was the invention that is credited to Vladimir Haensel of UOP with the introduction and use of platinum/alumina catalysts in catalytic reforming. A digression is necessary in order to share that some of us were fortunate to meet and be acquainted with the great Vladimir Haensel through our interactions in the North American Catalysis Society's annual meetings in the United States in the 1980s. The application of Haensel's platinum/alumina catalyst and catalytic reforming technology inventions led to major increases in gasoline yields, much reduced coke deposition, and hence longer cycles or times between catalyst regenerations.[5] UOP introduced a new catalytic reforming technology that was appropriately named the UOP Platforming process. The units were essentially fixed-bed regenerative reformers that operated with platinum/alumina monometallic catalysts. Catalyst regenerations, though less frequent relative to reforming operations before the introduction of monometallic platinum alumina catalysts, were now conducted after a total shutdown of the reformer for necessary equipment maintenance and catalyst regeneration and activation. Since catalytic reformers were the main source of hydrogen in the 1950s and 1960s, shutdowns of the catalytic reformers negatively impacted refinery utilization and profitability. It was, therefore, considered necessary to minimize significant losses in production and profitability associated with catalytic reformer outages for turnaround maintenance. Pipeline hydrogen was used if available by oil refiners to maintain production of the other oil refining process units. The frequent refinery production reductions due to frequent catalytic reformer shutdowns for catalyst regenerations, unit upsets, and routine maintenance spurred advancements in catalysis, equipment, and regeneration technologies to optimize catalytic reformer operations and profitability.

Several major process technology improvements occurred over time in the 1950s and 1960s that led to incorporating special reformer catalyst regeneration technologies. One such technology is incorporated in catalytic reforming process units that permit "cyclic" regeneration of catalysts. The major advancement permits the conduction of catalyst regeneration in a designated reactor that is aligned in a parallel piping arrangement with the other reactors in the process section of the catalytic reformer. This catalytic reforming process technology enables the oil refiner to operate the process reactors with hydrocarbon and hydrogen in the reactor process section while simultaneously regenerating catalyst in an oxidative environment in a parallel catalyst regeneration system. Selected reactors are alternately safely switched out of the main reforming process section with the use of motor-operated valves, and catalyst in the isolated or "swing" reactor is regenerated without shutting the catalytic reformer down. This mode of regeneration of isolated reactors while running the reactor side is known as cyclic catalyst regenerative reforming or simply the cyclic catalytic reforming process. ExxonMobil and Amoco, now BP, licensed the Powerformer and Ultraformer technologies, respectively. As will be reviewed later, continuous catalyst regenerative reforming technologies were later introduced by Honeywell UOP and IFP in the 1970s.

With respect to the history of the progression of catalyst innovations for catalytic reforming, the next major invention, in 1969, led to the introduction and use of a platinum modifier or promoter and the introduction of bimetallic platinum-containing catalysts. Rhenium was introduced as the promoter for the platinum alumina catalyst in the patent by Kluksdahl and Chevron.[(6)] Bimetallic platinum/rhenium catalysts are now used predominantly in fixed-bed catalytic naphtha reformers. Bimetallic and trimetallic platinum containing catalysts are reviewed in the next chapter of this book.

4.2 REFORMING PROCESS FUNDAMENTALS

Catalytic reforming is a process that is used in the upgrading of low-octane heavy naphtha feeds to produce high-value, high-octane liquid products or reformates and hydrogen. Since reformates contain significant percentages of aromatics in the benzene to C11 hydrocarbon range, benzene, toluene, and xylenes are sometimes separated out and used as feedstock in the petrochemical industries. The rest of the aromatic-depleted reformate, the raffinate, is used as a low-octane blending component for gasoline or as solvent. In addition, a number of catalytic reformers are designed primarily for the production of benzene, toluene, and xylenes and not specifically to produce reformate for motor gasoline blending. Such catalytic reformers with the specific processing objective to produce benzene, toluene, and xylene are referred to as benzene, toluene, and xylene reformers or simply BTX

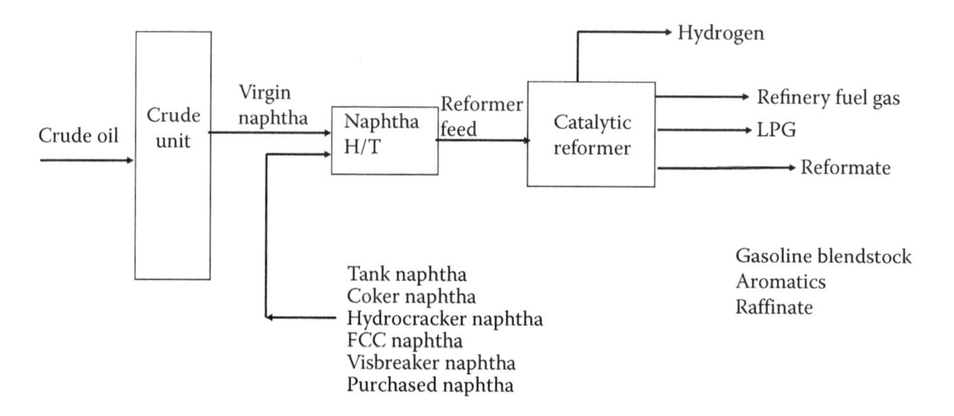

FIGURE 4.1 Catalytic reformer in a refinery.

reformers. In some petrochemical plants, xylenes and ethyl benzenes are isomerized to produce ortho-xylenes and para-xylenes.

A key feature of the catalytic reforming process is its versatility in the processing of a variety of naphthas with a wide range of paraffin and naphthene concentrations. Naphthas containing olefins and diolefins could be present in some of the unsaturated naphthas produced in hydrocarbon conversion units such as fluid catalytic crackers and cokers. Hydrotreated catalytic reformer feedstock could include straight-run (virgin) heavy naphtha from thc atmospheric crude oil distillation unit, tank farm, or other processing units such as the delayed coker, catalytic cracker, visbreaker, hydrocracker, and purchased or "opportunity" naphtha as shown in Figure 4.1. Naphtha feedstock must be hydrotreated to significantly reduce sulfur and nitrogen to less than 0.5 ppm and preferably to <0.2 wppm concentrations before processing the hydrotreated feed over the highly sensitive platinum-containing catalyst. Hydrocracker naphtha is usually fed to the catalytic reformer without undergoing additional hydrotreating. Bimetallic platinum/promoter alumina catalysts are very sensitive to metals and heteroatom contaminants in naphtha feeds. Naphtha feeds must be adequately hydrotreated to effect substantial reductions or elimination of metal contaminants such as silicon, arsenic and lead, and heteroatoms such as sulfur and nitrogen before processing over bimetallic platinum-containing catalysts.

4.2.1 Octane Numbers

A key objective of the catalytic reforming process is to increase the octane of low-octane naphtha by 25–50 octane numbers. The actual octane upgrade achieved depends on the reforming severity target, the starting octane of the naphtha feed, and the activity state of the reforming catalyst. A review of octane numbers and the octane numbers of relevant hydrocarbons in the naphtha boiling range follows. Octane number is defined as a measure of the antiknock quality of a hydrocarbon compound, which can then by extension be applied to the octane number of an aggregate or blend of hydrocarbons or fuel. Per a ranking system that was established for hydrocarbon compounds, the antiknock quality of straight-chain normal heptane is assigned a zero research octane number, and the highly branched iso paraffin, 2, 2, 4-trimethyl pentane or iso-octane, is assigned a 100 octane number. Iso-octane has a greater antiknock quality as a fuel relative to heptane and other hydrocarbon compounds with octane numbers less than 100. The octane number of a hydrocarbon or fuel is based on the result of a one-cylinder engine laboratory test in which the antiknock quality is compared to that of a blend of heptane and iso-octane per specifications that are based on the Standard ASTM Tests 2699 and 2700. The research octane number of a hydrocarbon or a fuel is measured at 600 rpm in a one-cylinder test engine and this has been determined to correlate best with engine performance under mild conditions at low speed in city driving. Another octane number measure, motor octane number, of a fuel correlates best with an engine performance test at 900 rpm. Motor octane numbers for fuels have been correlated with driving under heavy load conditions such as hill climbing and high-severity highway driving. A third octane number definition that is used commercially is the road octane number, which is the average of the research and motor octane numbers.

The road octane number, as represented by R, is used as a measure of the antiknock quality of gasoline fuels and is defined as

$$R = (RON + MON)/2 \text{ or as } R + M/2$$

The road octane number is also defined as the antiknock index or AKI.

In terms of research octane numbers or simply octane numbers for the purpose of comparing hydrocarbons, aromatic compounds have high octane numbers; for example, the octane number of benzene can vary in the range of 105–120 as shown in Table 4.1.[4] Within the family of paraffinic compounds, iso-paraffins have higher octane numbers relative to their paraffin isomers. Naphthenes have higher octane numbers relative to those of the iso-paraffins for hydrocarbon compounds with less than seven carbons. For compounds with seven carbons and greater, naphthenes have lower octane numbers relative to those of iso-paraffins. Of great interest is the fact that octane numbers for iso-paraffinic and aromatic compounds appear to increase and peak for hydrocarbons with eight carbon numbers and then drop off starting with hydrocarbons with nine carbons.

TABLE 4.1
Research Octane Numbers of Hydrocarbons

Carbon Number	Paraffins Iso/Normal	Naphthenes	Aromatics
5	92/62	102	NA
6	74/30	86	106
7	88/0	77	117
8	100/−19	70	115
9	85/−34	57	111
10	79/−53	50	109

TABLE 4.2
Research Octane Numbers of Hydrocarbons

Hydrocarbon	Carbon Number					
	5	6	7	8	9	10
N-Paraffins	62	25	0	−18	−18	−41
Iso Paraffins	85–92	73–104	42–112	22–100	0–100	0–100
Olefins	91–103	76–105	55–100	28–97	35+	35+
Naphthenes	101	83–91	67–92	47–87	18–43	NA
Aromatics	NA	100	120	102 118	>110	>110

Research octane number ranges are provided for paraffins, iso-paraffins, olefins, naphthenes, and aromatics in Table 4.2. It has been determined that the octane number for a blend of hydrocarbons often does not directly match the averaged octane numbers of the individual constituent hydrocarbons based on concentrations of the hydrocarbons. As a result, the octane number that is assumed to have been effectively contributed by a hydrocarbon after blending with other hydrocarbons is referred to as its blending octane number. The difference between the blending and research octane numbers of a hydrocarbon is usually ascribed to synergistic molecular effects between the hydrocarbons, and the blending effect differs significantly for hydrocarbons, as shown in Table 4.3 for selected notable hydrocarbons.(7)

TABLE 4.3
Octane Numbers of Selected Hydrocarbons[a]

Compound	Octane Number		Boiling Point, F
	Actual	Blending[b]	
	Paraffins		
Butane	93	113	32.0
Pentane	62	62	96.8
Hexane	25	19	156.2
Heptane	0	0	208.4
Iso-octane	100	100	210.7
	Naphthenes		
Cyclopentane	101	141	120.2
Methylcyclopentane	91	107	161.6
Cyclohexane	83	110	177.8
Higher alkylcyclopentanes and alkylcyclohexanes	70–80		190–270

[a] Satterfield, C., N., Heterogeneous Catalysis in Practice.

[b] Based on 20 vol. % of the compound in 80 vol. % of a 60:40 mixture of isooctane and heptane.

TABLE 4.4
Blending Octane Numbers of Selected Oxygenates

	Octane Number		
Compound	Actual	Blending[a]	Boiling Point, F
Methanol	92	134–138	148.0
Ethanol	97	128–135	178.0
Methyl tert-butyl ether (MTBE)	117	118	131.0
Ethyl tert butyl ether (ETBE)	115	119	163.4
Tert amyl methyl ether (TAME)	106	112	187.0

[a] The estimated "contributed" octane number of a compound in a blend of other hydrocarbons.

Blending octane numbers of naphthenes are much higher than their actual octane numbers, and those of the paraffins are much lower for paraffins with carbon numbers greater than 5.

Oxygenates such as ethers and alcohols are now used in gasoline to meet regulatory standards for fuels so as to minimize sulfur oxides (SO_x), nitrogen oxides (NO_x), and particulate pollutants in the environment. Additive ethers and alcohols provide necessary nascent oxygen for efficient combustion of fuel and for the conversion of heteroatoms in the catalytic converter to reduce automobile SO_x, NO_x, and particulate emissions. Varying amounts of concentrations of oxygenates are permitted in countries, and global gasoline standards are reviewed in Chapter 2. Oxygenates provide additional benefits as high-octane blending components, as shown by their high octane numbers in Table 4.4.

4.2.2 Process Objectives and Naphtha Feedstocks

The overall goal of the catalytic reforming process is to upgrade low-octane naphtha feeds to high-octane reformate. Heavy naphtha is an oil fraction with a broad boiling range of 160 F to 400 F composed of paraffin, naphthenes, and aromatic compounds in the C5–C12 carbon range. Based on the octane numbers of the hydrocarbons reviewed in the preceding section, it is necessary to convert a significant percentage of the normal paraffins in a naphtha feed to iso-paraffins and aromatics as well as converting naphthenes to aromatics. Process conditions for catalytic reformers are usually selected to minimize ring opening via thermal cracking and naphthene hydrogenation to form paraffins. The composition of a naphtha feedstock is relatable to the ease of reforming or conversion of its paraffins and naphthenes to iso-paraffins and aromatics. A simplified diagram of a fixed-bed catalytic reformer with three reactors, heaters, and a fractionating unit is shown in Figure 4.2. A platformer in the Khartoum Company refinery is provided as Figure 4.3.

4.2.2.1 Naphtha Composition

The feedstock to the reformer is pretreated in a naphtha hydrotreater to remove a variety of contaminants such as sulfur, nitrogen, oxygen, silicon, arsenic, iron

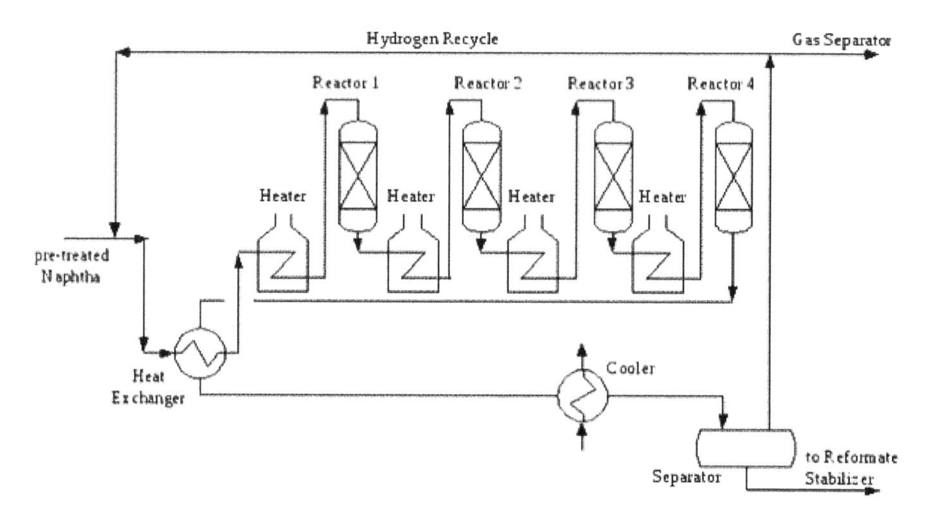

FIGURE 4.2 A simplified diagram of a catalytic reformer.

scales, and other metal contaminants. If unsaturated naphtha is included in the naphtha feed, the hydrotreater typically saturates the olefinic compounds. Naphtha feed as designated could comprise a variety of naphtha constituents, as indicated previously. It is important to note that the qualities of the constituent naphthas in a feedstock vary greatly with the degree of unsaturation of some of the hydrocarbons and concentrations of metal and heteroatoms of its constituent naphthas. Constituent

FIGURE 4.3 A catalytic reformer of the Khartoum company.

TABLE 4.5
Naphtha Feedstock Qualities

	WTI SR	Coker	FCC	Hydro/C	Solvent Refined Coal
API Gravity	54.0	57.0	54.0	41.0	38.5
ASTM Dist. F	200–400	150–350	150–270	170–350	160–375
Paraffin	49.0	34.0	27.0	45.0	38.0
Naphthenes	36.0	14.0	21.0	43.0	36.0
Aromatics	15.0	7.0	17.0	12.0	13.0
Olefin	0.0	45.0	35.0	0.0	13.0
N+2A	66.0	28.0	55.0	67.0	62.0
Sulfur, wppm	1500.0	3500.0	1000.0	0.5	7000.0
Nitrogen, wppm	1.5	100.0	40.0	0.5	2000.0

naphthas could add significant concentrations of specific contaminants' loads to the naphtha hydrotreater. Good knowledge of the properties of the feed mix and those of the constituent naphthas is required for efficient, timely, and successful troubleshooting of a catalytic reformer. Data in Table 4.5 have been reproduced from Chapter 2 to show the range of concentrations of contaminants and the distribution of paraffin, naphthenes, and aromatics in naphtha feeds to the naphtha hydrotreater. Such naphtha feeds are usually subjected to hydrotreating over base metal oxide catalysts at appropriate process conditions before feeding the hydrotreated products to catalytic reformers.

The properties of naphthas in Table 4.5 show that sulfur and nitrogen levels of a solvent-refined coal naphtha are prohibitively high, and this feed in a hydrotreating test conducted in a laboratory required multiple hydrotreating runs before achieving satisfactory reductions of sulfur and nitrogen concentrations of less than 0.5 ppm each. Coker naphthas usually have high sulfur and nitrogen and significant concentrations of diolefins. Due to significant concentrations of diolefins and heteroatom contaminants, coker naphthas are usually hydrotreated in a two-stage reactor system. The lead reactor is used mainly for saturating diolefins followed by hydrotreating in the second lag reactor. Hydrocracker and West Texas Intermediate naphthas have manageable concentrations of sulfur and nitrogen. as shown in Table 4.5. Fluid catalytic cracker naphtha usually contains high concentrations of sulfur and olefins. FCC naphthas are usually processed in dedicated gasoline hydrotreating units that can reduce sulfur concentrations of FCC naphthas and preserve the high octane of the naphtha. Sulfur concentrations of FCC naphthas depend on the degree of hydrotreating accorded the gas oil feeds before the FCC process units. Coker naphtha and other naphthas could also be profitably processed via coprocessing with catalytic cracker naphtha in dedicated gasoline hydrotreating units.

A good correlative factor has been developed that provides a measure of the quality of naphthas and the relative ease of processing or reforming of naphthas. The relative quality factor of a naphtha is defined in terms of the composition of the naphtha. An N+2A factor is determined and used. In this simple factor, N and A represent the concentrations of naphthenes and aromatics, respectively, in a naphtha. N+2A is used as a way of correlating naphtha feed composition to catalytic performance based on the expected performance of catalyst at defined process conditions. Based on the N+2A property and adequate hydrotreating of the naphthas in Table 4.5, coker naphtha with a low 28 N+2A would be a more difficult feed to reform relative to the other naphthas, as their N+2As are higher than 55. The hydrotreated coker naphtha shown in Table 4.5 would be the most difficult feed to reform relative to the hydrotreated naphthas of the others due to its low concentrations of naphthenes and aromatic compounds. Hydrocracker and West Texas Intermediate naphthas have low concentrations of sulfur and nitrogen. Sulfur and olefins in fluid catalytic cracker naphthas are typically high, and in order to preserve the high octane of FCC naphthas, processing in dedicated gasoline hydrotreating units offers a good processing option for the naphthas. Naphtha feeds are typically characterized as either paraffinic or naphthenic feeds, and that designation is dependent on the N+2A factor of either a specific naphtha or naphtha feed mix. Feeds that have high concentrations of paraffins and low naphthenes and aromatic compounds are referred to as paraffinic naphthas. Naphthas that have high concentrations of naphthenes and aromatics are referred to as naphthenic feeds. In Table 4.5, naphtha feeds with N+2A values of 55 and higher are naphthenic and the coker naphtha with its low N+2A of 28 is paraffinic.

The quality of naphtha as determined by the naphthene and paraffin content and defined by the N+2A factor has significant implications in reforming catalysis, as the catalytic reformer feed properties impact catalyst activity, selectivity, and stability performances. Catalytic reforming performance data in Table 4.6 from a catalytic reformer (Rheniforming unit) show much better activity and selectivity performances

TABLE 4.6
Effect of Naphtha Quality on Reforming Performance

Naphtha	Paraffinic	Naphthenic
Boiling Range, F	200–300	200–300
Reactor pressure	200	200
	Feed PNA, vol. %	
Paraffins	68.6	32.6
Naphthenes	23.4	55.5
Aromatics	8.0	11.9
N+2A	39.4	79.3
Sulfur, wppm	<0.2	<0.2
Nitrogen, wppm	<0.5	<0.5
	Performance	
Reformate Octane	99.0	100.0
Hydrogen, SCF/B	1200	1400
C1–C3 (SCF/B)	335	160
C5+ yield, vol. %	73.5	84.7
Aromatics, vol. %	67.9	69.9

for a naphthenic feed relative to a paraffinic feed.[8,9] As shown in Table 4.6, the N+2As of the paraffinic and naphthenic naphthas are 39.4 and 79.3, respectively. These key performance differences are reviewed in greater detail later in this book. For now and for the purpose of comparison, the desired liquid yield is defined as the C5+ yield. High C5+ liquid yields and high aromatics and hydrogen production are indicative of good catalytic reformer performance.

For the feeds shown in Table 4.6 and even at a higher product octane target of 100 for the naphthenic feed relative to 99 octane for the paraffinic feed, catalytic performance with the naphthenic feed resulted in 11.2 volume percent C5+ yield advantage, 200 SCF/B higher hydrogen, and 2 volume percent higher aromatics relative to the paraffinic naphtha.

4.2.2.2 Heavy Naphtha Boiling Range

A full-range naphtha consists of light and heavy naphtha fractions. Heavy naphtha is usually separated out at the naphtha splitter after the crude distillation unit or in a naphtha splitter after the naphtha hydrotreater reactors. Hydrotreated light naphtha could either be fed to a paraffin isomerization unit or used as a low-octane gasoline blend stock. Selected heavy naphtha boiling ranges are dependent on the refiners' overall operating strategy for oil-refined products. The recommended boiling range is usually 160 F to 380 F as determined by ASTM D86 distillation. It is well established that operating on a naphtha feed with a higher endpoint above 380 F in a catalytic reformer would lead to the formation of polynuclear aromatic compounds. In high-severity reforming operations, the rate of catalytic coke make would increase substantially. The low boiling point of 160 F is recommended so as to eliminate C5 and lighter hydrocarbons that are not ideal reformer feed compounds, as they typically crack to lighter hydrocarbons and gas in high-severity catalytic reforming operations. In low-severity catalytic reforming operations, the C5 and lighter hydrocarbons are unchanged. However, due to favorable diesel demand and prices, US oil refiners operate their catalytic reformers on naphthas with final boiling points that are about 350 F and lower so as to maximize diesel production. In the United States, the initial boiling points of feed naphthas have been increased and are higher than the recommended 160 F so as to minimize the concentrations of benzene precursor compounds in the feed to the catalytic reformer. The boiling point ranges for paraffins, naphthenes, and aromatics by carbon number are as listed in Table 4.7. Paraffins with carbon numbers of 4–6 are not reactive in catalytic reforming in the production of liquid products and, as discussed, in some high-severity operations are hydrocracked to produce refinery fuel gas and light hydrocarbons. At low operating severity, the low carbon compounds simply go through the reactors and undergo minimal catalytic conversions.

In summary, it is appropriate to emphasize that naphthas vary with respect to the crude oils from which they are derived. Naphthas also vary greatly with respect to the downstream processing units such as a coking, visbreaker, or cracking unit that produced them. These factors are clearly highlighted by the data in Table 4.8, which compare the properties of naphthas from a hydrocracker unit and from Mid Continent and light Arabian crude oils. As shown in Table 4.8, the naphtha from light Arabian crude oil is highly paraffinic, with N+2A quality of 33, and those of the hydrocracker

TABLE 4.7
Boiling Points of Reformer Feed Hydrocarbons

Carbon No.	Paraffins	Naphthenes	Aromatics	PNA[b]
4[a]	10–30	30–55		
5[a]	50–100	70–125		
6[a]	120–155	125–160	175	
7	175–210	160–220	230	
8	210–260	220–270	275–290	
9	260–305	270–315	305–350	
10	300–345	395	360–400	425
11	385	400	402	456–470
12	420	435	440	495–515
13	456	470		525

[a] These paraffins do not typically react to give liquid product. In high severity reforming operations, they hydrocrack to produce lower carbon number hydrocarbons.

[b] PNA is poly nuclear aromatics. PNAs are essentially coke precursors in naphtha reforming.

and Mid Continent crude oil are greater than 60. In addition, the octane numbers of the hydrocracker and Mid Continent naphthas are 62 and 55, respectively, and that of the Light Arabian naphtha is 33.

In general, paraffinic and naphthenic naphthas from crude oils typically fall in the broad ranges of properties listed in Table 4.9. The N+2A qualities for paraffinic naphthas fall in the range of 25–50 and those of naphthenic naphthas in the range of 50–80. Octane numbers of naphthas are important in catalytic reforming, as they dictate the severity of operations required to achieve reformate octane targets. Naphthenic naphthas have octanes in the range of 50–60 and paraffinic naphthas between 40 and 50.

TABLE 4.8
Qualities of Naphthas Vary for Crude Oils and Refinery Process Units

Naphtha	Hydrocracker	Mid Continent	Light Arabian
API Gravity, Degrees	51	56	66
ASTM D-86			
IBP	208	178	176
50%	284	246	215
FBP	394	330	298
N+2A. vol. %	79	65	33
RON, octane	62	55	41

TABLE 4.9
Paraffinic and Naphthenic Naphtha of Crude Oils

Naphtha	Naphthenic	Paraffinic
Crude, vol. %	15–25	15–25
API Gravity	50–60	50–60
Paraffins, vol. %	40–55	50–70
Naphthenes, vol. %	30–40	20–30
Aromatics, vol. %	10–20	5–15
N+2A, vol. %	50–80	25–50
RON, octane	50–60	40–50

4.3 KEY CATALYTIC REFORMING REACTIONS

Reviews of the impacts of naphtha compositions and naphtha quality factor indicated that the ease of reforming and expected product yields were dependent on the concentrations of paraffins, naphthenes, and aromatics in naphthas. A simplified measure of the naphtha quality of N+2A was deemed to be useful in correlating the expected performance of catalytic reforming in the processing of a naphtha. It was determined that in catalytic reforming, naphthenic naphtha feeds produced higher reformate, aromatics, and hydrogen products relative to those for paraffinic naphthas. In addition, paraffinic feeds produced more undesirable methane through butane gas yields as a result of "undesirable" reforming reactions. Several reactions occur during catalytic reforming of hydrocarbons. It has, however, been found to be more informative and useful to focus on five or six major reactions that significantly contribute to reformate, production, hydrogen yields, and catalytic coke. The five major reactions are naphthene dehydrogenation, isomerization of paraffins, paraffin dehydrocyclization, hydrocracking of paraffins, and hydrodealkylation of alkyl aromatics. In addition, some reviews are provided of the contributions of feed hydrocarbons and contaminants, and process conditions that promote coking reactions and higher rates of catalytic coke depositions on catalytic reforming catalysts. Condensation reactions and impact of sterically hindered hydrocarbons in the production of multiring aromatics and coke are reviewed in this chapter.

4.3.1 Naphthene Dehydrogenation

Naphthene dehydrogenation reactions are rapid, endothermic reactions that convert naphthenes to aromatics and hydrogen. The reactions occur with a net increase in the moles of gaseous products and are thermodynamically favored by high temperatures and low pressures. An example of naphthene dehydrogenation is provided in Figure 4.4 for the dehydrogenation of cyclohexane.

Since naphthene dehydrogenation reactions are rapid and consume substantial amounts of heat, heaters are required between reactors in the catalytic reforming processes to supply the necessary heat for sustaining hydrocarbon reforming reactions. Furthermore, since naphthene dehydrogenation occurs readily, it is customary to

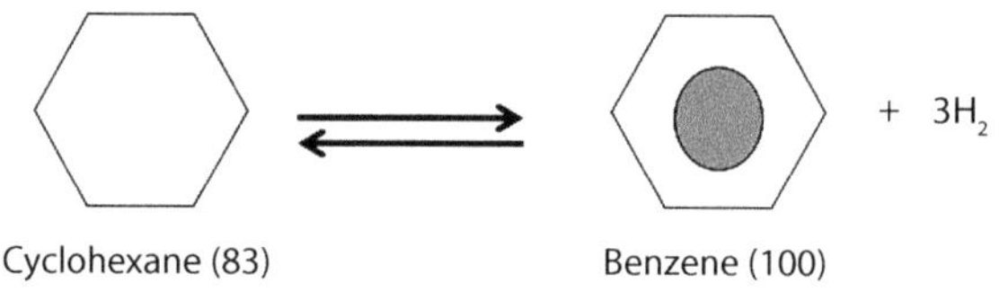

Cyclohexane (83) ⇌ Benzene (100) + $3H_2$

FIGURE 4.4 Dehydrogenation of cyclohexane to benzene.

size the reactors accordingly to take full advantage of the favorable kinetics of the reactions. As will be shown later, the first reactor is usually smaller than the second, third, and fourth reactors if it is a four-reactor catalytic reformer.

Naphthene dehydrogenation has been shown by many researchers to be a metal-catalyzed reaction, and in catalytic reforming, the metal of choice and that found to be most effective is platinum. The reactions are highly selective in producing aromatics and hydrogen.[8,9] It is the key reaction that produces most of the hydrogen in a catalytic reformer. Therefore, the production of low levels of aromatics and hydrogen could indicate subpar dehydrogenation/hydrogenation activities during reforming for a variety of reasons. Because naphthene dehydrogenation reactions are endothermic, another measure of loss of platinum activity in the first bed could be rapid loss of temperature in the lead reactor. These factors will be covered in greater detail in the section on key performance indicators and troubleshooting of catalytic reformer operations problems. Naphthene distribution also contributes significantly to octane upgrades, as observed in the reforming of low-octane naphthas in catalytic reformers. As shown in Figure 4.4, there are 17 octane number increases and 3 moles of hydrogen produced in the dehydrogenation of cyclohexane to produce benzene. Due to the rapid rates of reaction and octane boost of naphthenes, naphthenes are usually used as model compound feeds in research and development programs for developing new catalysts and technologies in catalytic reforming. Methyl cyclopentane is one of the hydrocarbons used as a model compound in catalytic reforming studies to assess the contributions of acidic and hydrogenation/dehydrogenation functionalities of catalysts.

4.3.2 Isomerization of Naphthenes and Paraffins

Naphthene isomerization reactions are critically important in the formation of aromatics. As indicated in the preceding section, naphthene dehydrogenation that produces aromatics is a facile reaction over platinum catalysts. The isomerization of paraffins to alkyl cyclopentanes requires the sequential involvement of both the dehydrogenation and acidic functionalities of the catalysts, and the formation of the intermediate compounds are considered critical steps in the formation of aromatics.[7,10,11]

The alkyl cyclopentanes produced are in turn isomerized to form other isomers of alkyl cyclopentanes and alkyl cyclohexanes. Isomerization of naphthenes are slow, slightly exothermic reactions that are sometimes accompanied by undesirable ring opening reactions. An example of an alkyl cyclopentane isomerization reaction is shown in Figure 4.5. The reactions usually yield small octane number gains. In the case of the isomerization of 1, 3 dimethyl cyclopentane to methyl cyclohexane, there is a gain of 8 octane numbers per mole of the reactant, as shown in Figure 4.5.

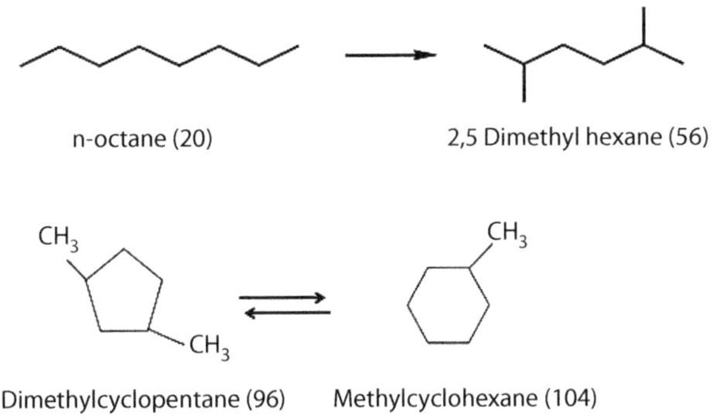

n-octane (20) 2,5 Dimethyl hexane (56)

CH_3 CH_3 CH_3

Dimethylcyclopentane (96) Methylcyclohexane (104)

FIGURE 4.5 Paraffin and naphthene isomerization.

Paraffin isomerization reactions occur to a minor extent in catalytic reforming. Due to greater number of competitive reactions, paraffin isomerization reactions are not as prominent as those that occur for C5 and C6 hydrocarbons in dedicated paraffin isomerization process units. Paraffin isomerization reactions are slower than naphthene isomerization reactions in catalytic reforming processes. Paraffin isomerization reactions are less selective relative to naphthene isomerization reactions, as significant undesirable reactions such as paraffin hydrocracking and hydrogenolysis occur. The octane boost associated with paraffins isomerization reactions is high. The isomerization of n-octane to 2, 5 dimethyl hexane leads to a 36 octane number gain per mole of the reactant converted, as shown in Figure 4.5, whereas the isomerization of dimethyl cyclopentane to methyl cyclohexane results in 8 octane number boost.

4.3.3 Paraffin Dehydrocyclization

Paraffin dehydrocyclization or paraffin aromatization, as the reaction is sometimes referred to, involves the conversion of paraffins over both the hydrogenation/dehydrogenation (platinum) and acidic sites of catalytic reforming catalysts to produce aromatics.[(7,10,11)] There are significant octane boosts for the conversion of low-octane paraffins to premium high octane number aromatic compounds. The aromatization of heptane to toluene is given in Figure 4.6.

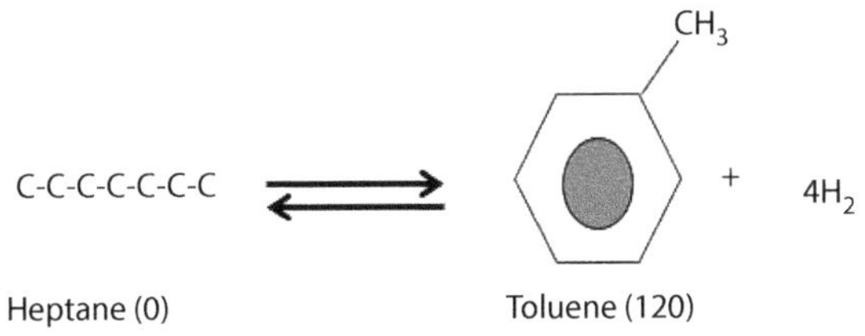

CH_3

C-C-C-C-C-C-C + $4H_2$

Heptane (0) Toluene (120)

FIGURE 4.6 Heptane dehydrocyclization.

Successful conversion of one mole of heptane to toluene contributes a huge 120 octane number gain. Similarly, conversion of other straight-chain paraffins to aromatics adds significantly to the octane numbers of reformates. Paraffin dehydrocyclization reactions are extremely slow and the reaction rates are about two orders of magnitude lower than those of naphthene dehydrogenation. A key objective of the catalytic reforming process is high conversion of paraffins via aromatization reactions to increase the aromatics concentrations of reformates. The catalytic reforming process is designed to maximize the degree of conversion of a high percentage of reformer feed paraffins within the reactors. One of the key factors in the design of the reactors is the need to provide sufficient residence time so a high percentage of slow paraffin aromatization reactions are enhanced and completed. Sixty to 70% of the catalyst load is usually charged into the last two reactors in a four-reactor catalytic reforming process unit. Aromatization rates have been determined by a number of researchers over the past 60 years.(12–15)

A paraffin dehydrocyclization reaction involves multiple intermediate steps with molecular species moving rapidly and sequentially between the metal and acidic sites. Based on the many pathways that the intermediate molecular species can take, various products other than the desired aromatics can occur. Mills and Mobil conducted a major research and development program that has enabled researchers and process technologists in catalytic reforming to develop sound basic working knowledge of the key mechanisms of paraffin aromatization reactions.(16) The specific mechanism provided by Mills et al. has provided insights for the author of this book in his inventions for catalytic process technologies and in solving myriad catalytic reformer performance challenges over the past 40 years, and this mechanism is shared with you.(16,17)

It should be noted per the paraffin aromatization mechanism provided by Mills and others that appropriate "levels" of metal and acid functionalities are required for dehydrogenation and isomerization or rearrangement of the straight-chain paraffin to form an alkyl cyclopentane intermediate, as shown in Figure 4.7. The formation of the alkyl cyclopentane intermediate is an established necessary first step in paraffin dehydrocyclization reactions to form aromatic compounds. Platinum sites are then

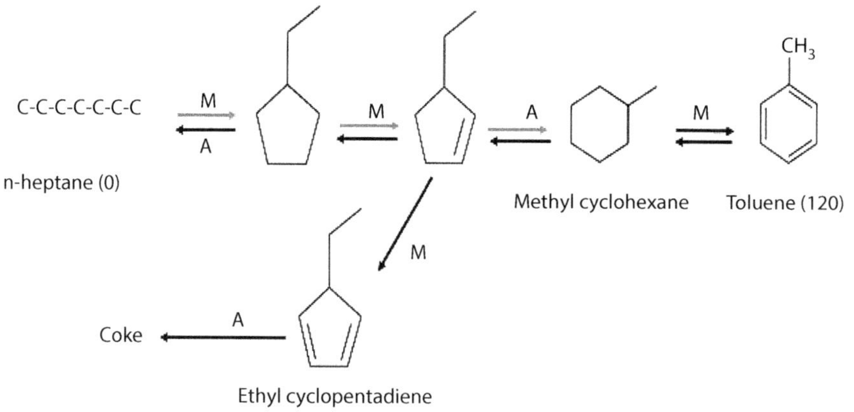

FIGURE 4.7 A simplified mechanism for paraffin aromatization reaction.

required to convert the alky cyclopentane to a cyclic olefin compound and acidic site to form alkyl cyclohexane and then aromatics. However, an undesirable pathway involves further dehydrogenation of alkyl cyclopentenes over metal sites to form alkyl cyclopentadienes. Further reactions of alkyl cyclopentadienes over acidic sites lead to coke precursors and coke, thus reducing liquid and aromatic compound yields and depleting catalytic activity via increased coke deposition on the catalyst.

A number of model compounds, such as hexane, heptane, methyl cyclopentane, cyclohexane, methyl cyclohexane, and benzene, and the paraffin aromatization mechanism above have been invaluable in reconciling bench scale and pilot plant data with expected commercial catalytic reformer unit performance for researchers and oil refining technologists of catalytic reforming technologies.

4.3.4 Paraffin Hydrocracking

Hydrocracking is the reaction of straight-chain paraffinic hydrocarbons with hydrogen to produce methane, ethane, propane, and butane. In most motor gasoline and aromatic reformers, hydrocracking is not a desirable reaction, as it leads to liquid product losses. Essentially, there is a significant loss of reformates and aromatics depending on the extent of paraffin hydrocracking reactions. However, the reactions could also be beneficial in reducing coking by the cracking that occurs for sterically hindered hydrocarbons. Hydrocracking leads to the elimination of bulky alkyl side chains on naphthenes and aromatics. In the case of hydrocracking of side chains on aromatic rings, the reaction is sometimes referred to as aromatic hydrodealkylation. Hydrodealkylation of aromatic compounds is covered in more detail in the next section. The hydrocracking of hexane is as provided in Figure 4.8.

C-C-C-C-C-C ⟶ C-C-C-C + C-C

Hexane (25) Butane (93) Ethane (118)

FIGURE 4.8 Hydrocracking of hexane.

Paraffin hydrocracking is moderately fast and the reactions are faster than those of paraffin dehydrocyclization. The reactions are catalyzed by metal and acidic functionalities of the catalyst. Reactions are exothermic and, if they occur in the lead reactors, lead to significant declines in the lead reactor temperature drop. Paraffin hydrocracking reactions typically occur mostly in the last two reactors.

In specific oil refined market environments with high demand for liquefied petroleum gas from naphtha, those specific markets have been supplied with specific, suitable catalysts for maximizing the production of propane and butanes from naphtha instead of liquid products.(17)

4.3.5 Hydrodealkylation of Alkyl Aromatics

Hydrodealkylation of naphthenes and aromatics is a desirable reaction for increasing C6 through C8 aromatic compound yields and reducing the rate of coke formation in catalytic reformers. Cracking of bulky alkyl side chains from naphthene intermediates

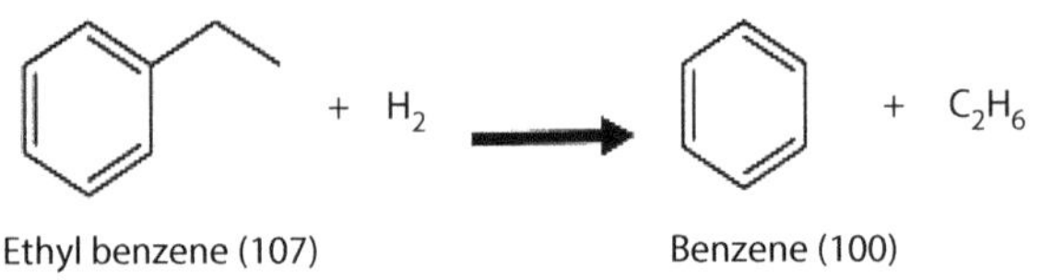

$+ H_2$ → $+ C_2H_6$

Ethyl benzene (107) Benzene (100)

FIGURE 4.9 Hydrodealkylation of ethyl benzene.

in the paraffin aromatization mechanism leads to a lower rate of coke formation and higher C5+ liquid yields. Depending on the metal and acidic functionalities of the catalysts and proccss conditions of the catalytic reformers, alkyl aromatic compounds react with hydrogen to produce benzene, toluene, and xylenes and alkanes. As an example, hydrodealkylation of ethyl benzene as shown in Figure 4.9 leads to the production of benzene and ethane. Hydrodealkylation reactions confounded oil refiners who had expected negligible benzene in reformates after eliminating benzene precursor hydrocarbons from heavy naphtha feeds to the catalytic reformer and instead found appreciable benzene in the reformate. The rates of aromatic hydrodealkylation reactions are slow and comparable to the rates of paraffin hydrocracking reactions. Hydrodealkylation of aromatics reactions are slightly exothermic.

Several major catalytic reforming reactions have been reviewed. Universal Oil Product has combined a number of these reactions in an overall scheme showing pathways for the reactions and the specific catalyzing actions of metal and acid functionalities, as shown in Figure 4.10.[3,19]

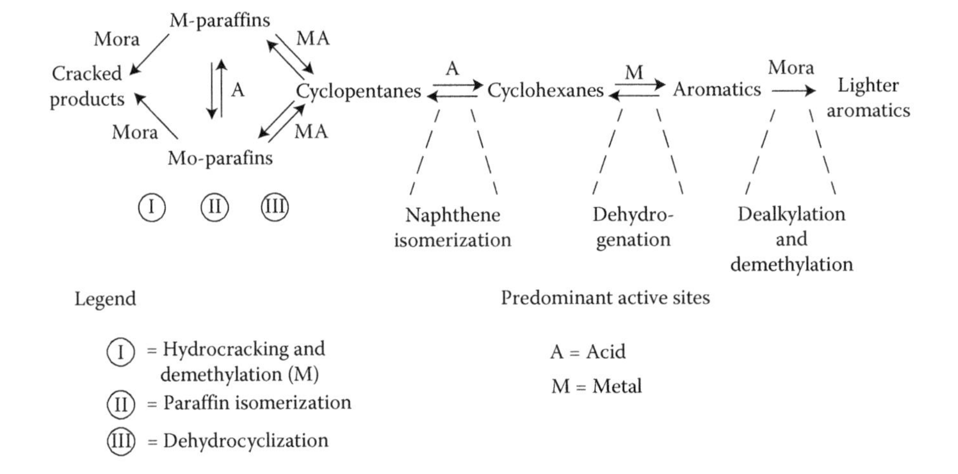

FIGURE 4.10 Generalized catalytic reforming reaction scheme.
From Poparad, A. et al., Reforming Solutions for Improved Profits in an Up-Down World, Paper AM-11-59. UOP LLC. Accessed July 2016. https://www.uop.com/?document=uop-reforming-solutions-for-improved-profits-paper&download=1; Oyekan, S. O., Personal Lecture Notes, 2010; Hawkins, G. B., Catalytic Reforming: Catalyst, Process Technology and Operations Review. Accessed September 16, 2017. https://www.slideshare.net/GerardBHawkins/catalytic-reforming-catalyst-process-technology-and-operations-overview.[3,19,20]

4.4 PROCESS VARIABLES OF CATALYTIC REFORMERS

The independent process variables in catalytic reforming are target product or reformate octane, reactor pressure, hydrogen to hydrocarbon molar ratio, liquid hourly space velocity, naphtha properties, catalyst life, and feedstock additives. Dependent variables are reformate and hydrogen yields, reactor temperature, catalyst deactivation due to coking or catalyst poisoning, and reformate quality. Catalytic reforming process variables are reviewed in this book.

Key process feedstock inputs such as paraffinic and naphthenic naphthas, boiling range of the naphtha, and the expected target octane for reformate help determine the key process variables for a catalytic reformer. Paraffinic naphthas would produce much lower yields of reformate and hydrogen, as reviewed earlier and in Table 4.6. Reforming of high-endpoint, naphthenic naphtha at appropriate process conditions typically leads to production of higher yields of reformates and coke relative to lower-endpoint naphthas. In general, high-quality naphthenic feeds, characterized by high N+2As, would require lower catalyst volumes and lower temperatures and produce higher yields of hydrogen relative to those of poor paraffinic feeds.

In addition to unit capacity, feed quality, and reformate target octanes, another determinant factor for selecting the process variables for a catalytic reformer is the mode of catalyst regeneration. Oil refiners usually select processing units for their grassroots and revamped refineries based on the markets that they are operating in, the turnaround maintenance schedules they plan to use, and availability of money and turnaround maintenance companies and equipment. Most of the top-tier oil performing refiners prefer to use "staggered" outage scheduling of process units and do not typically take complete shutdowns of their refineries for turnaround maintenance. For the ideal staggered process units' turnaround scheduling and for refineries with multiple catalytic reformers, shutting down one of the catalytic reformers to conduct equipment maintenance and catalyst regeneration is accomplished with minimal negative impacts on oil refining production planning. Flexibility in the modes of operating catalytic reformers is reviewed in Chapter 6 in the section on catalytic reforming process technologies. A second type of catalytic reformer is one that permits regeneration of a percentage of total system catalyst such as 20%–25% of total system catalyst via regeneration of one of the reactors while the other reactors in the process loop are conducting reforming of the naphtha feed. This mode of regeneration is referred to as the cyclic catalyst regeneration mode of operation. The latest key technology introduced to the industry in the early 1970s permits continuous regeneration of catalyst while reforming naphtha, and this mode of regeneration is referred to as continuous catalyst regeneration. The oil refiner's processing plans, turnaround maintenance frequency, hydrogen availability, capital, and complexity of the oil refineries usually dictate the catalytic reforming technology selected for use. In the past 20 years, most grassroots catalytic reformers have been continuous catalyst regeneration process units. Several semiregenerative catalytic reformers have been revamped to add continuous catalyst regeneration units and more extensive product separation systems for efficient hydrocarbon recovery and higher hydrogen purity. A broad range of process variables is provided in Table 4.10 which encompasses catalytic reforming technologies that incorporate one of the three distinct catalyst regeneration modes described earlier.[21]

4.4.1 Reactor Temperature

The reactor temperature is one of the important process variables in catalytic processes. For reactions in the catalytic reforming process, it has been suggested that an increase of reactor temperature of 30 F doubles the rate of reactions. However, the increase in the reaction rates could be negated within a short period of operating the catalytic reformer, as a 25 F rise in reactor temperature could also double the rate of coke deposition on catalytic reforming catalysts and lead to significant reformate and hydrogen yield losses due to increases in the rates of undesirable hydrocracking reactions. Thus, selection of reactor temperatures and operating severities for catalytic reformers must take into consideration catalyst regeneration frequency, the capabilities of catalyst regeneration systems, and reformer productivity targets.

In catalytic processes, weighted average bed temperatures (WABT) and weighted average inlet temperatures (WAIT) are used for correlating effective reactor temperatures to catalytic units' performances. In catalytic reforming, the reactor temperature is defined in a way to account for the unique use of multiple reactors and varying temperature drops in the reactors due to endothermic reactions taking place in the reactors. Weighted average inlet temperature is the preferred reactor temperature indicator in catalytic reformers.

The weighted average inlet temperature is calculated using the inlet temperatures and averaging them based on the fraction of catalysts in reactors. The WAIT formula for a four-reactor catalytic unit is as shown below.

$$
\begin{aligned}
\text{WAIT} = {} & (\text{weight fraction of catalyst in Reactor 1}) \times (\text{Reactor 1 Inlet Temperature}) \\
& + (\text{weight fraction of catalyst in Reactor 2}) \times (\text{Reactor 2 Inlet Temperature}) \\
& + (\text{weight fraction of catalyst in Reactor 3}) \times (\text{Reactor 3 Inlet Temperature}) \\
& + (\text{weight fraction of catalyst in Reactor 4}) \times (\text{Reactor 4 Inlet Temperature})
\end{aligned}
$$

Weighted average bed temperature is similarly defined as:

$$
\begin{aligned}
\text{WABT} = {} & (\text{weight fraction of catalyst in Reactor 1}) \times (\text{Reactor 1 Bed Temperature}) \\
& + (\text{weight fraction of catalyst in Reactor 2}) \times (\text{Reactor 2 Bed Temperature}) \\
& + (\text{weight fraction of catalyst in Reactor 3}) \times (\text{Reactor 3 Bed Temperature}) \\
& + (\text{weight fraction of catalyst in Reactor 4}) \times (\text{Reactor 4 Bed Temperature})
\end{aligned}
$$

The reactor temperature range listed in Table 4.10 is based on the weighted average inlet temperature.

Another indicator of catalytic performance in catalytic reforming units is the temperature drop determined for the reactors. Reactor temperature drops or delta temperatures (delta Ts or DTs) observed as a result of the net endothermic reactions occurring in the individual reactors and the total delta T are calculated and compared over time and cycles to monitor catalytic reformer performance. A history of such data can also be used for guidance in making decisions for the replacement of spent

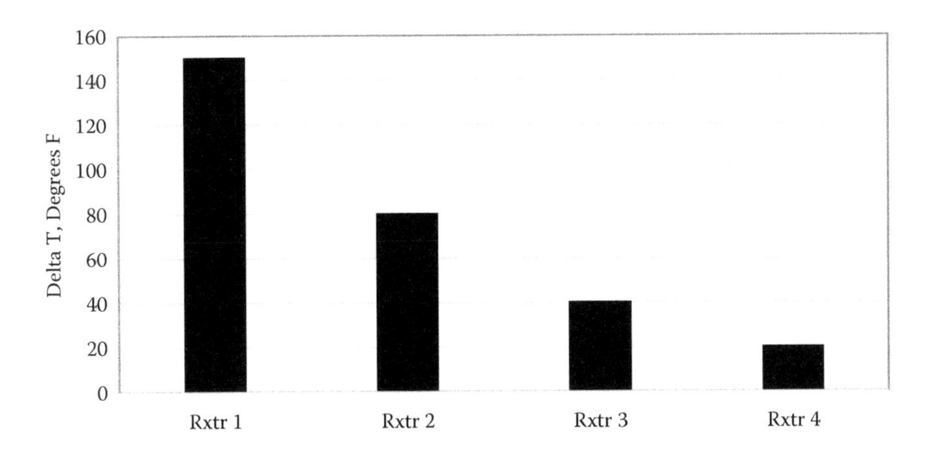

FIGURE 4.11 Reactor temperature deltas for a four-reactor catalytic reformer.

reforming catalysts. Reactor delta T is simply defined as the difference between the inlet and outlet temperatures of a reactor. An example of reactor DTs for a four-reactor fixed-bed semiregenerative catalytic reformer is provided in Figure 4.11.

A third temperature monitoring parameter for assessing catalytic reformer performance is delta WAIT. Delta WAIT is defined as the difference between the current overall reactors' weighted average inlet temperature and the start of run (SOR) weighted average inlet temperature. SOR WAIT is defined as the start of the run weighted average inlet temperature and is usually determined for semiregenerative catalytic reformers immediately after the initial catalyst activity equilibration at startup of a catalytic reformer. The activity term can also be applied with some in-depth understanding to fixed-bed cyclic regenerative and continuous catalyst regenerative reformers.

$$\text{Delta WAIT} = \text{WAIT} - \text{SOR WAIT}.$$

For catalytic reforming processes and especially for fixed-bed catalyst regenerative reformers, the increase in delta WAIT provides a good measure of catalyst deactivation. As a consequence, when a catalytic reformer delta WAIT increases by 30 or 40 degrees, it is at the end of the cycle of profitable reforming operations. That end-of-cycle indicator is usually the time to conduct turnaround maintenance and catalyst regeneration.

4.4.2 Reactor Pressure

During the pre-1970 era of catalytic reforming, high pressures in excess of 250 psig were required in order to moderate the rate of catalyst deactivation in the cyclic regenerative and especially in semiregenerative catalytic reformers. In order to successfully moderate the rate of catalyst deactivation for catalytic reformers, it is necessary to operate at moderate process severities and process conditions. High

TABLE 4.10
Range of Process Variables for Catalytic Reformers

Pressure, psig	35–700
Temperature, F	850–1000
LHSV, 1/hr	1–5
H_2/HC Molar Ratio	1.5–10
Reforming Severity, octane	80–107

hydrogen to hydrocarbon molar ratio recycle gas and low reformate octane targets were recommended in order to increase the times for processing operations or cycle lengths between catalyst regenerations. Additionally, appropriate naphtha feeds were used and naphthenic feeds were preferred. Additionally, high reactor pressures and pressures up to 700 psig were used for semiregenerative catalytic reformers. Based on the selectivity advantages that lower-pressure operations can contribute for improving liquid yields from naphthene dehydrogenation and paraffin dehydrocyclization reactions, technologies such as those based on continuous catalyst regeneration were developed and now permit operating catalytic reformers at very low pressures. Catalytic reforming is now conducted in units with reactor pressures as low as 35 psig. Reactor pressures are usually determined from the pressures of the low gas separators. Low-pressure operations lead to rapid coke deposition on the catalyst, and continuous catalyst regeneration operations enable the catalytic reformers to operate at essentially at fresh catalyst activity and selectivity. Pressure range for the three catalytic reforming processes is 35–700 psig, as listed in Table 4.10. Lower reforming operations are favorable for maximizing reformate and hydrogen yields. Reformate yields are as shown in Figure 4.12 for naphtha reforming operations at 100 and 300

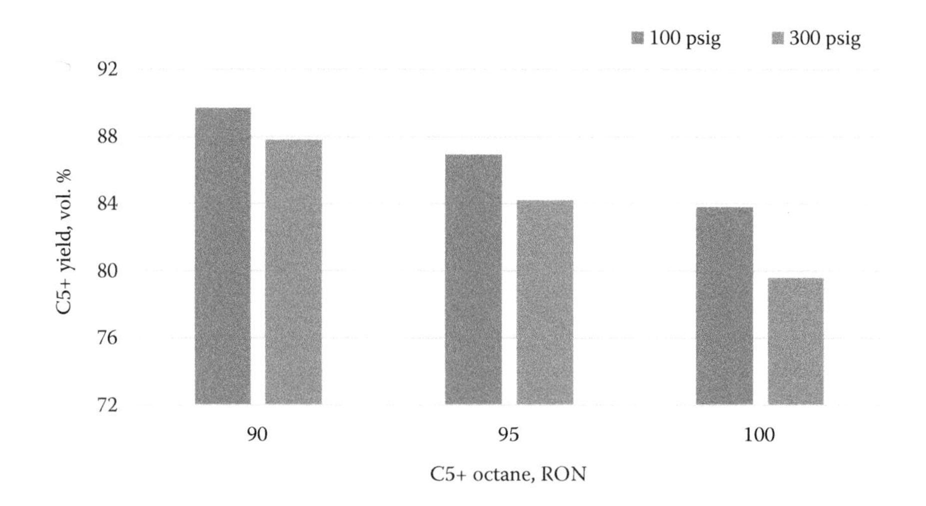

FIGURE 4.12 Effect of reforming pressure on reformate yield.

psig. The reformate yield benefit for the 100 psig operation was about 5 volume percent relative to the 300-psig operation at a reforming severity of 100 RON and on the same naphtha feed.[19,21]

4.4.3 Recycle Gas Hydrogen to Hydrocarbon Molar Ratio

In most hydroprocessing units, hydrogen is used as a reactant, and some of the residual hydrogen and light hydrocarbon gases recovered after the high-pressure separator are recycled to the reactor to help suppress coke formation and deposition on the catalyst. The higher the recycle gas and hydrogen purity, the higher the partial pressure of hydrogen in the reactor. High hydrogen partial pressures in the catalytic reformer have the same effects as high reactor pressure, as they aid in suppressing the rate of coke formation and deposition on the catalyst. High coke make leads to lower reformate, lower aromatics yields, and lower hydrogen production. In addition, for fixed-bed, semiregenerative catalytic reformers, much shorter cycle lengths or much shorter times between reformer shutdowns for catalyst regenerations would be expected for operations with low hydrogen partial pressures. The hydrogen recycle rate in catalytic reforming is measured as standard cubic feet of hydrogen per barrel of feed and as hydrogen to hydrocarbon molar ratio. The hydrogen to hydrocarbon molar ratio of the recycle gas is used predominantly for the catalytic reforming process. Catalytic reforming operations are usually conducted with recycle gas hydrogen to hydrocarbon molar ratios in the range of 1.5–10.0. The wide range encompasses the low hydrogen to hydrocarbon molar ratio of 1.5–3.0 used in continuous catalyst regenerative and cyclic catalyst regenerative units to the high 5–10 hydrogen to hydrocarbon molar ratios used in fixed-bed semiregenerative catalytic reformers.[4,21]

4.4.4 Reformate Octane

A reformate octane target is usually selected as the desired liquid product octane in the operation of a catalytic reformer consistent with gasoline blending requirements and hydrogen production. In the case of fixed-bed catalytic reformers, the octane targets are also selected so that the reformer can be operated over a long cycle of 12–18 months. This is necessary so as to minimize the frequency of reformer shutdowns for turnaround maintenance and catalyst regenerations and replacements. The octane target for a catalytic reformer is referred to as the octane severity of the process unit. A catalytic reformer operating at 101 octane relative to another operating at 99 RON is considered to be operating at a higher reforming severity. Higher-octane reforming operations of 101 octane would produce lower reformate yields relative to reforming operations at a lower 99 octane severity if the catalyst operates on the same naphtha feed. The octane number used as the reformer octane target is based on the research octane number, which was defined and reviewed in Chapter 3. It is determined via a single engine operated at a speed of 600 rpm with an air intake temperature of 60 F to 125 F in the oil refiners' laboratory.[22] Octane numbers of reformate in processing units can also be determined online via the use of the near-infrared (NIR) methodology. The reformate octane target is achieved in catalytic reformers by setting the reactor temperature that would produce reformate with quality that matches the target octane. The reactor temperatures can then be adjusted as required to produce reformates with different octanes as desired. Reformers operate at reformate severities that range from 80 to 107 octane, as shown in Table 4.10 due to the fact that catalytic reformers operate over a wide range of octane for either motor gasoline or aromatics production.[19]

4.5 GENERAL CATALYTIC REFORMING OPERATIONS

As previously indicated, catalytic reforming units are critically important in processing strategies and plans of oil refiners, as the units not only produce reformate for motor gasoline and aromatics for petrochemical plants, they are also the only units that produce hydrogen within the refinery. Catalytic reforming units produce almost 60% of the hydrogen required in oil refineries for use in the other hydrogen-consuming hydroprocessing units.

A simplified sequence of unit operations for a catalytic reformer is shown in Figure 4.13. The feed to a catalytic reformer is a hydrotreated heavy naphtha product from a naphtha hydrotreater. Hydrotreated naphtha should meet sulfur, nitrogen, metal, and other contaminant concentration specifications for efficient processing of naphtha over highly sensitive monometallic and bimetallic platinum-containing catalysts. Bimetallic platinum/rhenium and platinum/tin catalysts are more sensitive to sulfur relative to monometallic platinum catalysts.

Hot effluent product and gas from the last reactor of a multireactor system exchanges heat in the combined feed effluent (CFE) exchanger with naphtha feed. The heated feed is sent to the charge heater for further heat up to the specified lead reactor inlet temperature. Due to the fact that key reactions such as naphthene dehydrogenation and paraffin dehydrocyclization are endothermic, there are significant heat losses as determined by the temperature drops through the reactors. The greatest temperature drops occur in the lead reactors and drop off as represented in Tables 4.11 and 4.12 for three- and four-reactor systems, respectively. For the three- and four-reactor

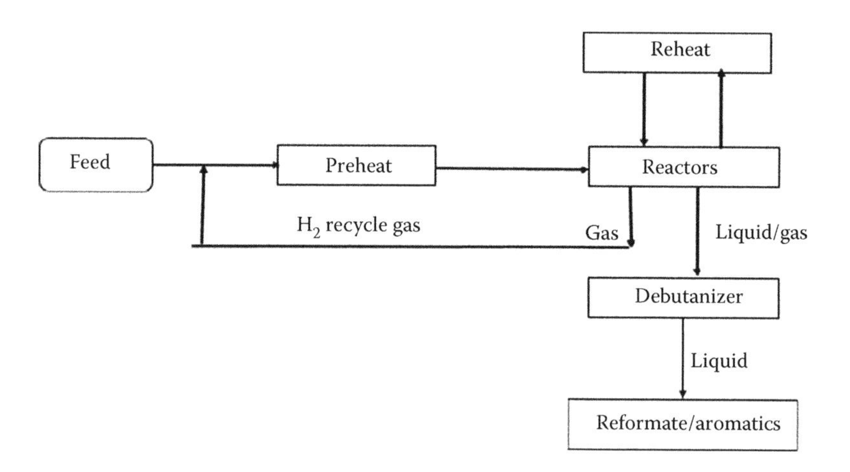

FIGURE 4.13 A simplified drawing of the flow scheme of a catalytic reformer.

TABLE 4.11
Reactor Temperature Deltas for a Three-Reactor Reformer

Reactor	Delta Temperature, F	Percent DT
1	211.0	51.3
2	121.0	29.4
3	79.0	19.3

Where: DT is delta temperature or temperature drop.
Percent DT = (Reactor DT/Total DT for the Reactors) × 100.

TABLE 4.12
Reactor Temperature Deltas for a Four-Reactor Reformer

Reactor	Delta Temperature, F	Percent DT
1	150.0	42.6
2	86.0	24.4
3	68.0	19.3
4	48.0	13.7

reforming systems in Tables 4.11 and 4.12, the total temperature deltas were 411 and 352 degrees, respectively.[19,21]

Reheat is required after each of the reactors to maintain desired reactor inlet temperatures for the next reactors. Some of the reheat is also necessary for semiregeneration and cyclic regeneration catalyst systems to compensate for additional deactivation of the catalyst in order to maintain desired reactor inlet temperatures. For continuous catalyst regeneration reformers, catalyst deactivation and losses via catalyst fines are compensated for by the addition of fresh catalyst as makeup catalyst.

Reactor product effluent is sent to the separator, where some hydrogen and light hydrocarbon gases are recycled to the reactors and the rest of the product is separated in the debutanizer or stabilizer to recover reformate, LPG, and refinery fuel gas.

4.6 CATALYTIC REFORMING PROCESS EQUIPMENT

There are several necessary pieces of equipment in catalytic reformers, and they are essentially similar to those that are used in the other oil refining hydroprocessing units. Some of the major equipment includes the reactors, heaters, heat exchangers, rotating equipment, and product separation section equipment. The product separation section typically includes gas compressors, separation vessels, and chloride guard beds. Some of this equipment, its operations, and its reliability are reviewed later in this book.

4.6.1 Reactors

Multiple reactors are used in catalytic reforming process units to compensate for the fact that the overall heat requirement for the set of reactions is significant, as a number of the key reactions are endothermic. Naphthene dehydrogenation reactions are highly endothermic and cause significant temperature drops as the reactions progress through lead reactors. The net endothermic heat requirements are usually relatable to the naphtha feed quality and target octane severity operations, and these data are used in the design of grassroots and revamps of catalytic reformers. Thus, the total heat of reaction for the catalytic reformer is dependent on the naphthene concentration of the naphtha feed and desired reformate octane. The total expected heat of reactions is used in determining and designing the number of reactors for a catalytic reformer. Individual reactors are sized to hold a percentage of the total catalyst load to permit completion of the necessary reactions to meet catalytic reforming process objectives. The number of reactors could be three, four, or five, with the catalyst load divided proportionately with the number of reactors. Thus, the percentages of catalyst loads in a three-reactor system could be 20%, 30%, and 50% and 10%, and 15%, 25%, and 50% for a four-reactor system. For a catalytic reformer with five reactors, the catalyst load percentages could be 5%, 10%, 15%, 20%, and 50%, respectively, for the five reactors. The last reactor usually holds the highest percentage of the catalyst so as to provide sufficient residence time for completion of paraffin dehydrocyclization reactions that are required to meet target reformate octanes.[4,19,21]

Specific reactors used are dependent on the process technology licensor as well as important requirements of the specific catalytic reforming technology selected. The materials of construction used in the manufacturing of a reactor are dependent on whether it is a cold- or hot-wall reactor. For the hot-wall reactor type, low 1.25 Cr-0.50 Mo, ASTM specification A-337 Grade 11 is used, and for cold-wall reactors, C-0.5 Mo, A-204 Grade A/B/C is used. Low Cr-Mo steels have high temperature strengths and are resistant to hydrogen attack. Reactor internals are usually assembled by using parts that are made from Austenitic 316, 321, and 347 steel.[23] Austenitic steels are used because of their weldability and have high corrosion resistance properties.

Most of the reactors used in catalytic reformers are vertical downflow, radial downflow, and radial upflow types. Reactors and catalyst loading schemes are selected so as to provide adequate catalyst containment and permit unobstructed gas flow and good flow distribution of reactants, intermediates, and products through catalyst beds.

4.6.1.1 Downflow Reactor

A downflow reactor is the type used in fixed-bed semicatalyst regeneration catalytic reformers, and it is similar to the earlier naphtha hydrotreater reactors. A sketch of a downflow fixed-bed reactor that has been loaded with catalyst and inert materials or inert balls is provided in Figure 4.14. Catalyst inert balls of different sizes are used to ensure catalyst containment and for good flow distribution through the catalyst bed.

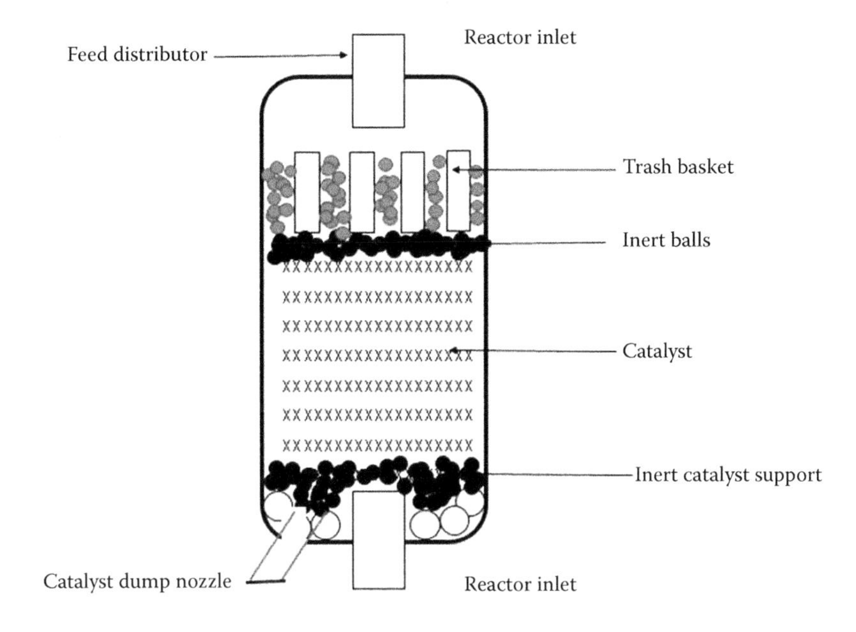

FIGURE 4.14 Downflow fixed bed reactor.

4.6.1.2 Spherical Reactor

A second type of reactor is the spherical reactor, which is used mainly in cyclic catalyst regenerative reformers such as in ExxonMobil's licensed Powerformer and Amoco's Ultraformer. Key features of the spherical reactors are similar to those of the downflow fixed-bed reactor except for the larger diameters of the spherical reactor, which could pose practical problems with leveling of catalyst and inert balls mixing during catalyst loading.

A diagram of a spherical reactor loaded with catalyst and inert balls is shown in Figure 4.15.

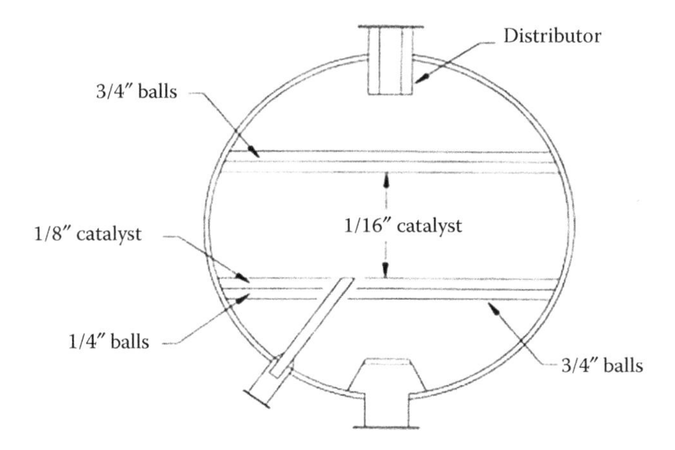

FIGURE 4.15 Spherical downflow reactor.

4.6.1.3 Radial Downflow Reactor

Radial downflow reactors are used extensively in fixed-bed semiregenerative reformers, and a diagram of the radial downflow reactor is as shown in Figure 4.16. In this reactor, catalyst is contained in the space between the perforated center pipe and outer screen through which the mixed reactants, intermediate compounds, and products flow radially and exit through the center pipe. Catalyst containment is achieved via the use of the outer screen, center pipe, and catalyst support. Radial downflow reactors are now more commonly used in grassroots fixed-bed semiregenerative catalytic reformers.

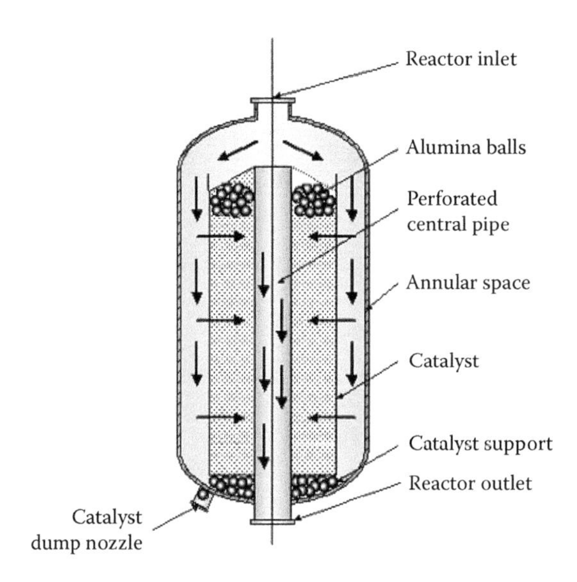

FIGURE 4.16 Radial downflow reactor.

The number of catalytic reformers using radial downflow reactors is increasing, as they are used in grassroots continuous catalyst regenerative reformers licensed by UOP and Axens. UOP-licensed CCR process technology typically uses the stacked radial reactors configuration, while Axens, licensor of the Octanizer CCR catalytic reformer technology, has the side-by-side reactor configuration for its CCR process technology as well as for its Octanizer semiregenerative catalytic reformers. Chevrons Rheniforming, Atlantic Richfield Oil Company (ARCO), Engelhard (now BASF) Magnaforming, and UOP's Platforming units also use the side-by-side reactor configurations for their semiregenerative catalytic reformers.[4,19,21] Badische Anilin und Soda Fabrik (BASF) acquired Engelhard in 2006.

4.6.1.4 Advances in Continuous Catalyst Regeneration Radial Reactor Design

There has been at least three generations of specially constructed radial reactors used in UOP's stacked reactors for CCR and CycleMax process units over the past 45 years. The first set of UOP's CCR process units used an outer basket that was

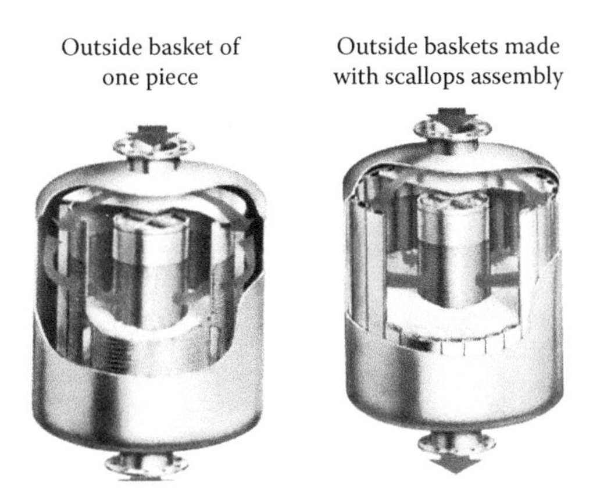

FIGURE 4.17 Radial reactors with outer basket and scallops.
Courtesy of Aqseptence Group—Johnson Screens.

made of one cylindrical piece and referred to as the outer basket of the reactor. Later improvements led to the use of an outer basket that was assembled from appropriately sized scallops arranged in a symmetric fashion around the center pipe screen. The catalyst bed is contained within the annular space between the center pipe and one outer basket or an outer basket created by aligning and installing a number of scallops around the center pipes. The two forms of radial reactors with the single outer basket and assembly of scallop outer basket are compared in Figure 4.17.[24]

There are a number of documented benefits for the use of scallops as the outer basket in reactors relative to reactors with the single-unit cylindrical outer basket. The scallop outer baskets permit high contact areas between the screens and catalyst beds, which leads to lower fluid velocities. The vertical slot orientation of the scallop reduces abrasion between the surfaces of the screens and catalyst and that leads to reductions in the concentrations of catalyst fines that enter catalyst beds. Reductions in the concentrations of catalyst fines in catalyst beds leads to good bed permeability and an even flow distribution. Another key benefit is the fact that repairs and maintenance of the scallops are much easier and less time consuming relative to those required for the massive one-unit outer baskets. There are potential benefits for reformate and hydrogen productivity, and profitability associated with the opportunities to increase catalyst loads and thereby reduce liquid hourly space velocities for constant octane severity operations as reactor outer baskets are replaced with the scallop outer basket assemblies.

There have been significant improvements in the type of materials used in the manufacture of scallops for the outer screen. The Johnson Screens company of Minneapolis, Minnesota, has developed three major types of designs for the scallop over the years. They offer perforated, Vee-wire, and Optimiser scallops and the necessary center pipes to match the type of material selected for the scallops.[24] As indicated in the picture credits, after over 100 years of business operations, the Johnson Screens company underwent two name changes after 2013. For those who may not have followed, the Johnson Screens company name was first changed to Bilfinger Water Technologies and then to the current Aqseptence Group—Johnson Screens.

FIGURE 4.18 Optimiser, Vee-wire, and perforated scallop designs.
Courtesy of Aqseptence Group—Johnson Screens.

The Optimiser scallop is as shown in Figure 4.18 with its trapezoidal cross-section in contrast to the semicircle cross section of the Vee-wire and perforated scallops. The correct installation of the reactor internals of the appropriate center pipe and scallop materials is vital for the reliable, profitable operations of a stacked CCR or CycleMax CCR unit. It is vital to select appropriate companies for the fabrication of the reactors and reactor internals and their installation due to the number of items that have to be incorporated into the reactors. It is also highly recommended that oil refining companies use the inspection services group of Honeywell UOP and other companies if feasible to ensure that the necessary reactor installation work for the CCR or CycleMax reactors is completed satisfactorily during commissioning of grassroots units and turnaround work for existing catalytic reformers. An example of the precision can be gleaned from the completed installation of Vee-wire scallops and other reactor internals, as shown in Figure 4.19.[24]

FIGURE 4.19 View looking up into Vee-wire scallops installation for a reactor.
Courtesy of Aqseptence Group—Johnson Screens.

FIGURE 4.20 Installation of optimiser scallops in a reactor.
Courtesy of Aqseptence Group—Johnson Screens.

The Optimiser scallops are much more rugged and reliable relative to the perforated and Vee-wire scallops and are now the internal outer screen assemblage of choice for larger reactors and high-capacity CCR and CycleMax reformers. Adequate inspection of ongoing and finished installation of the Optimiser scallops and associated reactor internals is highly recommended. The installation of the trapezoidal Optimiser screens is as shown in Figure 4.20.[24]

A variation of downflow radial reactors is upflow radial reactors in which the reactor effluent flows through the center pipe in an upflow direction and out from the same direction that the hydrocarbon feed and hydrogen entered the reactor. Upflow radial reactors are useful, as their usage aids in minimizing "dead zone" or "no-flow" sections and the amount of "heel" catalyst at the bottom of each of the stacked radial reactors. There are several advantages for upflow radial reactors. The advantages include minimization of the dead zones in reactors, reduction in the amount of heel catalyst, enhanced catalytic performance, and lower catalyst and platform costs.

4.6.2 Process Heaters

The catalytic reforming process is endothermic due to the highly endothermic naphthene dehydrogenation reactions. Heat transfer to the reactants and hydrogen is required before feeding the lead reactor and for heat transfer to the reactants and intermediate product mix between reactors. As indicated, the naphtha feed and hydrogen are heated initially through heat exchange with the product effluent from the last reactor in a combined feed heat exchanger. In between the reactors, the effluent from a reactor is heated to provide necessary heat to the process reactants/products. Direct-fired heaters are used to provide the necessary heat. The direct-fired heater is a heat exchanger that uses the hot gases of combustion to raise the temperature of a feed flowing through coils of tubes arranged within the heater. Natural gas, refinery fuel gas, waste gases, and residual oils are typically used as fuels to fired heaters or process heaters. Depending on the use, the fired heaters are also called furnaces and process heaters. A simplified sketch of the key sections of a fired heater is provided in Figure 4.21.[25] The drawing shows the stack, convection and radiant, breeching, and shield sections.

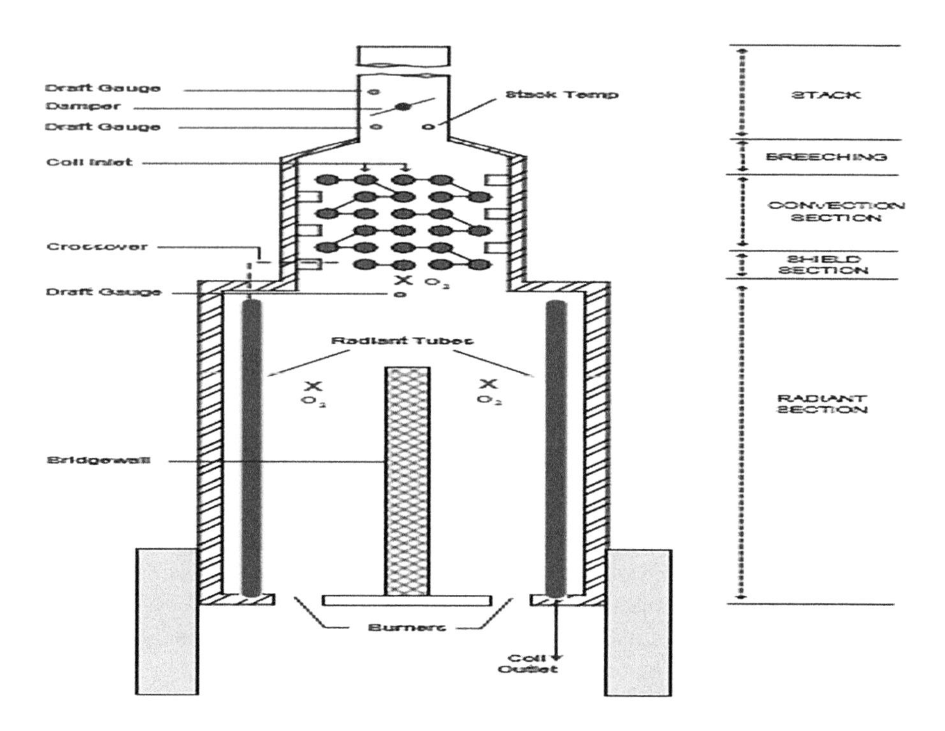

FIGURE 4.21 Sections of a process heater.
Courtesy of Ametek Process Instruments. From Ametek Process Instruments, Process Heaters, Furnaces and Fired Heaters. Accessed July 10, 2017, www.ametekpi.com.[25]

The complete combustion of natural gas leads to the production of carbon dioxide and water, as shown in the equation below with the evolution of about 14.4 mbtu/lb of methane.

$$CH_4 + 2O_2 \rightarrow CO_2 + 2H_2O \ (14.4\,\text{mbtu})$$

Incomplete combustion of the natural gas leads to the production of carbon monoxide and much less heat production, which then leads to greater firing due to insufficient oxygen.

$$2CH_4 + 2O_2 \rightarrow 2CO + 4H_2O \ (4\,\text{mbtu/lb})$$

Natural gas and essentially hydrocarbon fuel combustion products are water, carbon dioxide, carbon monoxide, sulfur oxides, and nitrogen oxides. To minimize the generation of excessive carbon monoxide and the energy required to heat up air, the excess oxygen in the stack gases should be greater than 2% and less than 4%. Process tubes in the radiant section containing either process feed or reactant/product mixes, as in the case of catalytic reformers, receive about 85% of the required heat in this section. The tubes in the radiant section absorb direct radiant heat from the burners and the refractory walls of the radiant section.

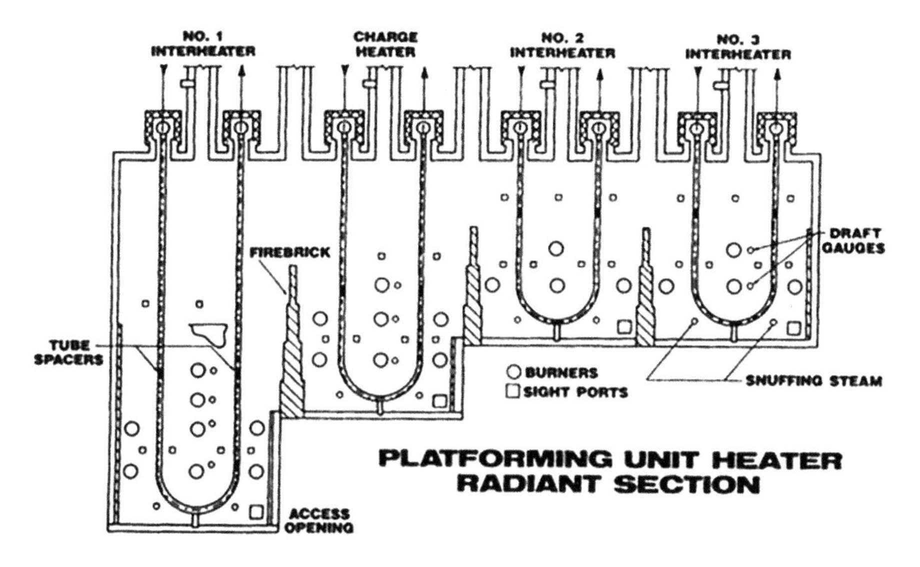

FIGURE 4.22 Platforming U-type heater radiant section.

Due to high fuel costs associated with the operations of fired heaters and the minimization of NO_x and CO_x emissions, complete combustion of fuels is desired, as shown by the amount of heat evolved, for example, in the combustion of methane relative to that for incomplete combustion. Different heater styles are in use, and they include those with U-tube, double-I, and single-I configurations.

A multicell fired process heater that was licensed and popularized by UOP for catalytic reformers is shown in Figure 4.22 with the firebricks separating the cells. Burners are located in the center space of the U-tube sections and next to the refractory walls as shown in Figure 4.22 to maximize radiant heat to the tubes containing process fluids.[3,25]

A picture of a process heater is provided in Figure 4.23 with the multicell radiant section for heating the reactor feed and the interreactor effluents.

UOP Platforming U-type multicell heaters have been shown to provide significant capital and operating cost benefits over individual heaters that were used as the charge and interreactor heaters in the older fixed-bed semiregenerative catalytic reformers. With larger multicell heaters, furnace light off startup and operational procedures must be carefully adhered to minimize damage to heaters and nearby process units. Prudent safe operation of multicell heaters can really unlock the capital and energy benefits expected with using them, especially for the highly endothermic catalytic reformers.

There are some general operational variable and key performance indicators that should be monitored for the safe utilization of the 9% Cr, 1% Mo metallurgy tube in the fired heaters. The key performance indicators include maintaining operations in the fired heaters such that the tube wall temperature (TWT) is less than 635°C or 1175 F and the bridge wall temperature (BWT) is less than 900°C or 1650 F. In addition, due to increased concerns with potential occurrence of metal carburization in heater tubes and consequent damages, it has been recommended that sufficient

FIGURE 4.23 Process heater for a catalytic reformer.

concentrations of hydrogen sulfide be present in the recycle gas by the licensors of the continuous catalyst regeneration catalytic reforming technologies.

In order to operate at the high 85+% fuel efficiency levels, the fired heaters should be monitored as often as required using good analytical instrumentations to measure firebox pressures, bridge wall temperatures, tube wall temperatures, stack temperatures, and oxygen and combustible concentrations. Burners should be properly designed and operated adequately at all times. Fuel gas filters and coalescers should be installed and operated as recommended to ensure delivery of clean fuels to the burner tips. Fired heater manuals' guidelines should be used to ensure trouble-free, reliable, and highly efficient fired heater operations.

Key factors that can influence heater efficiency include operating at high charge rates and increased tube fouling. Increased charge rates and tube fouling factors can lead to TWT and BWT limitations, increased fuel firing, and increased NO_x and CO_x emissions. Additionally, the factors also lead to shorter tube life. The external surfaces of the radiant tube section can be fouled with heavy oxidation scale and combustion products, thus limiting heat transfer to reactants and product mixes in the tubes. Fouling of the internal section of the tubes could also occur due to coke deposition. Oxidation scales that accumulate externally on heater tubes can be removed via cleaning by specialized heat transfer maintenance companies such as IGS-Cetek and others.[26] Applications of the IGS-Cetek high emissivity ceramic coating technology to cleaned external surfaces of heater tubes have been shown to lead to over 400 F reduction in tube wall temperature and consequent

FIGURE 4.24 Tubes after application of high emissivity ceramic coatings by IGS-Cetek. From Bacon, J., Latte, D. D., Increasing Process Efficiency via High emissivity Ceramic Coatings, www.scribd.com.[27] (Courtesy of IGS-Cetek.)

immediate catalytic reformer productivity benefits. The ceramic coatings provide a durable, protective thin-film layer on the external surfaces of the heater tubes, which prevents oxidation and corrosion of the tube metal. A set of process tubes that has been subjected to the cleaning and coating process by IGS-Cetek is as shown in Figure 4.24.[27]

4.6.3 Heat Exchangers

A heat exchanger is equipment that is used in the transfer of heat between a solid object and a fluid or between two or more fluids. Shell and tube, welded plate, plate and shell, plate fin, tube and tube, vertical heat exchangers, and waste heat recovery equipment represent a variety of some the heat exchangers in use. For most of the applications in catalytic reformers and especially for the combined feed and effluent exchangers, shell and tube and welded plate exchangers are predominantly used. As a consequence, a short review of the catalytic reformer heat exchanger and especially of the CFE will focus on shell and tube and plate heat exchangers.

4.6.3.1 Shell and Tube Heat Exchangers

These exchangers are essentially a bundle of tubes that contain the fluid that is to be cooled or heated, and the bundle of tubes is enclosed by a shell that contains the other fluids for the heat transfer. The second fluid flows over the tubes that are to be heated or cooled. Tubes could be plain, helical, longitudinally finned, or twisted.[28] Shell and tube exchangers are strong and reliable, and as a result, they are employed in oil refining processes where the operating pressures could be higher than 400 psig and at higher reactor temperatures. A detailed diagram of a shell and tube heat exchanger with description is provided in Figure 4.25.[29]

The ends of the tubes are connected to fixed tube sheet through holes. Baffles, shown in Figure 4.25, are used to help direct flow over the tubes and minimize

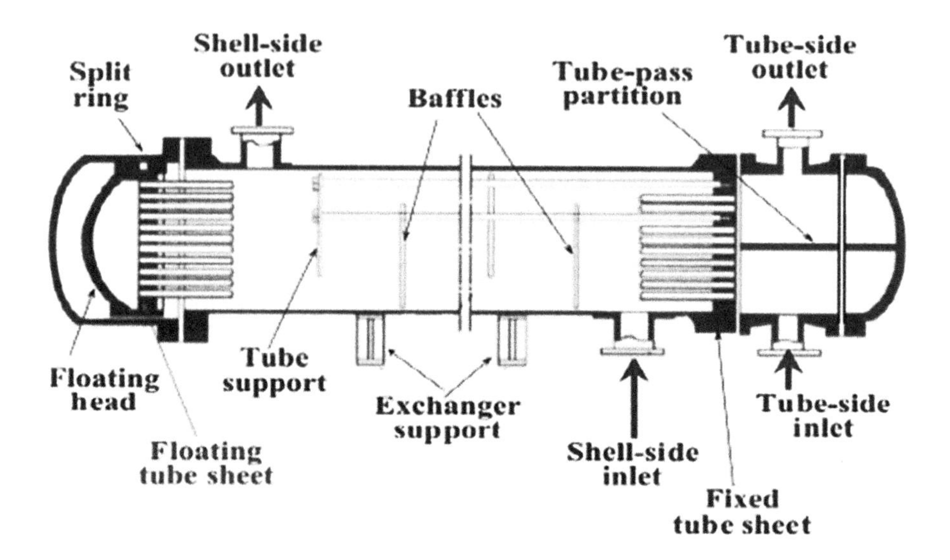

FIGURE 4.25 Shell and tube heat exchanger as per Patwardhan.
From Patwardhan, A. V., Momentum and Heat Transfer, Applications of the Principles of Heat Transfer to Design of Heat Exchangers, Institute of Chemical Technology. https://www.slideshare.net/KathiresanNadar/applications-of-the-principles-of-heat-transfer-to-design-of-heat-exchangers.[29]

vibration of the shell and tube exchanger. The baffles usually run in a perpendicular direction to the shell and confer some rigidity to the bundle of tubes. Another drawing showing the flow of one fluid over the tubes is provided in Figure 4.26.

A variation of shell and tube heat exchangers is the vertical combined feed effluent (VCFE) exchanger that is used as the combined feed effluent exchanger in recent continuous catalyst regeneration catalytic reformers. A schematic picture of a VCFE is provided in Figure 4.27. Vertical combined feed effluent exchangers

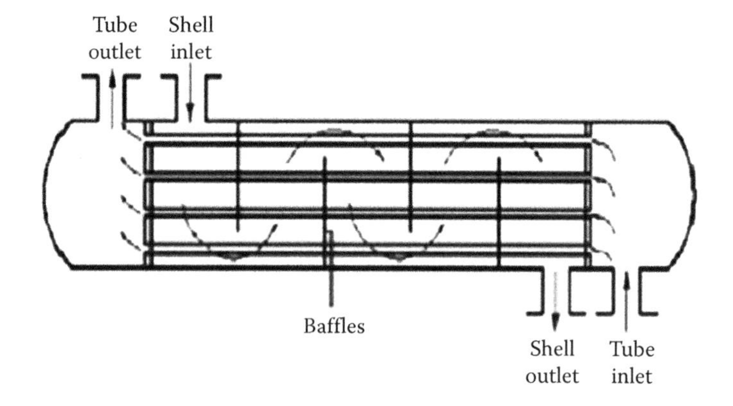

FIGURE 4.26 Shell and tube heat exchanger.

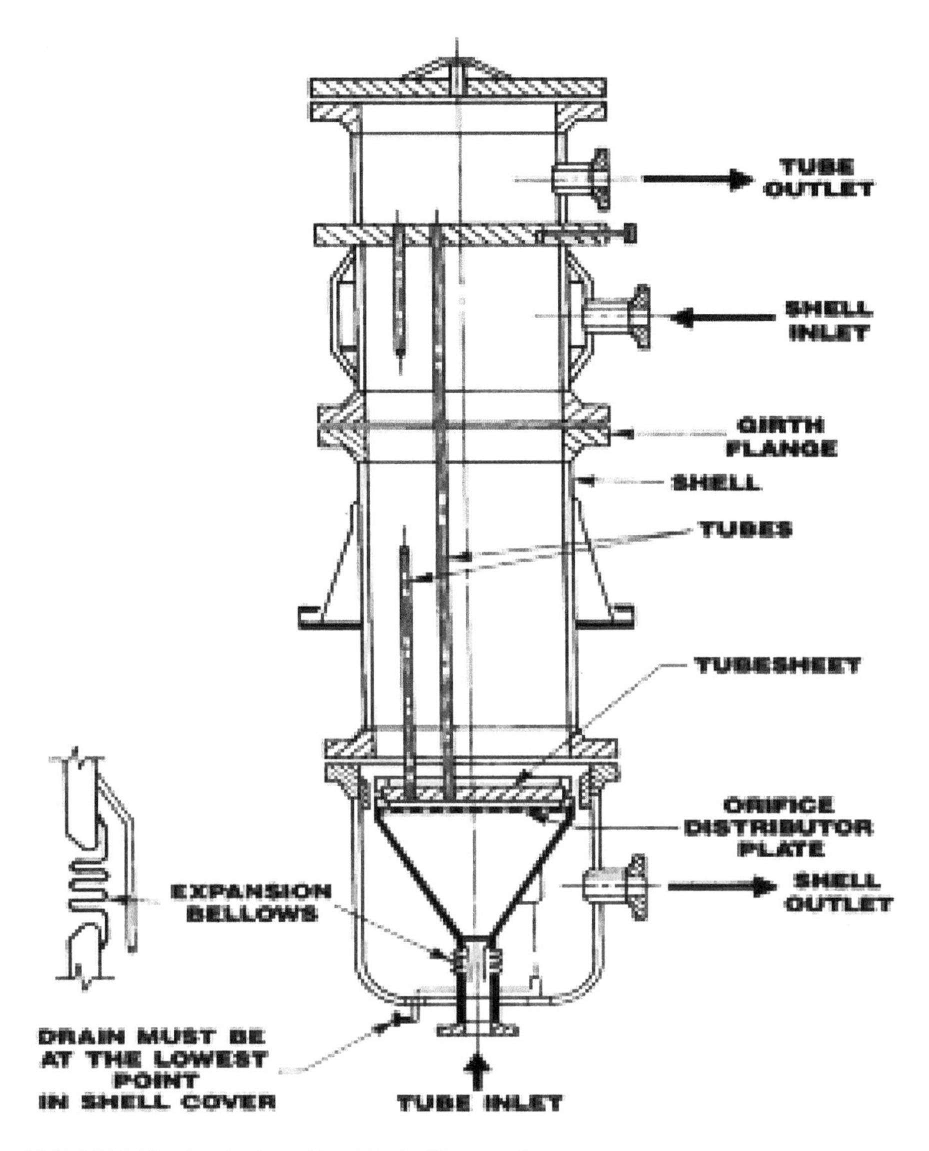

FIGURE 4.27 Vertical combined feed effluent exchanger.

are sometimes referred to as Texas towers. They are usually installed as one to four vertical exchangers in parallel. Texas tower exchangers are beneficial in terms of capital costs, energy efficiency, and smaller plots of land required for installing them in oil refineries relative to the equivalent energy banks of horizontal shell and tube exchangers. From past maintenance studies reports of successful cleaning and decoking of Texas towers by Tube Tech International, each of the towers could contain up to 3100 tubes. Each tube could have a diameter of 12–14 mm and a length of 21 meters or about 70 feet. Thus, the Texas tower tube bundle and structure is usually massive and tower cleaning at turnarounds is usually a major project. Vertical combined feed effluent exchangers pose a major challenge in terms of ease of conducting maintenance on the VCFEs, especially when the shell and tube sections are contaminated with extensive coke (shell side) and inorganic chloride salts and polycyclic aromatic compounds in the tubes. A number of maintenance service companies are proficient and available to clean Texas towers to meet the requirements of oil refiners.[30]

4.6.3.2 Plate Heat Exchangers

Plate heat exchangers are usually composed of many thin, separated plates that have large surface areas and small spaces between the alternate plates that enhance heat transfer between fluids. In special designs popularized by Alfa Laval, the plates are stacked and welded at the perimeter and formed into a bundle. Within this bundle of welded plates, hot and cold fluids flow separately between alternate plates. Welded plate heat exchangers are used in oil refining, especially in catalytic processes that operate in low to medium pressures relative to the robust shell and tube heat exchangers that operate in medium- to high-pressure processes. Alfa Laval tout their welded plate exchanger as having the highest thermal efficiencies in its class. According to Alfa Laval, the welded plate heat exchanger design is compact and has lower total emissions and lower capital and operating expenses relative to shell and tube heat exchangers. Over the past two decades, enhancements have been made to welded plate exchangers to enable them to be amenable to faster and cheaper cleaning relative to earlier welded plate heat exchangers.[31] The alternating flow of cold and hot fluids is shown in Figure 4.28. The bundle of welded plates that constitutes the heat exchanger is then enclosed in a huge shell to protect the bundle, as shown in Figure 4.29.

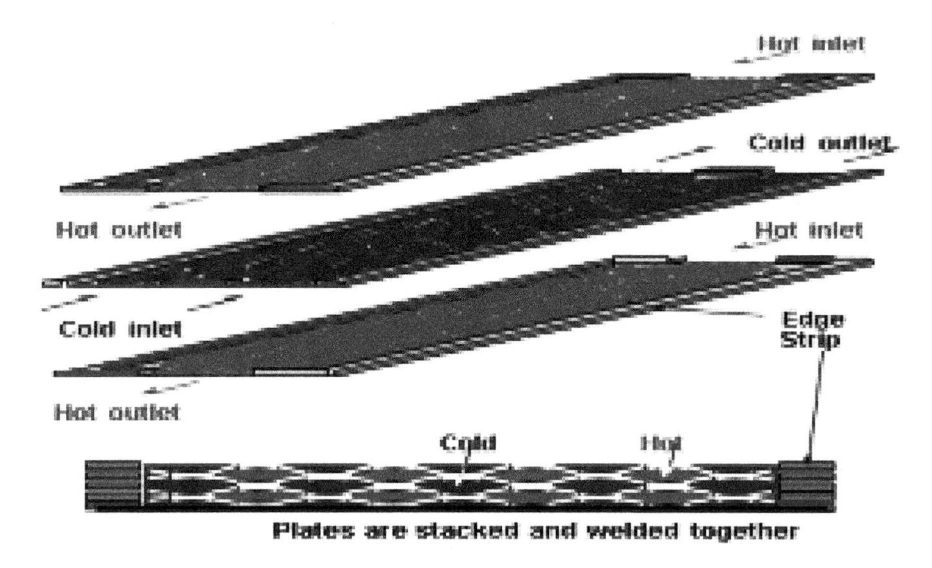

FIGURE 4.28 Alternate plates of flow of hot and cold fluids.

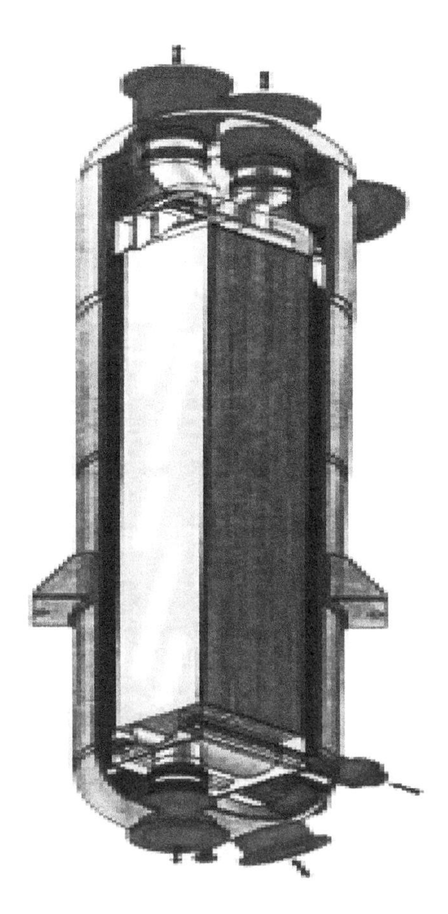

FIGURE 4.29 Bundle of welded plates in a shell enclosure.

4.6.4 Process Gas Compressors

Process gas compressors are used in catalytic reformers and for various applications in oil refining and petrochemical process units as makeup and recycle gas compressors. In a catalytic reformer, hydrogen-rich gas from a gas separator is returned to the suction of the recycle hydrogen gas compressors. The gas is compressed and recycled to the lead reactor as shown in Figure 4.30 to serve as a reactant and to provide hydrogen for reducing the rates of coke formation and coke deposition on the catalyst in the reactors. In addition to the recycle gas application, hydrogen gas compressors are utilized in the product separation section to boost gas pressures and achieve more effective separation of high-value liquid hydrocarbons while simultaneously increasing the purity of export hydrogen from catalytic reformers and especially from CCR reformers. Another benefit for the use of gas compressors for producing high-purity net hydrogen gas in excess of 90% is that in CCR catalytic reformers, a small portion of the high-purity hydrogen is used beneficially to achieve and sustain good catalyst circulation and excellent catalyst reactivations. The use of high-purity

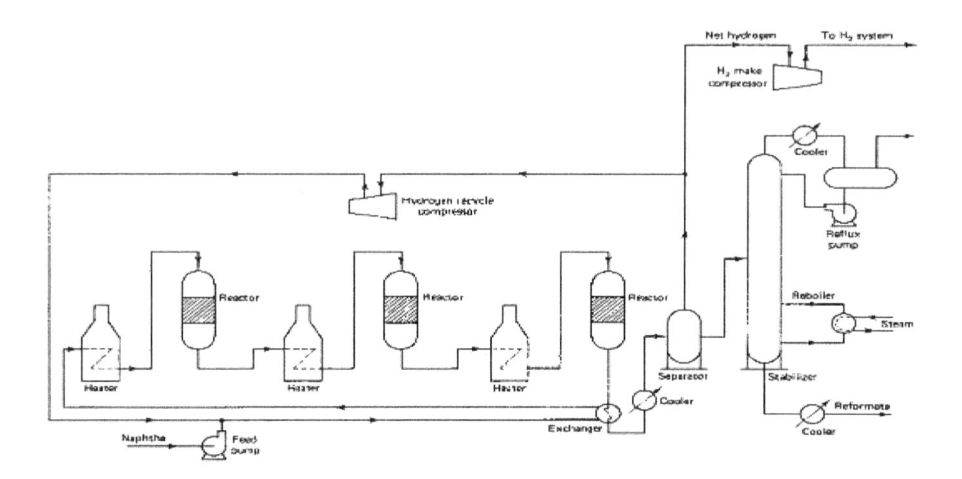

FIGURE 4.30 A semiregenerative catalytic reformer with a H_2 makeup compressor. From Compressed Air and Gas Institute, Gas & Process, www.cagi.org.[32]

hydrogen and good operation of gas coalescers are required to help minimize catalyst circulation challenges in CCR catalytic reformers.

The centrifugal and reciprocating gas compressors are the main types of gas compressors used in catalytic reformers. A centrifugal compressor is shown in Figure 4.31. A third type of process gas compressor that is now gaining acceptance is the screw compressor.[33] Reciprocating compressors are positive displacement machines in which the compressing and displacing element is a piston having a reciprocating motion within a cylinder. As the piston compresses the gas, it raises

FIGURE 4.31 A centrifugal compressor. Courtesy of Kobelco.
From Kobelco, Kobe Steel Group, Centrifugal Compressors, Accessed September 10, 2017, www.kobelcocompressors.com.[33]

FIGURE 4.32 4850 Kw net gas reciprocating compressor for a CCR catalytic reformer. From Kobelco, Kobe Steel Group, Centrifugal Compressors, Accessed September 10, 2017, www.kobelcocompressors.com.[33]

the pressure of the gas to the required specification. Centrifugal compressors are continuous flow machines in which one or more of the rotating impellers accelerate the gas in a radial direction. The resultant increase in velocity of the gas is then converted to pressure. Screw compressors are also positive displacement machines that use rotating twin rotors as pistons that compress the gas in a rotor chamber. Compression is achieved continuously by the rotation of the twin rotators.[34]

Centrifugal compressors are usually used for recycling gas in catalytic reformers and other low pressure–type process units. For high-pressure hydrogen service of greater than 1500 psig such as hydrocracking, reciprocating compressors are used for makeup hydrogen, and centrifugal compressors are typically favored for recycle service. Reciprocating compressors are also used in net hydrogen gas services and in the product separation of CCR catalytic reformers to produce higher-purity hydrogen and maximize reformate hydrocarbon production. A Kobelco reciprocating gas compressor is shown in Figure 4.32.

To ensure good performance and reliability of process gas compressors, the machines have to be monitored and maintained as often as required. Some of the usual reliability challenges are fouling of the compressors with inorganic chlorides and other contaminants, imbalance of centrifugal compressor rotors and premature fouling, and rapid deactivation of catalytic reformer catalysts. Abnormal rates of catalyst deactivation in reformers, as is well known, can significantly and negatively impact catalytic reformer and refinery productivity and profitability.

The recommendations in this book include using the services of specialized professionals such as rotating equipment and heater specialists, corporate consultants, and external refining consultants to ensure that oil refiners can sustain top-tier performance of their refinery process units. Such corporate reliability experts can help coordinate purchase of appropriate equipment and services, establish excellent equipment monitoring programs, and participate in all phases and plans by oil refiners for grassroots and revamp refinery projects, where their expertise is highly required.

REFERENCES

1. US EIA, Number and Capacity of Petroleum Refineries, June 22, 2016. Accessed September 16, 2017, www.eia.gov/petroleum/refinerycapacity/refcap16.pdf
2. Ielsen, R. N., Advances in Catalytic reforming, Process Economics Program, Report 129B, SRI Consulting, October, 2006. Menlo Park, California. Accessed September 16, 2017. https://ihsmarkit.com/pdf/RP129B_toc_173313110917062932.pdf
3. Poparad, A., Ellis, B., Glover, B., Metro, S., Reforming Solutions for Improved Profits in an Up-Down World, Paper AM-11-59. UOP LLC. Accessed July 2016. https://www.uop.com/?document=uop-reforming-solutions-for-improved-profits-paper&download=1
4. Oyekan, S. O., Catalytic Reforming Lecture Notes, AIChE Course, 1992.
5. Haensel, V., Donaldson, G. R., Platforming of Pure Hydrocarbons. *Ind. Eng. Chem.* 43(9), 2102–2104, September 1951.
6. Kluksdahl, H. E., Reforming a sulfur-free naphtha with a platinum-rhenium catalyst. US Patent 3,415,737.
7. Satterfield, C. N., *Heterogeneous Catalysis in Practice.* McGraw-Hill, 1980.
8. Sinfelt, J. H., Rohrer, J. C., Kinetics of the Catalytic Isomerization: Dehydroisomerization of Methylcyclopentane. *J. Phys. Chem.*, 65, 978, 1961.
9. Krane, H. G., Groh, A. B., Schulman, B. L., Sinfelt, J. H., Reactions in Catalytic Reforming of Naphtha. *Proc. 5th World Petroleum Congress, New York Section III*, 39, 1959.
10. Gary, J. H., Handwerk, G. E., *Petroleum Refining*, Marcel Dekker, New York.
11. Gates, B. C., Katzer, J. R., Schuit, G. C. A., *Chemistry of Catalytic Processes*, McGraw Hill, New York.
12. Hettinger, W. P., Keith, C. D., Gring, J. L., & Teter, J. W., Hydroforming reactions: Effect of Certain Catalyst Properties and Poisons. *Ind. Eng. Chem.*, 47, 719–730, 1955.
13. Sinfelt, J. H., Hurwitz, H., Rohrer, J. C., Role of Dehydrogenation activity in the Catalytic Isomerization and Dehydrocyclization of Hydrocarbons. *J. Catal.*, 1(5), 481–483, 1962.
14. Sinfelt, J. H., Hurwitz, H., Rohrer, J. C., Kinetics of n-Pentane Isomerization Over Pt-Al2O3 Catalyst. *J. Phys. Chem.*, 64(7), 892–894, 1960.
15. McVicker, G. B., Collins, P. J., Ziemiak, J. J., Model Compound Reforming Studies: A Comparison of Alumina-Supported Platinum and Iridium Catalysts. *J. Catal.*, 74(1), 156–172, 1982.
16. Mills, G. A., Heineman, H., Milliken, T. H., Oblad, A. G., Houdriforming Reactions, Catalytic Mechanism, *Ind. Eng. Chem.*, 45(1), 134–137, January 1953.
17. Oyekan, S. O., Swan, G. A., Catalytic Reforming Process. US Patent 4,436, 612, 1984.
18. McClung, R. G., Bonacci, J. C., Yarrington, R. M., Reformer Catalysts Provide Flexibility to obtain Yield-Octane Requirements, Engelhard (now BASF) Publication.
19. Oyekan, S. O., Personal Lecture Notes, 2010.
20. Hawkins, G. B., Catalytic Reforming: Catalyst, Process Technology and Operations Review. Accessed September 16, 2017. https://www.slideshare.net/GerardBHawkins/catalytic-reforming-catalyst-process-technology-and-operations-overview
21. Oyekan, S. O., Personal Notes on Catalytic Reforming.
22. Peyton, K. B., *Fuel Field Manual.* McGraw Hill, 1997.
23. Buchheim, G. M., Danis, J. I., *API RP 571, Damage Mechanisms in the Refining Industry Training Course*, Marathon Ashland Petroleum, Catlettsburg, Kentucky, November 16–18, 2005.

24. Evans, T., Johnson Screens Presentation on Reactor and Regenerator Experiences and Recommendations, New Orleans, September, 2011. https://www.aqseptence.com/app/en/keybrands/johnson-screens/
25. Ametek Process Instruments, Process Heaters, Furnaces and Fired Heaters. Accessed July 10, 2017, www.ametekpi.com.
26. Bacon, J., Target: Fired Heater – An Often Neglected Necessity, Cetek Notes for Soni O. Oyekan.
27. Bacon, J., Latte, D. D., Increasing Process Efficiency via High emissivity Ceramic Coatings, www.scribd.com.
28. Albanis, T., Kladopoulou, E., Ljubicic, B., Fazzini, M., A Simple Twist of Fate, Hydrocarbon Engineering, February 2013. Hellenic Petroleum Company, S.A. Greece and Koch Heat Transfer Company, S.R.A, Italy. Accessed September 17, 2017. http://www.kochheattransfer.com/attachments/article/97/ASimpleTwistOfFate.pdf
29. Patwardhan, A. V., Momentum and Heat Transfer, Applications of the Principles of Heat Transfer to Design of Heat Exchangers, Institute of Chemical Technology. https://www.slideshare.net/KathiresanNadar/applications-of-the-principles-of-heat-transfer-to-design-of-heat-exchangers
30. Tube Tech International, Vertical Combined feed Effluent Exchangers, www.tubetech.co.
31. Reverdy, F., Alfa Laval Packinox Heat Exchanger, August 2005. Accessed July 10, 2016. https://www.alfalaval.com/globalassets/documents/industries/energy/crude-oil-refinery/ppi00202en.pdf
32. Compressed Air and Gas Institute, Gas & Process, www.cagi.org
33. Kobelco, Kobe Steel Group, Centrifugal Compressors, Accessed September 10, 2017, www.kobelcocompressors.com
34. Ohama, T., Kurioka, Y., Tanaka, H., Koga, T., Process Applications Where API 619 Screw Compressors Replaced Reciprocating and Centrifugal Compressors, Accessed July 15, 2016, www.kobelcocompressors.com

5 Catalytic Reforming Catalysts

5.0 INTRODUCTION TO CATALYTIC REFORMING CATALYSTS

A typical catalytic reformer catalyst is comprised of platinum deposited on an alumina support, one or two metals added as promoters or modifiers, and chloride. The platinum moiety on the alumina usually provides the dehydrogenation/hydrogenation functionality, and the alumina and chloride provide the acidic isomerization functionality. Catalytic reformers can arguably be rated as one of the major catalytic process conversion units in oil refineries that fully utilize the range of capabilities and efficacy of multifunctional or bifunctional catalysis. Multifunctional catalysis is required for the selective production of desirable high-value products and for longer durations of time between turnaround maintenance periods. Major catalytic reforming process objectives are that the process achieve optimal production of primary products such as reformate and hydrogen, and that it minimize the production of undesirable secondary byproducts. In addition, it should produce desirable reformate and hydrogen at low rates of catalyst coking or catalyst deactivation. These objectives are achievable through the use of excellent catalysts in the processing of hydrocarbons in the appropriate boiling range of heavy naphtha fractions. Several key reactions are catalyzed by platinum containing catalysts in naphtha reforming. Catalytic reforming catalysis requires that a "good balance" of hydrogenation/dehydrogenation and acidic isomerization functionalities of the catalysts is maintained continuously for optimal naphtha reforming. The variety of reactions catalyzed by the bifunctional catalysts include paraffin dehydrogenation, paraffin hydrocracking, paraffin isomerization, naphthene hydrogenation and ring opening, naphthene dehydrogenation, naphthene isomerization, naphthene hydrodealkylation, aromatic saturation and ring opening, hydrodealkylation of alky aromatics, complex multistep paraffin aromatization, and hydrocarbon condensation reactions. In addition, thermal reactions also occur due to processing of the naphtha feed at reactor temperatures that are higher than 900 F that are required for the production of high-octane reformates. Hydrocracking and fluid catalytic cracking of gas oil processes in oil refining are also classified as processes that require significant applications of catalysis. For gas oil hydrocracking processes, two to three catalysts are used for hydrodemetellation and hydrotreating to help remove metals and heteroatom compounds that could negatively impact hydrocracker cracking catalyst performance and stability. The main cracking catalysts are usually followed by a "finishing" hydrotreating catalyst. Finishing hydrotreating catalysts are recommended to minimize the reformation of mercaptans. In fluid catalytic cracking units, several specific catalytic additives are used in addition to the main cracking catalyst. Additives are added to promote the production of light olefinic products and higher-octane FCC naphtha. Additionally, catalytic additives enable oil refiners to meet environmental constraints with respect to concentrations of nitrogen oxides (NO_x) and sulfur oxides (SO_x) in regeneration effluent gas, and minimize excessive catalyst fines make via the use of carbon monoxide promoter additives.

Due to the complex nature of catalytic reforming catalysts, significant amounts of research and development work have been conducted over the past seven decades with the goals of optimizing the dehydrogenation/hydrogenation and acidic isomerization functionalities of the catalysts, and these have led to literally thousands of catalysts and process inventions for catalytic reforming processes.[1–5,8–21] Catalytic reforming processes and catalyst technologies continue to contribute cost-effective options for oil refiners in the production of high-value transportation fuels.

5.1 REFORMING CATALYST NOMENCLATURES

A review of the nomenclatures for the present and past catalytic reforming catalysts can aid in possibly identifying the number and type of active catalytic metals on catalysts and the specific catalytic reforming technologies wherein optimal benefits could be realized by the use of appropriate catalysts. Changes in the number of catalysts and technology provider offerings due to acquisition of product lines from competing catalyst provider companies have reduced the number of catalyst suppliers and designations for catalytic reforming processes.

Some of the more popular catalyst designations are identified for the major catalyst and technology suppliers. The catalysts that are currently used by oil refiners are reviewed. Catalyst providers usually work with oil refiners so as to aid them in their efforts to select appropriate high-performance catalysts for their catalytic reformers. Usually, catalyst and process technology suppliers are invited by an oil refiner to submit performance projections for catalysts based on the specific requirements of the oil refiner for its catalytic reformers. Performance projection data, and catalyst properties, catalyst prices are usually included in the bid package provided by each of the catalyst suppliers.

Catalysts from companies are given different designations by the suppliers so as to market the good features of their catalysts and differentiate them from similar catalyst products from other competing catalyst suppliers. Past catalyst designations by Axens have included the CR-series for motor gasoline catalytic reformers, the AR-series for aromatic production, and, more recently, the PS-series for platinum/tin and the PR-series for platinum/rhenium catalysts. Axens retained the PS-series nomenclature and other catalyst designations used by the Criterion Catalyst Company after acquiring Criterion's catalytic reforming business in 2012. Prior to Criterion's exit from the catalytic reforming catalyst business, its nomenclatures included the P-series for platinum monometallic catalysts, PS-series for platinum tin, and PR-series for platinum-rhenium catalysts. Axens also has the high density Symphony™ PS-series and Symphony™ CR-series catalysts. Engelhard catalyst designations were the RD-series and E-series catalysts used before the formation of the Acreon partnership between the Procatalyse and Engelhard reforming catalyst businesses in the early 1990s. Acreon and Institut de Francais Petrole (IFP) were later combined into Axens. ExxonMobil in the 1980s and 1990s used the KX-series for the designation of its catalytic reforming catalysts, which are used primarily in fixed bed semi-regenerative and cyclic regenerative catalytic reformers.

Universal Oil Products and now Honeywell UOP use the R-designation for their catalytic reforming catalysts and I-designation for paraffin isomerization catalysts. Sinopec offers the PRT-series for semiregenerative reformers and the RC-series and GCR-series of catalysts for CCR units.[32] Hydrogenation/dehydrogenation and acidic isomerization functionalities are incorporated in the sets of catalytic catalysts reviewed. The degrees of incorporation of the metal and acid functionalities, their distributions, and their physical and chemical characteristics lead to the observed differences in the activity and selectivity performances of catalysts in catalytic reforming units.

5.2 DEHYDROGENATION/HYDROGENATION FUNCTIONALITY OF REFORMING CATALYSTS

Based on the pioneering studies of Vladimir Haensel and Universal Oil Products, platinum and palladium were initially identified as metals that could provide desired dehydrogenation/hydrogenation functionalities and minimize nonselective cracking of hydrocarbons relative to what was achieved with molybdenum oxide catalysts, as reviewed previously. The predominant catalyst in the mid-1940s was based on molybdenum oxide as the active component. Molybdenum oxide–based catalysts had provided significant improvements in the conversion of straight-run naphtha feeds to high-octane reformate relative to previous thermal cracking processes. In the late 1940s, innovative work by Haensel and UOP determined the range of platinum concentrations based on the cost of platinum and site accessibility that was appropriate for good activity and selectivity of catalytic reforming catalysts. Platinum was selected and it is now the preferred metal to provide the appropriate level of hydrogenation/dehydrogenation functionality for catalytic reforming catalysts.[1–5] While oxides of molybdenum, cobalt, nickel, and tungsten are used in hydrogenation services such as in hydrodesulfurization, hydrodenitrogenation, and hydrodeoxygenation reactions, they lack the dehydrogenation functionality component required for dehydrogenation of paraffins and naphthenes to generate the necessary intermediates in paraffin aromatization. Similarly, palladium, which is used in the selective hydrogenation of diolefins in alkylation feed pretreatment, is not as effective as platinum.

As a result of huge investments made over the past 50 years, significant advancements have been made in catalysts and catalytic reforming processes to meet demands for increased gasoline and aromatic production. These efforts have resulted in the development of high-performance catalysts that have excellent attrition resistance and high chloride retention characteristics. Good properties of current catalysts enable oil refiners to achieve high reformate and hydrogen productivities relative to catalysts of the pre-1950 period.

5.3 ACIDIC ISOMERIZATION FUNCTIONALITY

Dehydrogenation/hydrogenation and acidic sites of catalytic reforming catalysts have to be supported on the surfaces of one or more suitable inorganic oxide compounds. Typically, the preferred inorganic oxide is an aluminum oxide or alumina. A catalyst support that can aid in optimizing the distribution and dispersion of the metallic hydrogenation/dehydrogenation and acidic isomerization sites is usually preferred for catalytic reforming catalysis. The other desired characteristics of the support are that it its porous, has a high surface area, and provides facile accessibilities for the transport of reactants, intermediates, and product molecules to and from active catalytic sites.

A catalytic reforming catalyst support material could be a composite of a porous inorganic oxide, molecular sieve, and appropriate binder. Inorganic oxides suitable for use as catalyst support material can include alumina, magnesia, titania, zirconia, chromia, zinc oxide, thoria, ceramic, porcelain, bauxite, silica, silica-alumina, silicon carbide, clays, and crystalline zeolitic aluminosilicate materials.[18] A variety of crystalline alumina forms such as alpha, eta, gamma, and theta alumina are used. Several excellent studies have been conducted on the phase transition of bayerite, gibbsite, and boehmite, and their results are as shown in Figures 5.1 and 5.2.[6,7] Boehmite and gibbsite are different forms of aluminum hydroxides that occur naturally in bauxite deposits.[26] Boehmite is the monohydrate form of alumina (AlOOH) and gibbsite is the trihydrate form (Al $(OH)_3$) and the fact that the boehmite is aluminum monohydrate is highly significant, as the gamma alumina produced from boehmite has a high concentration of residual hydroxide groups.

As shown in Figures 5.1 and 5.2, boehmite undergoes phase transition at calcination conditions of 450°C (842 F) to gamma alumina. If heated further to 750°C (1382 F), gamma alumina is converted to delta alumina, and at temperatures greater than 1000°C (1932 F) to the theta phase or theta alumina. At temperatures in excess of 1200°C (2192 F), a low-surface-area alpha alumina becomes the final

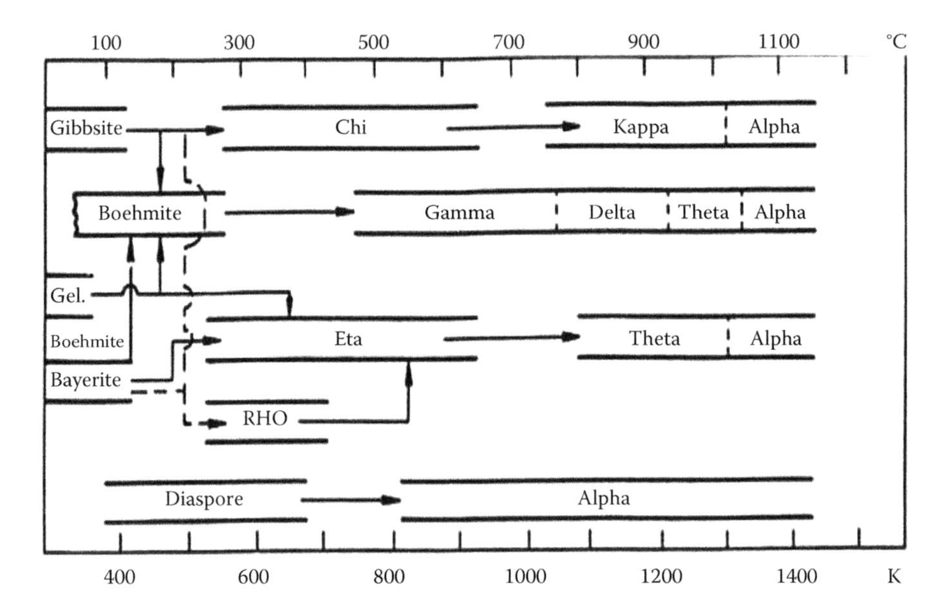

FIGURE 5.1 Phase transition for gibbsite, bayerite, and boehmite.
From Wefers, K., Misra, C., Oxides and Hydroxides of Alumina, Alcoa Technical Paper 19, Aluminum. Accessed February 15, 2018. http://epsc511.wustl.edu/Aluminum_Oxides_Alcoa1987.pdf[6]

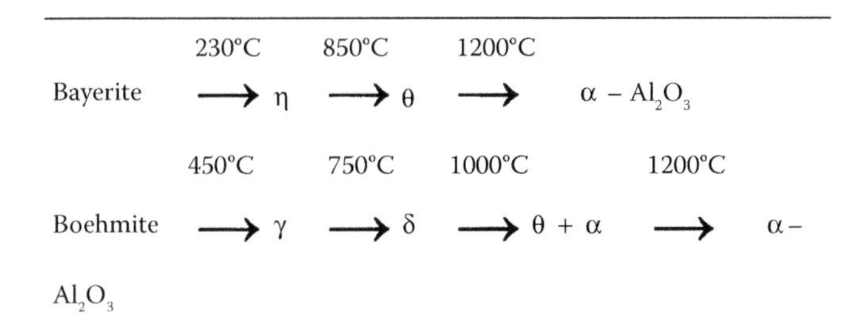

FIGURE 5.2 Phase transition for bayerite and boehmite.
From Chase, G. C. et al., Metal Oxide Fibers and Nano fibers: Method for Making same and there off, EP 2267199 A2.[7]

alumina product. Alpha alumina is usually a very hard, solid particle with a surface area that could be as low as 2 m^2/gm.

Eta alumina that was used as catalyst support for catalytic reforming catalysts about 40 years ago could be produced as shown starting with bayerite, as the material can undergo phase transition at 230°C (446 F) to produce the high-surface-area eta alumina. The surface area of eta alumina is typically in the range of 350 to 450 m^2/gram. It was discovered in hydrothermal stability studies that eta alumina has the disadvantage of exhibiting lower hydrothermal stability relative to gamma alumina. Hydrothermal stability is defined as a measure of the stability of the catalyst and alumina structure and properties at high temperature steaming conditions that are similar to processing conditions during catalyst regenerations in catalytic reforming.

Gamma alumina is now the preferred alumina and has effectively replaced eta alumina, which was used as catalyst support by some catalyst providers between 1948 and 1970. Since 1970, gamma alumina is now used predominantly as alumina support due to its superior hydrothermal stability and surface area retention characteristics relative to those of eta alumina. The gamma alumina support that is used in catalytic reforming actively participates in several of the chemical reactions in conjunction with chloride ions as the acidic isomerization functionality or acidic sites in catalytic reforming. As a result, a significant number of extensive studies have been conducted on alumina types and alumina acidity. The acidic nature of the alumina chloride moiety on the catalyst is substantially reviewed in a later section of this chapter.

5.4 PLATINUM CATALYTIC REFORMING CATALYSTS

The platinum/chloride/alumina catalyst was the main catalyst of choice in semiregenerative and cyclic catalyst regenerative processes between 1949 and 1967. Platinum-only catalysts are catalysts that have platinum concentrations in the range of 0.3 to 0.8 wt. %, 0.7 to 1.1 wt. % chloride, and the balance of about 98 weight percent is the alumina support. Platinum-only catalysts are referred to as monometallic catalysts. Platinum/chloride/alumina catalysts are used mainly in fixed-bed semiregenerative and cyclic regenerative reformers, due in part to their low sulfur sensitivity in the processing of naphtha feeds with up to 20 wppm sulfur. For catalytic reforming operations in which the monometallic catalyst is exposed to high feed sulfur concentrations of greater than 20 wppm and nitrogen concentrations greater than 2 wppm, catalyst regenerations are not recommended, as fresh catalyst-type activity and product selectivity are typically not recovered. Monometallic catalysts are still being used in a number of cyclic regenerative catalytic reformer applications and for some semiregenerative catalytic reformers.

5.5 MULTIMETALLIC CATALYTIC REFORMING CATALYSTS

Kluksdahl's catalyst invention in 1967 ushered in the current era of platinum/rhenium catalytic systems, which have enabled oil refiners to operate semiregenerative catalytic reformers for much longer periods of time or longer operating cycles relative to those for monometallic catalysts. In addition, multimetallic reforming catalysts usually produce higher yields of reformate and hydrogen products relative to monometallic catalysts.[8,9] Catalytic research and development studies in the 1965 to 1985 period led to the introduction of promoter metals or platinum "modifiers" such as rhenium, tin, iridium, germanium, and so on. Multimetallic catalysts exhibited superior activity and product selectivity performances relative to those of monometallic catalysts. Though platinum/tin and other multimetallic catalysts were introduced after 1967, several excellent studies have been conducted for enhanced improvements in catalytic performance and better understanding of the sensitivity of bimetallic platinum/rhenium catalysts and their expected catalytic performance in operations in the reforming of naphtha feeds containing 0.1 to 1.0 wppm sulfur.[8–20]

Furthermore, for bimetallic platinum/tin and platinum/rhenium catalysts, some researchers have suggested that the addition of a second "modifier" or "promoter" leads to improvements in the dispersion of the platinum metal sites. Other researchers have suggested that, depending on the second metal type, improvements in the activity and product selectivity of platinum bimetallic catalysts could be due to significant reductions in the rates of catalytic coke make.[8–20]

As indicated earlier in the chapter, gamma alumina is the preferred alumina support. The surface area of fresh catalysts containing gamma alumina and supplied by catalyst providers is typically in the range 180 to 230 m^2/gm.

In terms of the nomenclature in use, platinum/rhenium catalysts containing equivalent concentrations of platinum and rhenium are usually referred to as "equimolar" or "balanced" catalysts. They are referred to as equimolar since the atomic weights of platinum and rhenium are 195.1 and 186.2 pounds per pound mole, respectively. High-rhenium catalysts are, therefore, defined as catalysts in which the rhenium concentration is higher than that of platinum. Some process technology licensors and catalyst provider companies claim that "promoter" metals are added, usually for performance enhancement of the hydrogenation/dehydrogenation functionality. Over the past decades, catalyst provider companies have also moderated the acidic isomerization functionality via alumina calcination and promoter metal insertion.[11–21] Catalysts in which more than one promoter metal has been incorporated are usually referred to as trimetallic catalysts. A few more salient features of multimetallic catalysts will be reviewed in the next chapter. In summary, monometallic platinum, bimetallic platinum/rhenium, and platinum/tin catalysts are the catalysts used in fixed-bed semiregenerative and cyclic catalyst regenerative reformers. Platinum-tin catalysts are used exclusively in continuous catalyst regenerative reformers.

5.6 CATALYST MANUFACTURING PROCESS

A brief overview of the catalyst manufacturing process for catalytic reforming should provide some understanding of key stages that are used in the production of alumina support and the fresh catalyst. As indicated in the preceding section, a reforming catalyst is composed of at least 97 weight percent of alumina, 0.2 to 0.6 weight percent platinum, 0.2 to 0.6 weight percent promoter metal, and about 1.0 weight percent chloride. The special manufacturing steps required to produce cylindrical and spherical extrudates of alumina are vital in the catalyst production process.

The alumina manufacturing process starts with the production of the alumina gel, which in one procedure involves the acidification of aluminum tri-hydroxide or the Bayer hydrate. In another variation, alumina is produced by reacting aluminum with an alcohol. The first process involving acidification of aluminum tri-hydroxide is credited to UOP and the second process to Condea.[65] It is at this stage that changes in the structure of the alumina such as surface area, pore distribution, and acidity are effected by the use of appropriate temperatures, pH, and time.

Temperatures as high as 450°C (842 F) may be required to transform the alumina to the gamma phase and impart necessary crush strength, acidity, and porosity to the alumina.

The next stage in catalyst production is the alumina formation or agglomeration. Alumina is formed to specific sizes, shapes, types, and crush strengths to meet the required physical and chemical properties for the catalyst support material. Three alumina-forming procedures are utilized: disc pelletizing, extrusion, and oil-forming technologies. Extrusion and oil-drop technologies are used mainly in the forming of alumina support for catalysts that are used extensively in oil refining hydroprocessing units. A paste of the alumina is made by mixing pseudoboehmite or boehmite, essentially alumina, with water and a peptizing agent such as acetic acid. The paste is extruded to form cylindrical, trilobe, quadrilobe extrudates that could be 3/16 of an inch in length and 1/16 inch in radius. In the oil-forming or oil-drop alumina-forming technology, a sprayer is used to create droplets of the aqueous alumina paste. The droplets move vertically downward at moderate rates through a long column of oil in appropriate oil conditions conducive to forming spherical pellets. Superficial tension exerted on the droplets leads to spherical pellets or beads after moving through the high column of oil. Calcination of the alumina extrudates or pellets is conducted to harden them before use in the next platinum, promoter metal, and chloride impregnation of the shaped alumina.

After the formation of the shaped alumina pellets or "extrudates," platinum, promoter metals, and chloride impregnations are conducted so as to deposit the necessary calculated concentrations of platinum, promoter metals, and chloride on the alumina. Precursor metal compounds such as chloroplatinic acid, perrhenic acid, ammonium perrhenate, hydrochloric acid, and stannous chloride are used. Depending on the catalyst to be manufactured, the appropriate set of precursor metal compounds, hydrochloric acid, and impregnating agents are used. Hydrochloric acid and carbon dioxide have been used as impregnating agents to facilitate uniform distributions of platinum, promoter metals, and chloride. After the metal and chloride impregnation step, drying and calcination procedures are conducted at moderate temperatures to complete the catalyst production.

5.7 CATALYST PROPERTIES

Successful operations of cost-effective, profitable catalytic process units are usually dependent on a good understanding by the operators of the physical and chemical properties of catalysts, and their projected catalytic performances before purchasing the catalysts for use in catalytic process units. This requirement is more critical for the catalytic reforming process that has so far been shown to require precisely manufactured platinum-containing catalysts and specific multifunctionality requirements for naphtha processing. Catalytic properties should be monitored for the fresh catalyst when received by the oil refiner for crush strength, alumina phase, surface area, bulk density, porosity, metal and chloride contents, metal and chloride distribution, and metal contaminant levels. Attrition resistance is an important physical property that should be given a significant amount of attention for fresh catalysts that have to be used in continuous catalyst regenerative reformers. Use of a catalyst with subpar attrition resistance characteristics would require more dedicated process unit monitoring to address catalyst-induced challenges. The challenges could include the inability to maintain steady-state continuous catalyst circulation. Maintaining catalyst and steady-state catalyst circulation rates are highly vital for realizing the full profitability potentials of oil refiners' continuous catalytic reforming units and refineries. Other critical properties that should be monitored on a variety of appropriate frequencies for CCR catalysts include surface area, platinum dispersion, metal contaminants, spent and regenerated catalyst coke, spent and regenerated catalyst chlorides, chloride retention, and catalyst fines make. A detailed review would be accorded to challenges associated with the use of average attrition resistance catalysts and other negative profitability-impacting factors that oil refining operators have to contend with in the operation of CCR units.

In addition to assessing the amount and quality of the fresh catalysts purchased from a catalyst supplier, catalyst samples should be taken where online catalyst samplers are available during the catalyst regeneration phases for fixed-bed catalytic reformers. Furthermore, catalyst samples should be taken during planned dump, screen, and unloading of the catalyst in fixed-bed catalytic reformers. Good physical and chemical property data from the catalyst inspection reports for the sampled catalysts should provide the basis for a good assessment of the state of the catalyst with time. The catalyst data could then be used in making necessary operational changes during troubleshooting and optimization to improve the operation of catalytic reformers. Samples should be taken of the spent catalysts from catalytic reforming reactors before sending the dumped, spent catalysts to storage and subsequently for precious metal reclamation. For CCR units, it is also highly recommended that catalyst samples be taken frequently during operations, preferably daily for coke and chloride, and analyzed in a timely manner so as to enable the oil refiner to make necessary timely operational changes with respect to catalyst circulation and catalyst regenerator and reformer operations. Such timely adjustments would significantly improve the CCR unit's operations and profitability. Additionally, the reliability of the CCR process units can be proactively managed with available catalyst inspection data. Thus, a comprehensive catalyst monitoring program is highly recommended in order to facilitate the conducting of timely fresh catalyst inspections, required inspections of on-stream catalysts, and timely catalyst samplings. Extensive catalyst and unit monitoring programs are required for effective management of catalytic reformer operations and of catalysts and platinum.[37]

5.7.1 Physical Properties

Oil refinery programs that emphasize continuous use of catalysts with good physical and chemical properties in oil refining processes, and especially in catalytic reforming, improve oil refiners' chances of achieving profitable catalytic reforming unit operations and cost-effective precious metal management programs. The physical dimensions of a reforming catalyst cylindrical extrudate such as length and diameter, and diameter in the case of spherical particles, are usually provided in descriptions of catalysts in the catalyst bid process. Selected key physical properties that should be reviewed periodically include bulk density, crush strength, porosity, surface area, and attrition resistance. These properties vary in their specific importance and impacts on the catalytic and reliability performance of the catalytic reformer. They also have relative impacts at the different stages in the life of the catalyst and in the operation of process units, and especially for the CCR reformer. The physical properties listed impact catalyst loading and on-stream catalytic performance. Physical structure of the catalyst could impact catalytic sites accessibility, transport of reactants, intermediate compounds, and products within catalyst pores and through catalyst beds.

5.7.1.1 Bulk Density

Bulk density is defined simply as the mass of the catalyst divided by the fixed volume of the catalyst. This catalyst property is useful for the calculation of the mass or volume of a catalyst and inert materials required to fill reactors, and the information is necessary both for planning and as a check during catalyst loading of the reactors. Sometimes, the loading density of the catalyst is provided by the catalyst provider and could be slightly different from the bulk density. Other density measures used in oil refining include sock-loaded and dense-loaded catalyst densities. Catalyst sock- and dense-loading densities are usually provided by catalyst providers. The use of special catalyst loading technologies could permit loading up to 17% greater mass of catalyst into a fixed reactor relative to use of the sock-loading procedure.[22–24] Sock-loading densities reported for catalytic reforming catalysts have been in the range of 35 to 42 pounds per cubic foot, and dense-loading densities are in the range of 47 to 50 pounds per cubic foot.

In addition to the comparative projected performance data and other catalyst factors provided by catalyst and technology provider companies as their quotes for the catalyst bid process, bulk densities of the catalysts are used in calculating the total cost of each of the offered catalysts to fill the reactors. Furthermore, oil refiners can process higher rates of feed through the reactors for greater profitability using higher bulk density catalysts, and this could offer significant benefits in reduced catalyst costs and catalytic performance in catalytic reformers.

5.7.1.2 Crush Strength

Good mechanical strength of a catalyst is a highly desirable property. A physically inferior catalyst as determined by its low crush strength could pose significant catalyst challenges for the oil refiner. A catalyst with low crush strength could create excessive catalyst fines during catalyst loading into reactors and operations in catalytic reforming reactors. Crush strength or crush resistance is a measure of how resistant the catalyst particles or extrudates are to getting crushed and forming catalyst fines. Single-catalyst pellet crushing strength tests, ASTM D4179 and ASTM D6175, are used for cylindrical extrudates and spherical pellets, respectively, for oil refining process catalysts. Other methods, ASTM D7084 and SMS-1471, are used for the determination of bulk crushing strength of catalysts. Standard test methods based on the single-pellet crush strength and used for catalysts in catalytic reforming are the ASTM D4179 for cylindrical extrudate particles and ASTM D6175 for spherical catalysts.(25) The unit of crush strength is pounds of force per 1/8th of an inch pellet length. Average crush strengths for catalysts used in fixed-bed catalyst regenerative reformer applications have ranged from 5 to 8 pounds of force per 1/8th of an inch pellet for cylindrical extrudates and from 5 to 13 pounds of force per spherical pellet or spheres of catalysts used in continuous catalyst regenerative reforming processes.

Catalysts that have high crush resistances are highly valuable in catalytic processes where maintenance of the structural integrity of the catalyst extrudates is a necessary requirement for reliable steady state operations over extended periods of months and years. The benefits of maintaining excellent catalyst containment in reactors are of greater significance for catalytic process units when excellent routine and turnaround maintenance programs are conducted for the reactor internals such as for the center pipes and scallops. Effective catalyst containment within reactors adds to the cost effectiveness of platinum management programs, as platinum and other precious metal–type losses are greatly minimized. This is because losses of catalyst and catalyst fines are minimized in the reactors and for the catalytic reformer.

5.7.1.3 Catalyst Porosity

In addition to desired performance attributes such as activity, selectivity, and stability that an excellent heterogeneous catalyst should possess, the catalyst should also have good crush resistance, as reviewed in the preceding section, and appropriate average pore size and porosity. Pore size and pore size distributions are determined from physical adsorption isotherms of nitrogen as the adsorbate.(43,44) For large pores, mercury porosimetry is used, and in most cases, the nitrogen adsorption and mercury porosimetry determinations are in reasonable agreement. The pore sizes of catalyst carriers or supports are classified based on the size of the pores. Pores that have average radii of less than 20 Angstrom are referred to as micropores, intermediate pores or mesopores have pore radii of 20 to 500 A, and macropores are pores with average radii in excess of 500 A. Catalysts that are used in oil refining for the processing of naphtha fractions have mesoporous alumina support with average pore radii in the range of 20 to 500 Angstrom.(26) The average pore sizes also correlate directly to the porosity of the alumina and degree of carrier or support porosity for catalytic reforming catalysts. Effective diffusion and transport of reactants, product intermediates and products within the pores and to active sites of the alumina should be facile for selected catalysts. The porosity of a catalyst can be calculated from data generated from gas absorption and mercury penetration measurements. An excellent review of the available methods for calculating porosity of a catalyst from the analytical data is provided by Davies and Antos.[27] The average pore volumes of catalytic reforming catalysts are generally in the 0.2 to 0.8 cubic centimeters per gram range. Spherical catalysts used in CCR reforming process units have been reported to have pore volume between 0.65 to 0.80 cubic centimeters per gram and average pore diameter in the 80 to 200 A range.

Changes occur in the porosity of the catalysts with time on oil and number of catalyst regenerations, especially for spherical CCR catalysts. The smaller pores collapse with time as a result of the effects of catalyst regenerations and metal and sulfur contaminant buildup. As the smaller pores are eliminated, the average pore sizes of the alumina support increase gradually with time, as shown in Figure 5.3.

Some of the changes in pore sizes of the same catalyst loaded in different CCR units could reflect some slight variations in catalyst qualities and differences in catalyst regenerator operations. High frequency of catalyst regenerations that reflect short catalyst cycles would have some negative impacts on the physical state of the catalyst and may lead to significant changes in pore sizes. Pore size changes are compared for essentially different productions of CCR catalysts in Table 5.1.

The data in Table 5.1 show that small changes occurred in pore sizes and pore volumes, emphasizing the excellent retention of the essential physical characteristics of the catalysts over several years of catalyst regeneration operations.

5.7.1.4 Surface Area

Catalyst carriers or support provide surfaces for deposition of active metal and acidic sites. They also provide a surface for the optimization of the contributions of the active catalytic sites and modifications of the contributions of the metal and active sites for improved catalytic performance. The innovative ideas applied in the modification of dehydrogenation/dehydrogenation and acidic isomerization sites have led to performance improvements of catalytic reforming catalysts. A key physical

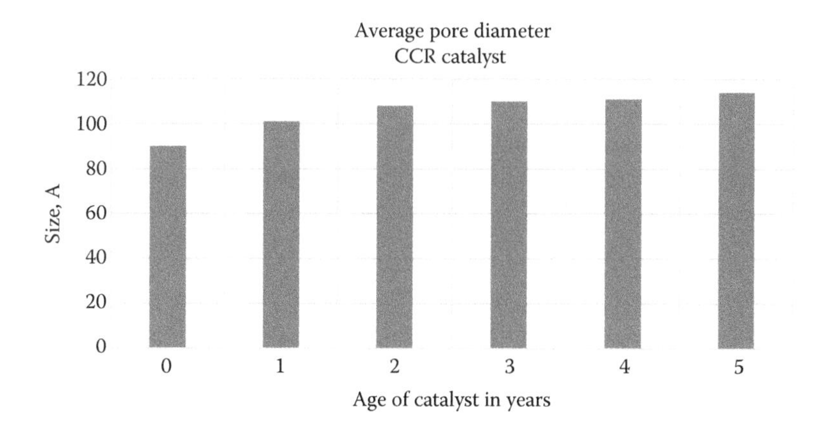

FIGURE 5.3 Increase in catalyst pore size with age of catalyst.

TABLE 5.1
Changes in Catalyst Pore Sizes with Time

Unit	Catalyst	Catalyst Age, years	Hg Pore Volume, cc/gm	Pore Diameter Change, A
1	A	0.1	0.73	0
2	A	1.5	0.75	0
3	A	2.1	0.75	16
4	A	3.2	0.74	11
5	A	3.3	0.74	33
6	A	4.4	0.74	8
7	B	10.2	0.72	10

property of catalyst support that is typically tracked is the surface area. The surface area of a catalyst can be determined via either the Langmuir or Brunauer, Emmett, and Teller (BET) calculation methods from gas adsorption data on catalyst carriers or supports. For BET surface area determinations, nitrogen is used as the adsorption gas and argon and krypton are used for low surface area determinations.[26,28,29]

Gamma alumina is the preferred alumina as a result of its higher hydrothermal stability relative to eta alumina, as indicated in a preceding section chapter of this book. The surface area retention advantage of gamma alumina was established relative to eta alumina by comparing the surface areas of the aluminas after various hydrothermal treatment studies.[30] As indicated earlier, there have been numerous improvements made over the past decades. Modifications made to the physical structure of gamma alumina support as reported by Unmuth and Fleming led them to establish 275 to 300 m^2/gm surface area alumina support as possessing ideal properties for high-performance catalytic reforming catalysts in the 1980s.[31]

Numerous significant improvements have been made with respect to improving hydrothermal stabilities or surface area retention characteristics of catalytic reforming catalysts, especially for those used in CCR units. These improvements are due to higher temperatures or special hydrothermal treatment of the gamma alumina. Higher temperature calcinations have led to lower-surface-area alumina, and optimal surface areas of fresh catalytic reforming catalysts are now in the range of 180 to 210 m^2.gm. Additionally, higher temperature calcinations of the gamma alumina have led to catalytic reforming performance benefits due to optimization of the contributions of the metal and acidic functionalities in the naphtha reforming process. Essentially, higher temperature calcination of the alumina leads to moderation of the acidic functionality of the reforming catalyst and to higher yields of reformate and hydrogen.

During catalytic processing operations, especially in catalytic processes that require continuous regenerations of catalysts, the surface area of the catalysts will decline due to a number of factors. Nonuniform heat treatment of the catalytic particles leads to variations in the alumina phases of catalyst particles such that after some years and numerous regeneration cycles of operation, the catalyst particle mix could consists of various percentages of gamma, delta, theta, and alpha alumina particles. Surface areas of representative composite catalyst samples taken would then reflect the averaging of the surface areas. In addition, some catalyst structural collapse would eliminate a portion of the smaller pores, as reviewed previously. A third factor is the effect of fresh catalyst added to replace catalyst fines, as the surface area of catalyst in a CCR reformer would reflect the averaging of the fresh and old catalyst surface areas.

Surface area changes and to some extent the differing degrees of alumina phase transitions occurring in the catalyst support as indicated can be tracked in catalytic processing units that have catalyst sampling equipment. Catalyst samples taken at defined frequencies such as monthly can provide necessary surface area and other physical and chemical characteristic data for the catalyst. Typical catalyst surface area monitoring could lead to the production of surface area curves as shown in Figure 5.4.

The use of a database for the tracking of the surface area characteristics of the catalyst and other timely catalyst information should provide a good basis for cost-effective management of the catalytic reforming process, especially for the CCR unit. Surface areas of fresh or start-of-run (SOR) CCR catalysts and the rate of surface area losses after subjecting the catalysts to more than 250 catalyst regeneration cycles are provided in Table 5.2.

Some of the CCR catalysts in Table 5.2 show excellent surface area retention, and these are due to the excellent hydrothermal properties of the CCR catalysts and effective utilization of process unit management principles. Some of the CCR catalysts listed in Table 5.2 were in operation for more than 10 years and had been subjected to over 700 regeneration cycles. A few of the catalysts exhibited poor hydrothermal stabilities, and those catalysts also had lower crush resistance qualities.

The use of surface area as a key criterion for catalyst replacement is now partially obviated by the excellent hydrothermal stabilities of recent CCR catalytic reforming catalysts. Other performance criteria such as catalytic activity, product selectivity, catalyst poisoning, and catalyst deactivation continue to be factors used in making timely decisions for catalyst replacement. Another catalyst performance replacement criterion of vital importance for reliable and profitable operations of catalytic

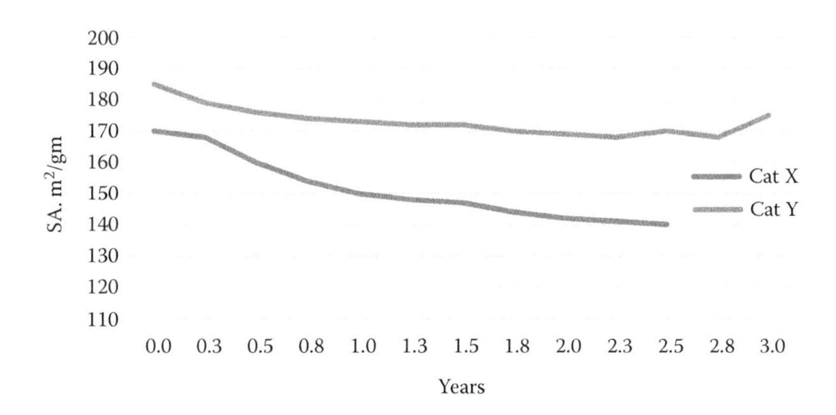

FIGURE 5.4 CCR catalyst surface area.

TABLE 5.2
Surface Area Retention Data for CCR Unit Catalysts

Catalyst	Fresh Surface Area, m^2/gm	Last Surface Area, m^2/gm	Age of Catalyst, years	Surface Area Loss, %/year
A	210.0	170.0	7.0	2.7
B	190.0	161.0	6.0	2.5
C	190.0	158.0	2.5	6.7
D	200.0	160.0	10.0	2.0
E	200.0	154.0	10.0	2.3
F	190.0	164.0	5.3	2.6
G	210.0	183.0	3.0	4.3
H	175.0	141.0	1.7	8.9

reformers and more specifically CCR units' operations is catalyst attrition resistance, and that is the topic in the next section.

5.7.1.5 Attrition Resistance

A good attrition resistance characteristic is a critical property of catalysts that are used in fluid catalytic cracking units, continuous catalyst regenerative processes, and other oil refining and petrochemical catalytic process systems. The process unit and catalyst systems that require continuous catalyst movement within and between reactor zones in the catalyst regeneration section of processing units usually benefit from using catalysts with excellent attrition resistances. Thus, oil refiners are advised that for catalytic processing units that require continuous circulation of catalysts through zones as indicated, catalysts that have excellent structural integrity should be used. Crush strength and attrition resistance properties of the catalysts are good indicators of expected fines make within the process units. The continuous catalyst regeneration reforming process is highly demanding due to the need to use spherical catalysts with excellent structural and attrition resistant properties. Catalysts in the CCR reforming type units are continuously subjected to abrasion by contact with rough metallic surfaces, contacts of moving catalyst spheres in the annular zone of reactors, and catalyst contacts with reactor internal elements and piping. Spherical catalysts are also subjected to high temperature coke burn operations in the catalyst regeneration systems that can also lead to alumina phase transitions, leading to the formation of softer theta alumina and catalyst fines. In order to meet the requirement of achieving optimal containment of the catalyst and therefore containment of platinum and promoter metals within the CCR unit, catalyst fines make must be significantly minimized in the overall CCR-type units. One of the major negative impacts of poor attrition-resistant catalysts is that the catalyst fines they generate cause poor reactor and regenerator operations due to pluggage of the catalyst containment screens within the reactors and regenerators.[33]

Furthermore, and possibly the most important reason to use a high-performance catalyst with excellent attrition resistance, is that high catalyst fines make and inefficient catalyst fines removal could lead to the inability to move catalyst particles continuously from the reactors to the regenerators. Many catalyst circulation challenges are due to excessive catalyst fines that could accumulate in the catalyst circulation system and impede catalyst circulation. Downtimes due to catalyst circulation problems associated with excessive catalyst fines make negatively impact CCR catalytic reformer and refinery profitability. Refinery profitability would be even more negatively impacted if the CCR unit were the sole supplier of hydrogen for the oil refiner's other hydroprocessing units. Additionally, due to different catalyst attrition resistance characteristics, it is highly recommended that dissimilar catalysts not be used as a mix or as makeup catalysts in CCR unit–type operations without a full assessment of the potential consequences of increased catalyst fines make on process unit operability.

Test methods used for measuring attrition resistances of catalysts include the jet-cup, air-jet, and a modified attrition drum (MAD) test consisting of two concentric rotating drums that was developed by Ashland Petroleum.[33–36] The space between the two concentric rotating drums is set at 11 mm, and that accommodates about seven layers of catalyst spheres. Drums are rotated in opposite directions to permit duplicating catalyst-to-catalyst and catalyst-to-metallic surface contacts, thereby simulating catalyst interactions in CCR-type conditions in the reactors, piping, and regenerators. Catalyst fines generated in the MAD test usually display similar fines type distribution relative to those generated in commercial CCR process units. Comparative data for catalysts fines distributions for the MAD test and commercial CCR units are shown in Figure 5.5.

Use of a suitable attrition resistance test method in monitoring catalyst fines make by catalysts is recommended for making catalyst selection decisions and in the management of catalyst fines–induced challenges in CCR reforming–type units that operate with catalyst circulation and continuous regenerations.

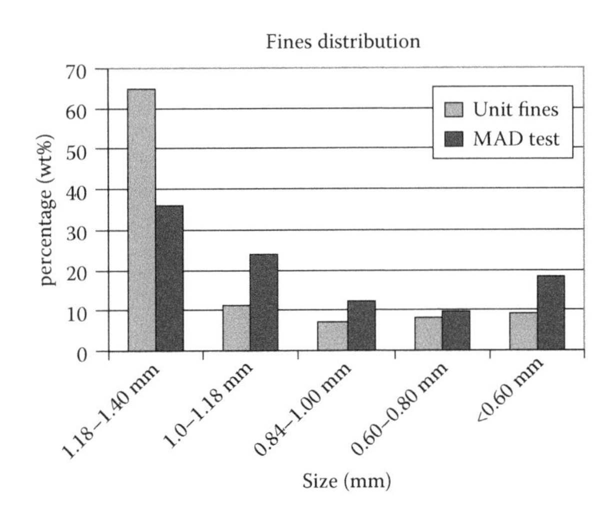

FIGURE 5.5 Comparison of MAD test and CCR unit fines.
From Doolin, P. K., Zalewski, D. J., Oyekan, S. O., Continuous Regeneration and Continuous Reforming Issues, in *Catalytic Naphtha Reforming* by Antos, G. J. and Aitani, A. M., Marcel Dekker, 2004.[33]

Equipment and technology advancements could aid in moderating some of the negative impacts of high catalyst fines make on catalyst circulation. However, as indicated, those technology advancements would most likely not moderate the negative impacts of a catalyst with a high propensity for making catalyst fines and creating other losses associated with catalyst, platinum and promoter metals, CCR unit operations, and profitability.

5.7.2 Chemical Properties

The typical basic chemical properties that are supplied by catalyst providers with their shipment of catalytic reforming catalysts include the nominal metal concentrations, chloride concentration, and sometimes the concentrations of other trace metals on the catalysts. These basic properties are used by oil refiners to verify if the catalyst as provided by the catalyst supplier meets agreed-upon chemical properties. Depending on the analytical assets of the oil refiners' laboratories and staff, technical capabilities, and expertise of key corporate personnel of the oil refining company, other sets of chemical properties could be determined by the oil refiners. Oil refiners are usually interested in assessing relevant properties of catalysts supplied that can provide projected naphtha reforming performance. The list of the chemical properties could then be expanded to include metal and chloride distributions, acidic isomerization sites, and reference baseline metal concentrations of the catalyst to permit assessing metal contaminant accumulation on the catalyst during its use. A third is to infer from the chemical properties and possibly from the results of previously conducted evaluation programs of potential catalysts the expected performance of the catalytic reforming catalyst. Relevant associated brief analytical tests for some of the chemical properties are reviewed in the section on catalyst management challenges in Chapter 6. There are a variety of useful analytical instrumentation and tools for characterizing oil refining catalysts and especially catalytic reforming catalysts. Elemental surface analysis of catalysts is usually conducted using electron microscopic tools that include scanning transmission electron microscopy (STEM), Auger spectroscopy, x-ray photoelectron spectroscopy (XPS), and x-ray fluorescence (XRF). A comprehensive description of the analytical tools and usage is provided by Bradley, Gattuso, and Bertolacini.[66] For molecular species on the catalysts, temperature programmed desorption (TPD), infrared (IR) spectroscopy, and other molecular chemisorption for platinum dispersion determination based on carbon monoxide and hydrogen as adsorbates are recommended.

5.7.2.1 Catalytic Reforming Catalyst Metals

The key active metal that provides the dehydrogenation/hydrogenation functionality on all the catalytic reforming catalysts is platinum. As indicated in the preceding sections, monometallic catalysts are used mostly in fixed-bed semiregenerative and cyclic regenerative reformers, and the platinum concentrations could be in the range of 0.3 to 0.8 wt. %. Platinum/rhenium bimetallic catalysts are also used in fixed-bed semiregenerative and cyclic regenerative reformers in low-sulfur naphtha reforming processing services. The concentrations of the platinum are typically in the range

TABLE 5.3
Catalytic Reforming Catalysts Metal Compositions

Catalyst Type	Designation	Platinum, wt. %	Rhenium, wt. %	Tin, wt. %	Chloride, wt. %
Pt	Monometallic	0.3–0.8	None	None	0.6–1.0
Pt/Re	Bimetallic	0.2–0.4	0.2–0.75	None	0.6–1.3
Pt/Sn	Bimetallic	0.2–0.3		0.2–0.4	0.6–1.3

of 0.20 to 0.3 wt. % and those of rhenium are in the range of 0.2 to 0.75 wt. %. Feed naphtha sulfur is usually specified at less than 0.5 wppm for platinum/rhenium catalyst reforming operations. For the equimolar, skewed, and higher-rhenium Pt/Re catalysts, feed sulfur is sometimes set at less than 0.2 wppm. Platinum/tin catalysts are used in all catalytic reforming processes. Platinum concentrations are usually between 0.2 and 0.3 wt. % and tin concentrations are in the range of 0.2 to 0.4 wt. %. The concentrations of chloride in the catalysts range from 0.6 to 1.3 wt. %. Alumina support concentration in the catalyst is usually about 98 weight percent. The concentrations of metals in platinum, platinum/rhenium, and platinum/tin catalysts are summarized in Table 5.3.

Equimolar Pt/Re is the industry nomenclature for a catalyst with the same weights of Pt and Re, as the atomic weights of platinum and rhenium metals are 195 and 186, respectively. Skewed and higher rhenium refer to Pt/Re catalysts with weight ratios of Re/Pt that are greater than 1.

For most platinum/rhenium and platinum/tin catalysts, special metals are used as platinum and alumina modifiers. The modifier metals can include tungsten, phosphorus, indium, and iridium. Some oil refiners purchase preactivated and presulfided platinum/rhenium catalysts, and for those catalysts, the sulfur content after activation could be in the range of 0.03 to 0.10 weight percent. The next set of metals that are usually in very low concentrations are those that may exist as "trace," and these are usually on the order of weight parts per million in concentrations. These trace metals should be determined for the fresh catalyst, and the fresh catalyst contaminant concentrations could be used later as reference concentrations in the tracking of feed contaminants that could later accumulate on the catalyst during naphtha processing.

5.7.2.2 Catalyst Chloride

Chloride concentrations for catalytic reforming catalysts are listed in Table 5.3. It must be stated, though, that the effectiveness of the catalyst chloride is highly dependent on the concentration of the chloride and the interaction of the hydroxyl groups of the alumina surface and chloride. It is this combination of the alumina hydroxyls and chloride that provides the acidic isomerization functionality of catalytic reforming catalysts. As reviewed in Section 5.3, calcination of gamma alumina support at moderate temperatures that are lower than 750 F could reduce the number of hydroxyl groups and thereby moderate the acidic isomerization functionality of the catalysts. In addition, special thermal and chemical treatments in some catalytic reforming catalyst manufacturing procedures that are applied to gamma alumina can also enable "insertion" of appropriate acidic functionality modifiers. The implications of acidic functionality moderation are given an additional review in Chapter 6 with respect to its impact on the performance of catalytic reforming catalysts.

During naphtha processing, it is also recommended that the catalytic reformer reactors be operated with a water to hydrogen chloride molar ratio of 15 to 25 so that the catalyst chloride can be maintained in the desired range of 0.8 to 1.0 wt. % on catalyst in fixed-bed semi-regenerative catalytic reformers.(37) For CCR catalytic reformers, catalyst chlorides are more easily maintained at desired targets by timely responses to chloride changes. Frequent, and possibly daily, spent and regenerated catalyst chloride data generated via laboratory analysis are typically used for making desired chloride management changes for optimal performance of the CCR units.

5.8 CATALYST EVALUATION PROGRAM

Catalyst evaluation programs can yield great benefits for catalyst manufacturers and their customers with respect to good determination of physical and chemical characteristics, as well as the activities and product selectivity of test and production catalysts. Fresh catalyst samples are analyzed to ensure that the batches of catalyst manufactured meet target physical and chemical specifications. In the catalytic naphtha reforming process, spent and regenerated catalyst samples can be analyzed to determine if catalyst regenerations are successful and use the data in the troubleshooting of catalytic reforming catalyst performance. Analytical data collected over time for spent and regenerated catalyst samples and catalytic performance can also be used to determine the end of catalyst life and plan for catalyst replacements.(37)

The costs associated with the acquisition of catalysts for oil refining processes are usually small fractions of total operating costs. However, as reviewed in this book, making good catalyst selections and running effective, timely catalyst management programs can greatly enhance the reliability and profitability of oil refining processes. In hydrotreating processes, physical characteristics, catalyst performance, and stability are emphasized. In fluid catalytic processes, good physical characteristics and catalytic performance are desired despite the fact that the performance of the main active fluid catalytic cracking catalyst performances could be moderated by the "cocktail mix" of the active catalysts with several catalyst additives in the FCC units. For catalytic reforming processes and depending on the specific reforming technologies used by oil refiners, catalyst investment could be substantial based on the cost of the catalyst and associated platinum. Hence, the review of physical properties of catalytic reforming catalysts such as crush strength and attrition resistance is crucial, and catalyst evaluation programs that generate reliable inspections data should be a top priority. As an example, catalyst and acquisition cost for a 75 MBPD CCR platformer operating with 400,000 pounds of catalyst containing 0.3 wt. % platinum is about $6 million for catalyst acquisition, and at a high platinum price of $2000 per troy ounce, the initial platinum cost would be in excess of $35 million. The oil refiner could elect to lease the required platinum and pay the costs associated with leasing to a platinum supplier. In CCR unit operations, platinum and catalyst losses can occur through the loss of catalyst fines due to poor physical characteristics of catalysts and significantly increase of the cost of catalyst management.

5.8.1 Catalyst Testing

A number of the major oil refining companies have central corporate groups of highly experienced technical personnel and associated laboratories to support effective catalyst evaluation programs for all of the catalytic processes in their oil refineries. Smaller oil refining companies are sometimes financially constrained and have to rely on catalyst supplier companies for catalyst selection decisions and process support. Those oil refiners' options for optimizing their process units, as expected, are highly limited with respect to making independent assessments of available catalyst offerings, technologies, and services. Typically, good catalyst evaluation programs are maintained and coordinated by several appropriate technical personnel operating with appropriate analytical equipment and test units for evaluating catalysts. Their evaluations and ratings may involve assessing physical and chemical characteristics of the catalysts and ranking catalyst performances through model feed tests in small units and naphtha feed processing in pilot plant–type units.

5.8.1.1 Model Feed Test Program

Model compounds such as cyclohexane, methyl cyclopentane, and heptane, which are representative of key hydrocarbon compound types in naphtha feeds, are used in simple, small test units as feeds to evaluate and screen catalysts, study reaction kinetics and reaction mechanisms, and for other fundamental studies. Cyclohexane is used for an assessment of the platinum or dehydrogenation/hydrogenation functionality of the catalyst, as the reaction is relatively facile. Due to the high product selectivity to benzene, the extent of the cyclohexane dehydrogenation reaction can aid in studies with respect to adequacy of platinum activity, platinum poisoning, and moderation of platinum activity by other promoter metals added. For good assessments of relative contributions of the acidic isomerization and platinum functionalities, methyl cyclopentane and heptane are used.[38–40] Test catalyst samples should be activated in a dedicated catalyst activation unit and precautions should be taken to minimize cross-contamination of catalyst samples in the activation process. If feasible, activated catalysts should be analyzed to determine if the concentrations of chloride, sulfur, metal contaminants, and carbon are as required. Another set of dedicated test units should be used for catalyst evaluation using the selected model hydrocarbon feed. Platinum dispersion of the activated catalysts could also be determined before conducting catalyst test runs. After conducting test runs on similarly activated and characterized catalyst samples, the test data are then compiled and analyzed and the evaluated catalysts are ranked in terms of physical and chemical characterization data and catalytic performance.

It should be stated that the preceding reviewed model feed test program is suitable for use in exploratory catalyst research and development programs to identify good leads for the development of novel catalytic reforming catalysts.

Comparative physical and chemical characterization data should be compiled for all the catalysts that are being considered for selection, and the datasets and catalyst performance data should be used in selecting promising leads for developing novel catalysts in the more definitive pilot unit test programs.

5.8.1.2 Naphtha Feed Test Program

The more representative naphtha feed test unit program can provide catalyst performance data that are similar to what can be expected in oil refiners' catalytic reformers operating at similar pilot test operating conditions. Key elements of test methodology reviewed in the section on a model compound test program are applicable, including detailed fresh catalyst characterization and the activation of catalysts in dedicated catalyst activation units so as to eliminate cross-contamination of test catalysts during catalyst activation. Third, another major requirement that could substantially aid in the generation of reliable pilot unit test data is the use of a naphtha feed preparation and additive system. In addition to using reference hydrotreated naphtha for the pilot unit test programs, hydrotreated naphtha could be further treated by drying over materials such as 3A and 4A molecular sieves to reduce water and other adsorbents. The dried naphtha feed should be thoroughly mixed and additives such as organic sulfur and inorganic chloride added to required levels before using it as the feed in catalyst evaluations. For satisfactory naphtha feed drying and circulation, dedicated feed treatment and circulation units are recommended.[41,42]

Key requirements for a successful catalyst evaluation program include use of test unit reactors that are operated either strictly adiabatically or isothermally so as to generate data that would enable technical personnel to determine accurate intrinsic reaction rates and product selectivity.[45] The second requirement is that good analytical laboratory support is available that can provide much-needed timely catalyst characterization data and liquid and gas analysis. Third, the oil refiner or catalyst company should have capable technical personnel with the necessary skills, knowledge, and expertise to operate pilot plants reliably and conduct excellent analysis of catalytic reforming data.

A simplified flow diagram of a fixed-bed pilot unit is shown in Figure 5.6. The pilot unit typically consists of a single reactor, once through or recycle gas operation as

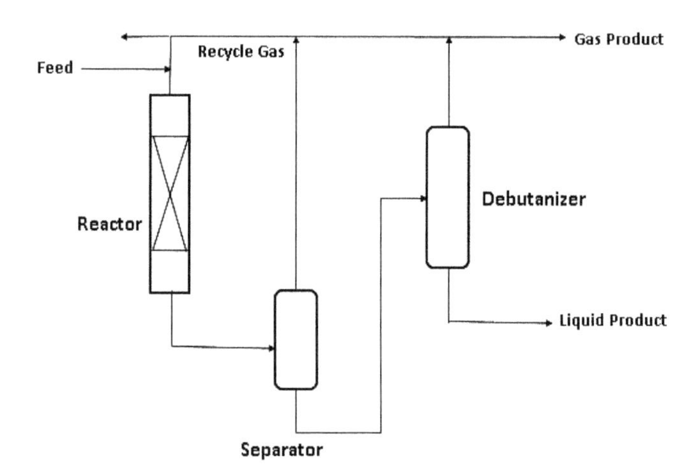

FIGURE 5.6 A simplified drawing of a catalytic reforming pilot unit.
From Oyekan, S. O., Catalytic Naphtha Reforming Lecture Notes in Catalytic Processes, in *Petroleum Refining* by Chuang, K. C. et al., AIChE Today Series, 1992.[37]

the one shown above; a gas/liquid separator; and the stabilizer for removing residual gas from the reformate. The pilot plant test unit shown in Figure 5.6 was used to generate test data on catalysts and naphtha feeds that were similar and consistent with catalytic performance data generated by catalytic reformers on selected catalysts. In order to study the effect of catalyst regeneration on hydrogenation/dehydrogenation and promoter metals, a catalyst regeneration program is usually conducted. It is highly recommended that catalyst regeneration studies be conducted to eliminate catalysts that require specific complex procedures for achieving full recovery of catalyst activity and product selectivity. Catalyst supplier companies are fully aware and strive to establish the regenerability of their catalysts during the development phases of their catalysts.

For the effective evaluation of specific catalysts for CCR processes, a few of the oil refiners have CCR-type pilot units that give them the advantage of generating pilot plant catalyst data that are more representative of expected catalyst performance and provide useful information on the structural integrity of the catalyst particles in catalytic reforming operations.

Two types of catalytic reforming catalyst evaluation methodologies are used in studies, the yield/octane and constant octane tests. The constant octane test is also referred to as the catalyst aging and catalyst stability test. Variations of yield/octane and constant octane test programs have been successfully developed and used over the past 70 years by catalytic reforming technologies and catalyst provider companies and by some of the oil refining companies.

5.8.1.3 Catalyst Activation

A consistent catalyst activation procedure improves the chances of generating reliable, accurate, representative catalytic data from the model feed or naphtha feed pilot units for test catalysts. Catalysts are activated to ensure evaluation of the samples at reasonably similar conditions to those in commercial reforming units. The pretreatments used are similar to those recommended for use by catalyst manufacturers. It is recommended that sufficient representative amounts of each catalyst be compiled via activation of batches of the catalyst. The composite of each activated catalyst can then be accumulated from the individual batches of the activated catalyst after individual sampling and characterization of the batches. This exercise permits retaining representative activated samples of the same catalyst. A sufficient composite of activated catalyst would provide flexibility in managing test runs, conducting several test runs, and use in catalyst regeneration studies.

The activation procedures could include variations of the following.[37]

a. Dry and purge catalyst to remove water at about 700 F to 800 F in flowing nitrogen.
   The heat-up rate should be selected and preferably maintained at between 50 to 100 degrees and not to exceed 100 degrees per hour rate so as to minimize undue breakage of catalyst particles.
b. Conduct catalyst rejuvenation or redispersion of the platinum and promoter metals at 950 F to 980 F.
   This is initiated by introducing 2 to 5% oxygen/nitrogen gas into the reactor at 800 F and raising the pressure to 200 to 300 psig. After about 2 hours,

the temperature is raised to 950–980 F over a 5-hour period followed by an air soak at the selected target temperature for 5 to 15 hours. During this period, an organic chloride is added to facilitate redispersion of the platinum oxide and promoter metal oxides. The metals rejuvenation conditions that favor achieving high platinum redispersion are high temperature greater than 900 F and less than 1000 F, sufficient oxygen, and about 15 psia oxygen partial pressure in the reactors, with chloride addition and sufficient time.[11,37]

c. Reduce the platinum oxide and promoter metal oxides of the catalysts to active forms at about 900 F in hydrogen.
   After the rejuvenation step, air is purged out and hydrogen introduced at 700 F. The reduction can be started at a temperature lower than 700 F and the temperature of the reactor contents raised to 900 F to complete the reduction of the of the platinum oxide and promoter metal oxides to their active metal states. The reduction temperature can also be raised to complete the catalyst reduction at about 950 F. A two-stage reduction procedure can also be applied for multimetallic catalytic reforming catalysts as required.[11]
d. Platinum and platinum/rhenium catalyst sulfiding.
   For platinum and platinum/rhenium catalysts, sulfiding of the catalyst is recommended. Hydrogen sulfide or an organo-sulfur compound is added, and sulfiding can be conducted at temperatures between 700 F and 900 F in hydrogen. Platinum/tin catalysts have a relatively simple activation procedure, as catalyst sulfiding is not required. Platinum/tin catalysts that do not require catalyst sulfiding are ideal catalysts for CCR reforming operations. This ease of the reactivation feature and high initial start-of-run reformate and hydrogen selectivity make platinum/tin the catalysts of choice for CCR reforming operations.
e. Conduct detailed catalyst characterizations—physical and chemical properties.
   After the catalyst sulfiding process for platinum and platinum/rhenium catalysts, the reactor is cooled and catalyst activation is completed for that batch of catalyst. The activated batch of catalyst is sampled, analyzed, and characterized with respect to its chemical properties, such as sulfur, chloride, and carbon. In addition, physical properties such as surface area, porosity, crush strength, and attrition resistance could also be determined for the fresh and activated catalyst.As indicated, several activated batches of a catalyst can be made into a composite and full physical and chemical characterizations can also be conducted on the composite, as indicated in the section above.

5.8.1.4 Yield/Octane Test

The yield/octane test is also referred to as an activity/selectivity test, as the test provides initial catalyst activity and reformate and hydrogen selectivity data. A yield/octane test run can be completed within 4 days of naphtha processing in a pilot plant unit. The space velocity or the reactor temperature can be modified to generate product yield data at different reformate octanes. For ease of data analysis, changing the reactor temperatures is favored over changing space velocity. Reformates, C5+ hydrocarbon liquids, hydrogen, and gas products are then determined for a number of reformate octanes. Low-severity test conditions utilized in yield/octane test runs are usually selected to minimize catalyst deactivation so as to generate fresh catalyst-type performances for facile comparisons. Test conditions usually include high reactor pressures and high hydrogen to recycle gas ratios.[37]

5.8.1.5 Constant Octane Stability Test

The constant octane test is a pilot plant test that is conducted on a catalytic reforming catalyst at constant reformate octane with the objective of simulating the performance of the catalyst in an oil refiner's catalytic reformer. The test is referred to as catalyst aging or catalyst stability, as the test catalyst is evaluated at high-severity conditions that provide information on the rate of catalyst deactivation or aging of the catalyst. Basically, a reformate octane is selected and the catalyst is tested at a start-of-run temperature that leads to production of the target reformate octane with a naphtha feed in the test run and at appropriate test conditions. Typically, the catalyst is tested at higher-severity operating conditions relative to those that were used in order to minimize catalyst deactivation in the yield/octane test. In the constant octane test methodology, the reactor temperature is increased to compensate for catalyst deactivation so as to achieve target reformate octane. In the naphtha feed test, the initial reactor temperature and reformate or C5+ yields represent the initial activity and product selectivity. The reactor temperature increase required to maintain constant octane correlates with catalyst deactivation and hence stability of the catalyst.

Consistent and representative pilot plant catalytic performance data can be generated using the recommended catalyst evaluation program provided in this book. Variations of this recommended catalyst evaluation program are also practiced by research and development centers of some of the global oil refining companies to great advantage with respect to operating their refineries reliably and profitably. Similar catalyst evaluation programs are also used by catalytic reforming technologies and catalyst supplier companies in catalyst development studies and catalytic reformers' optimization programs.

5.8.1.5.1 Platinum and Platinum/Rhenium Catalyst Evaluation

An excellent example of the use of a good catalyst test program involved the studies that were conducted to compare the performances of commercial platinum monometallic, Pt/Al_2O_3, and platinum/rhenium bimetallic, $Pt/Re/Al_2O_3$, catalysts conducted in the 1980s.[37] The inspections of the paraffinic naphtha feed used in the catalyst studies are provided in Table 5.4. The paraffinic naphtha endpoint was 395 F, and this is drastically different today with the current processing of low naphtha endpoint feeds in the 340 F to 360 F range due to oxygenate blending and diesel/gasoline price differential in some of the global fuel markets. Inspections of the activated platinum and platinum/rhenium catalyst samples are given in Table 5.5.

Pilot plant test conditions were set at 255 psig pressure, 4.0 WHSV, recycle gas ratio of 3 moles of hydrogen to 1 mole of the naphtha feed, and reformate target research octane number of 98. Test runs were conducted on a representative paraffinic naphtha feed, and its inspections are provided in Table 5.4.

A measuring criterion used for defining catalytic reformer operation cycle lengths and used sometimes by some oil refiners is a 30- or 40-degree increase in the reactor

TABLE 5.4
Paraffinic Naphtha Feed Inspections

	Naphtha Feed Inspections	
Gravity	API	57.8
Sulfur	Wppm	0.13
	ASTM D-86 Distillation, F	
IBP		208
50%		275
95%		340
FBP		395
	PNA Analysis, vol.%	
P/N/A		68/14.7/17.3
N+2A		49.3

weighted average inlet temperature. In a pilot plant test run, the reactor temperature rise end of the cycle criterion can be applied and a 40-degree increase is used in this review as a measure of the cycle length for a catalyst in pilot unit test runs. Based on that criterion, reactor temperature increase of 40 degrees and approximately 975 F was reached at 70 hours of test run for the Pt catalyst, whereas based on the 40-degree increase, a 970 F temperature was reached for the Pt/Re catalysts after 230 hours on oil. Thus, the Pt/Re/Al_2O_3 catalyst, as expected, maintained its activity for 3.3 times as long as the Pt/Al_2O_3 catalyst. The bimetallic catalyst also had a lower required reactor temperature of 930 F to meet target octane and higher initial activity at start of the test run relative to the monometallic catalyst as shown in Figure 5.7.

Reformate or C5+ yields are compared in Figure 5.8 for the Pt and Pt/Re catalysts. The curves show the expected greater C5+ yield selectivity of the Pt/Re catalyst relative to the Pt catalyst. Based on a criterion of 2 volume percent decline of C5+ yield as a measure of the end of a catalytic reformer operating cycle from start of run C5+ yield, the cycle for the Pt/Re catalyst is not established in Figure 5.8 because at

TABLE 5.5
Inspections of Activated Pt/Al_2O_3 and Pt/Re/Al_2O_3 Catalysts

	Activated Catalyst Samples	
Catalyst	**Pt/Al_2O_3**	**Pt/Re/Al_2O_3**
Platinum, wt. %	0.30	0.30
Rhenium, wt. %	0.00	0.30
Chloride, wt. %	1.00	1.10
Carbon, wt. %	0.03	0.03
Sulfur, wt. %	0.07	0.10

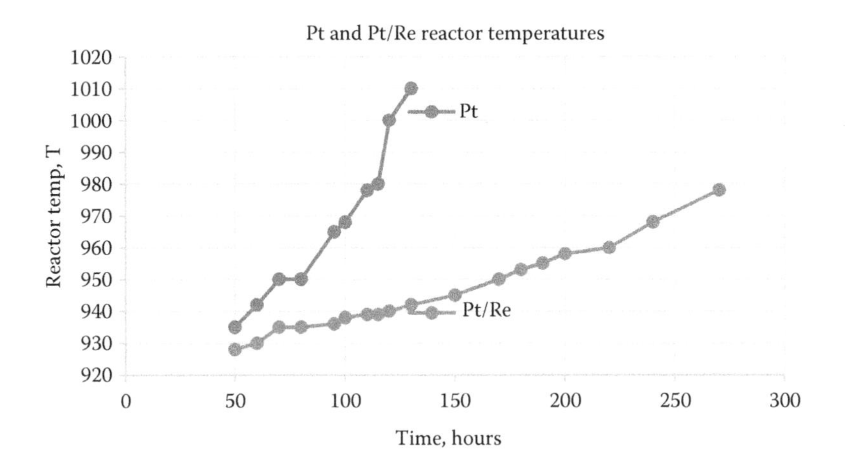

FIGURE 5.7 Platinum and platinum/rhenium catalysts activities.
From Oyekan, S. O., Catalytic Naphtha Reforming Lecture Notes in Catalytic Processes, in *Petroleum Refining* by Chuang, K. C. et al., AIChE Today Series, 1992.[37]

the end of the test run, the C5+ yield was approximately equivalent to that of the start of run, whereas the platinum monometallic catalyst had already suffered a 2 volume percent decline within 60 hours of its test run.

It is now known and accepted that $Pt/Re/Al_2O_3$ bimetallic catalysts exhibit higher reformate or C5+ yields relative those of Pt/Al_2O_3, monometallic catalysts. In addition, bimetallic catalysts typically exhibit excellent C5+ yield stability relative to that of monometallic catalysts in fixed-bed catalytic reforming units.

5.8.1.5.2 Platinum/Rhenium Sulfur Sensitivity

The next major catalyst study that was conducted in pilot plants and using special catalyst planning and test methodology as described in this book was by Oyekan and McClung at Engelhard, now BASF, in their 1982 to 1986 definitive studies on the

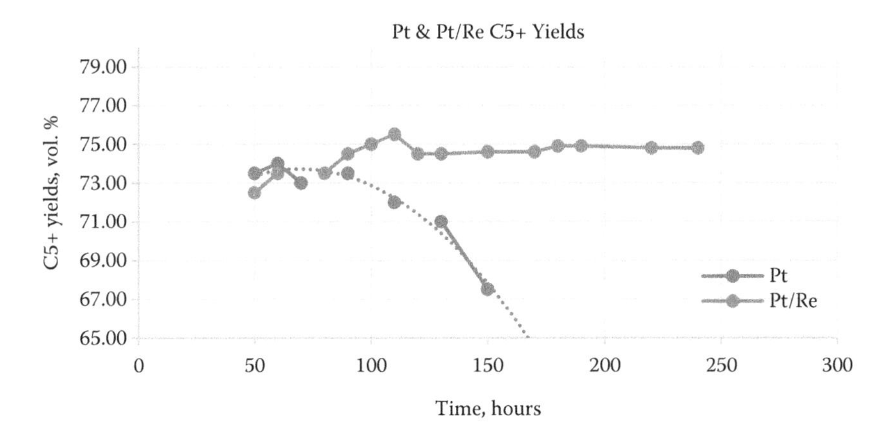

FIGURE 5.8 Reformate yields for Pt/Al_2O_3 and $Pt/Re/Al_2O_3$ catalysts.

sulfur sensitivity of Pt/Re catalysts.[42] In the1980s, negative catalytic performances were experienced in reforming high naphtha feed sulfur over Pt/Re catalysts. A major pilot plant study was undertaken to better understand the factors that were leading to lower catalytic performances. It was necessary to conduct Pt/Re catalysts studies, as oil refiners were using high rhenium, skewed, unbalanced Pt/Re catalysts in their semiregenerative and cyclic regenerative fixed-bed reformers with mixed catalytic performances. Catalysts that have higher weights of rhenium relative to platinum are referred to as high-rhenium Pt/Re or skewed Pt/Re catalysts, as reviewed previously. Oil refiners were experiencing poor performances with the high-rhenium Pt/Re catalysts in their catalytic reformers, especially with the high-rhenium catalysts that were processing high-sulfur naphtha feeds due to inefficient naphtha hydrotreater operations and hydrotreater naphtha product sulfur upsets. High-sulfur naphtha feeds are those that contain greater than 0.5 wppm of sulfur. Thus, there was a global need to understand the impacts of high-sulfur naphtha feeds on bimetallic Pt/Re catalyst activity and selectivity performances.

There were some prevailing hypotheses with respect to the performances of Pt/Re catalysts relative to platinum-only catalysts that should be reviewed. As was observed in the preceding section on Pt and Pt/Re catalysts, sulfided Pt/Re catalysts exhibited superior activity and better reformate selectivity relative to their platinum monometallic analogs. Some researchers ascribed the better performances of the Pt/Re catalysts to preferential chemisorption of sulfur on the rhenium.[46,47] Others ascribed different hypotheses for the better activity and selectivity stability performances of the Pt/Re catalysts relative to their monometallic catalysts.[48–50] Wagstaff and Prins suggested some alloying effect between the platinum and rhenium or bimetallic aggregation and that was responsible for the better platinum performance. Bertolacini and others suggested that the addition of rhenium to the platinum catalysts led to reduced coking and catalyst deactivation.[10,51,52]

Of great importance with respect to a good analysis of the pilot plant test data from the sulfur sensitivity studies is the need for some understanding of the platinum and rhenium states on activated catalysts and on the reducibility of rhenium. The interesting work of Johnson[53] established that platinum oxide is reduced to its zero-valent state and rhenium to the plus-4 state in the reduction step for oxidized platinum and promoter metals in the Pt/Re bimetallic catalyst. Other catalyst researchers suggested that rhenium could be reduced to the zero valent state depending on the pretreatment procedures used and the moisture content of the catalyst reduction environment.[49,54–57]. Electron spin resonance (ESR) and carbon dioxide (CO) chemisorption studies by Sachtler enabled him to suggest that the zero-valent and quadric-valent rhenium species coexist on the surface of $Pt/Re/Al_2O_3$. In summary, it can be inferred from the studies that Pt^0, Re^0, Re^{+4} exist on the reduced catalyst with the quadric-valent rhenium sites existing at about 5% of the rhenium sites after catalyst activation.[37,42]

5.8.1.5.3 *Platinum Rhenium Catalyst Sulfur Sensitivity Studies*

At the outset, good naphtha feed preparation and catalyst activation programs were considered vital for a successful pilot plant test program on platinum/rhenium catalyst sulfur sensitivity. In the naphtha feed program, hydrotreated naphtha was further

TABLE 5.6
Naphthenic Naphtha Test Feed

Naphtha Feedstock Inspections		
Gravity	API	54.3
Sulfur	wppm	0.3, 0.13
ASTM D-86 Distillation, F		
IBP		158
50%		292
95%		401
FBP		432
PNA Analysis, vol.%		
P/N/A		46/29/25
N+2A		79

hydrotreated over sulfur guard adsorbents to generate essentially a "zero"-sulfur naphtha. The feed was then treated in the feed circulation and drying system and an organosulfur compound was added to meet naphtha feed specifications for catalyst test runs. Two naphtha feeds were used and data presented in this book were generated via the use of the naphthenic feeds in Table 5.6. The treated naphtha feed was divided into two feeds and sulfur was added at the 0.3 and 1.3 wppm levels into the feeds, respectively. Pilot plant test runs were conducted at accelerated aging conditions of 255 psig, 4.0 WHSV, 3/1 H2 to HC molar ratio recycle gas, and at target 99 RON for the reformate. Test runs used equimolar Pt/Re catalyst labeled as A and a skewed Pt/Re catalyst labeled as B.[42]

The C5+ yield data for two different runs on the standard equimolar Pt/Re, catalyst A, are shown in Figure 5.9 for the 0.3 and 1.3 wppm sulfur naphthas. Using the criterion of 2 volume percent decline in C5+ yield as a measure of a catalytic reforming cycle, the cycle length for the 0.3 wppm sulfur naphtha operation is over five-fold longer than that for the 1.3 wppm sulfur naphtha operation.

The type of superior performance benefits that were determined in the pilot plant test studies in the reforming of low-sulfur naphtha feeds over platinum/rhenium catalysts can also be realized in oil refineries by the use of a either a liquid or vapor phase sulfur guard adsorbent unit. A vapor phase sulfur guard unit can be installed in the recycle gas section of the reforming process unit. The sulfur guard can remove hydrogen sulfide from the recycle gas and reduce hydrogen sulfide in the reactors section. A guard bed unit can also be installed to operate as a liquid phase sulfur adsorbent unit to guard against naphtha hydrotreater operation upsets that can increase sulfur compounds and hydrogen sulfide in the naphtha feed to the catalytic reformer.[58]

As discussed in the preceding section, cycle lengths were determined applying the 40-degree temperature increase for the activity data and a C5+ yield decline of 2 volume percent for product selectivity. Based on a 2 volume percent C5+ yield decline, the cycle length advantage is about five-fold for the reforming of 0.3 wppm

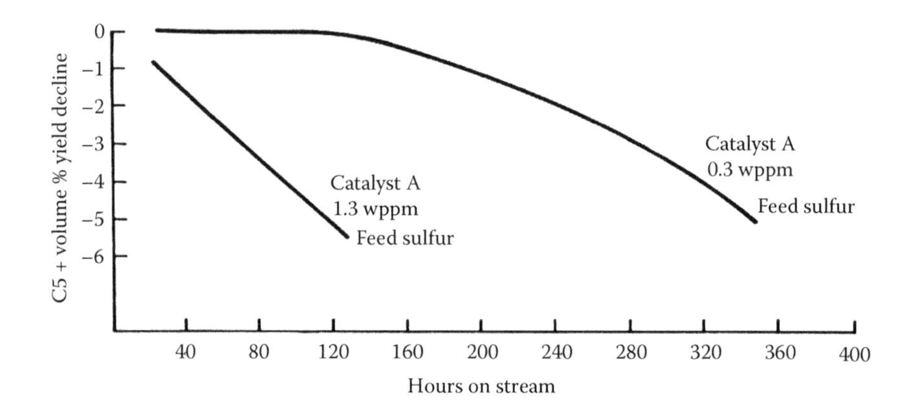

FIGURE 5.9 C5+ yields for equimolar catalyst A.
From McClung, R. G., Oyekan, S. O., Sulfur Sensitivity of Catalysts in Naphtha Reforming, *AIChE Spring National Meeting*, New Orleans, 1988.[42]

sulfur naphtha relative to reforming 1.3 wppm sulfur naphtha over the equimolar catalyst A, as shown in Figure 5.9. Similar results were also determined for cycle length advantage for the equimolar catalyst for low-sulfur naphtha reforming, as shown in Figure 5.10.

A high-rhenium catalyst, Catalyst B, was tested on the same 0.3 wppm sulfur naphthenic naphtha feed that was used earlier for the equimolar catalyst, Catalyst A, test run. Data from the high-rhenium catalyst test run for the 0.3 wppm sulfur naphtha feed were compared with those for the balanced catalyst for the 1.3 wppm sulfur naphtha feed. As expected, the high-rhenium catalyst B performed excellently in the reforming of low-sulfur low-naphtha feeds, as shown in Figure 5.11.

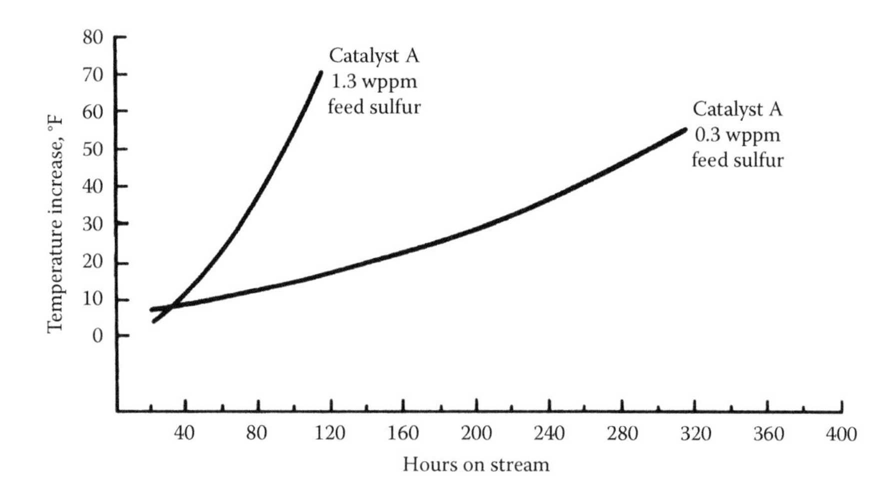

FIGURE 5.10 Temperature increases for 99 RON operations for catalyst A.

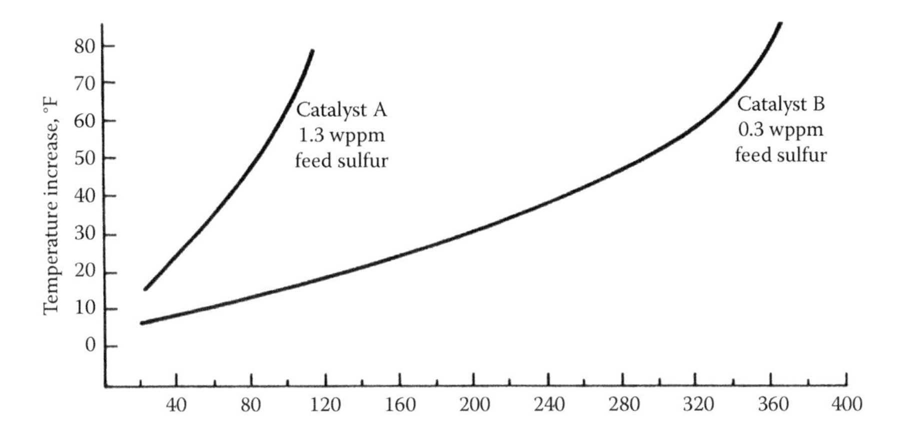

FIGURE 5.11 High-rhenium catalyst test run data.

In the 1980s, a correlation was developed for predicting the expected relative cycle lengths for reforming naphtha feeds with widely varying sulfur concentrations in the range of 0.1 to 2.0 wppm over equimolar Pt/Re and high-rhenium or skewed Pt/Re catalysts. The correlation curves are as shown in Figure 5.12 for the balanced, equimolar Pt/Re and high-rhenium Pt/ Re catalysts.

The two successful catalytic reforming catalysts studies reviewed as examples required significant planning and use of an excellent catalyst evaluation program. These studies incorporated a catalyst characterization program, catalyst activations in

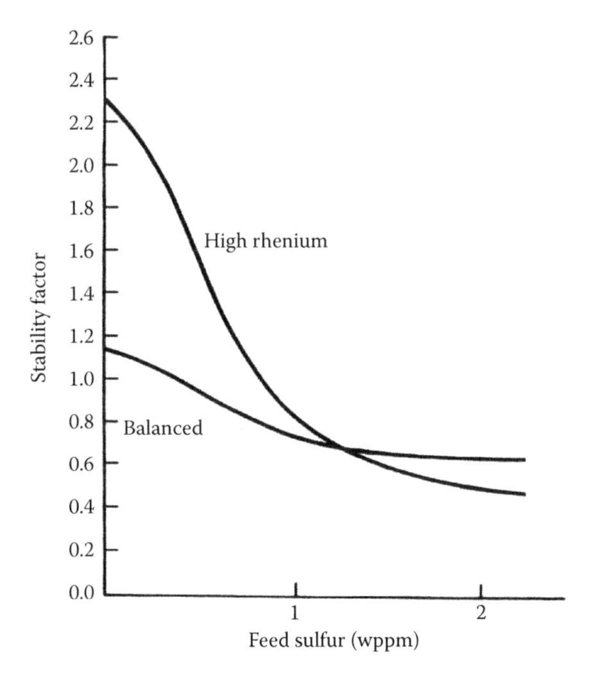

FIGURE 5.12 Stability factor correlations for Pt/Re catalysts.

dedicated activation units, and good naphtha feed preparation systems. The studies also used test naphtha feeds that had excellent quality consistency for over several months of the feed sulfur sensitivity studies. Additionally, test run data were not compromised by cross-contamination of catalyst samples during activations, and adequate moisture/chloride control was maintained during test runs. The studies have been shared to aid other research and development centers in their quest to produce meaningful and reliable pilot plant–type test data for their catalytic reforming businesses.

5.9 ALUMINA HYDROTHERMAL STABILITY TEST

A test that has experienced some resurgence in its usage is the hydrothermal stability test for the reforming catalyst alumina support.[(59)] The catalyst support or carrier is typically gamma alumina and the hydrothermal stability test is designed to help determine the stability of gamma alumina supported catalysts in reforming and regeneration operations over a wide range of temperatures and moisture. Data from such studies provide an assessment of structural and phase changes that reforming catalysts are subjected to during the coke burn and oxy-chlorination steps in catalyst regenerations. Retention of vital physical properties such as surface area, porosity, and pore sizes are directly measured. There are other key factors for using a gamma alumina carrier with excellent surface area retention characteristics. Other than providing physical surfaces for metal and acidic isomerization sites, the alumina surface also participates jointly with chlorides to provide acidic isomerization functionality. Studies by a number of researchers show that the structure of gamma alumina is traditionally considered to be a cubic defect spinel type, and the experimental unit is as shown in Figure 5.13.[(60)] It was indicated that the defective nature derives from the presence of trivalent Al cations in the spinel-like structure.

Other researchers have discussed the nature of the acidic sites on gamma alumina; some have suggested that there are relatively more Lewis acidic sites to Bronsted acidic sites, and the Bronsted sites were estimated to be about 5% of total acidic sites.[(61,63)] Of greater interest is what happens as some of the gamma alumina particles begin to undergo phase transitions to possibly delta, theta, and alpha particles, which was reviewed briefly in a preceding section. Some of these transformations lead to surface area losses, increased pore sizes, and drastic reductions in the concentrations of accessible Lewis acid sites.[(62,63)] These reductions in surface area, hydroxyl groups, and accessible Lewis acid sites lead to lower chloride retention characteristics for catalytic reforming catalysts.

The negative implications of lack of chloride retention by the catalyst in cyclic regenerative and continuous catalyst regenerative are staggering. First, with loss of hydroxyl groups due to surface area losses, the catalyst can no longer hold onto the chloride added, and this leads to higher utilization of organic chlorides. Higher chlorides lead to higher hydrogen chloride (HCl) in the reforming system. Excess HCl released from the catalyst increases the rate of ammonium salt fouling, acid-induced corrosion in product separation sections of the reformer, and higher HCl loads for the chloride guard beds for treating the net hydrogen gas. An inefficiently operated chloride guard bed system under the conditions of high HCl in the net hydrogen would lead to exacerbation of chloride-induced reliability challenges in the

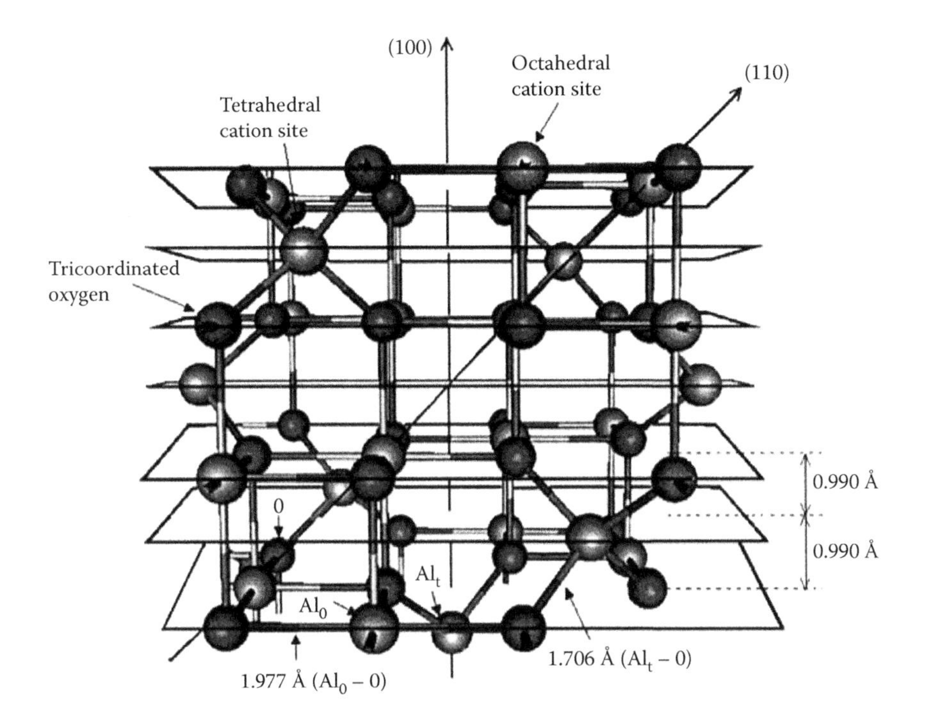

FIGURE 5.13 Cubic defect spinel-type structure of gamma alumina.
From Trueba, M., Trasatti, S. P., *Eur. J. Inorg. Chem.*, 3393–3430, 2005, © Wiley-VCH Verlag GmbH & Co. KGaA, 69451 Weinheim, Germany, 2005.[60]

refinery. In addition, due to the lower chloride retention of the catalyst, the regenerator vent gas chloride systems would have to be monitored to ensure meeting refinery maximum attainable control technology (RMACT 2) vent gas HCl and volatile organic compound specifications. Major activity and product selectivity debits would also occur for a catalyst with low surface area and poor chloride retention, characteristics especially for catalytic reformers that require frequent or continuous catalyst regenerations.

Therefore, based on the scenario described for catalysts with poor hydrothermal stabilities, it is important, if feasible, for the oil refiner to have a good assessment

TABLE 5.7
Equimolar and Skewed Pt/Re Catalyst

	Catalyst A	Catalyst B
	Equimolar Pt/Re	High Rhenium Pt/Re
Platinum, wt. %	0.35	0.35
Rhenium, wt. %	0.35	0.78
Chloride, wt. %	1.00	1.00

of the hydrothermal stability of catalysts that are being considered before making selections. The hydrothermal stability test is relatively simple, as the key parameters measured are surface area and porosity. Chloride retention and other physical characteristic changes for the catalyst could also be monitored. Hydrothermal stability test conditions are typically selected so that they are similar to hydrothermal conditions that coked catalysts are subjected to during catalyst regenerations. The catalyst is usually subjected to heating in an air/nitrogen gas mix at temperatures between 1100 F and 1200 F and reasonable amounts of water and hydrogen chloride. Catalyst samples can be taken over different times of steaming and surface areas determined for them in order to cover four or five time periods and produce a graph of the surface area of a catalyst as a function of time. The percentage of surface area decline from that of the fresh catalyst surface area provides a good measure of the hydrothermal stability of the alumina and catalyst. Hydrothermal stability test data for three CCR unit catalysts reported as percentages of the surface areas retained are listed in Table 5.8. As a reference and for easy interpretation of the surface area retention data in Tables 5.8 and 5.9, time zero surface area is the fresh catalyst surface area and, since it has no surface area loss, is defined as having 100% retention. The other surface areas are percentages of the fresh surface area.

Similar data were obtained for the three catalysts during operations in CCR reforming units. The surface area retention data could have been moderated a bit based on the different rates of fresh catalyst addition to the CCR units, and this is a function of the rate of catalyst fines make and CCR regenerator system upsets. Surface area retention data for the three catalysts taken out of CCR units over a period of 2 years are reported in Table 5.9.

TABLE 5.8
Laboratory Hydrothermal Stability Test Data

	Catalyst A	Catalyst B	Catalyst C
Time, Hrs.	SA, %	SA, %	SA, %
0.0	100.0	100.0	100.0
40.0	93.8	86.3	81.4
150.0	90.0	83.3	77.0

TABLE 5.9
Surface Area Retention of Catalysts in CCR Units

	Catalyst A	Catalyst B	Catalyst C
Time, Yrs.	SA, %	SA, %	SA, %
0.0	100.0	100.0	100.0
1.0	93.5	87.6	93.7
1.5	92.3	87.0	90.2
2.0	91.3	82.9	89.2

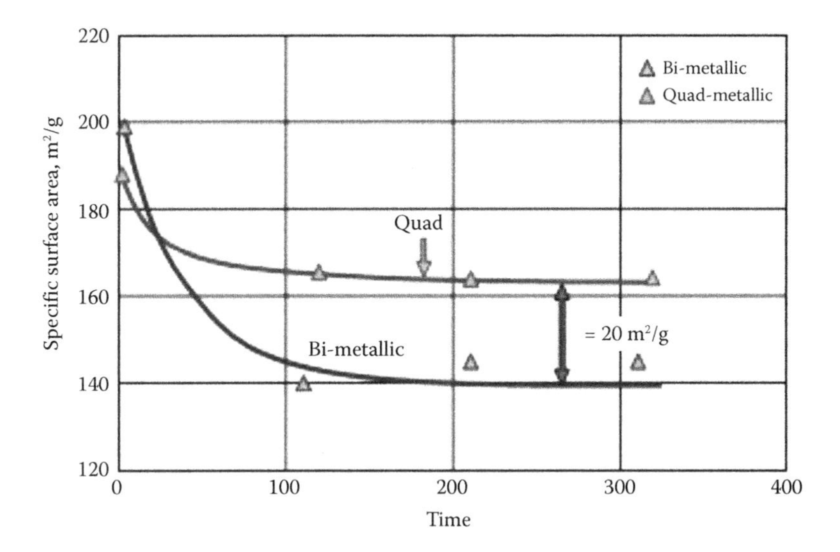

FIGURE 5.14 Surface area retention of current catalysts.
From Le-Goff, P. Y., Lopez, J., Ross, J., Redefining Reforming Catalyst Performance: High Selectivity and Stability, *Hydrocarbon Process.*, 47–52, September 2012.[64] (Courtesy of Axens.)

On a positive note, the hydrothermal stabilities of current catalysts for continuous catalyst regeneration reforming units appear to be good with respect to their surface area retention characteristics. Surface areas appear to decline and stabilize in the high 140 to 160 m^2/gm range versus less than 125 m^2/gm surface areas experienced with the first generation of catalysts for CCR units. Hydrothermal stability for two catalysts is shown by the curves in Figure 5.14.[64]

Greater benefits can still be realized by ensuring that a thorough precise evaluation program of catalysts of interest is conducted if the oil refiner has the assets such as pilot plant units, capable technical personnel, and time to plan and conduct such a study. The need to do this is obvious, as such a study can lead to high reliability and profitability in the operation of an oil refiner's catalytic reforming units and refineries.

REFERENCES

1. Curtis, G. F., Haensel, V., US Patent 2478926 A, 1949.
2. Haensel, V., US Patent 2666620A, 1949.
3. Haensel, V., US Patent 2566521, 1951.
4. Haensel, V., US Patent 2658,028, 1953.
5. Haensel, V., Donaldson, G. R., Platforming of Pure Hydrocarbons. *Ind. Eng. Chem.*, 43(9), 2102–2104, September 1951.
6. Wefers, K., Misra, C., Oxides and Hydroxides of Alumina, Alcoa Technical Paper 19, Aluminum. Accessed February 15, 2018. http://epsc511.wustl.edu/Aluminum_Oxides_Alcoa1987.pdf.

7. Chase, G. C., Espe, M. P., Evans, E. A., Ramsier, R. D., Rapp, J., Reneker, D. H., Tuttle, R. W., Metal Oxide Fibers and Nano fibers: Method for Making same and there off, EP 2267199 A2.
8. Kluksdahl, H. E., US Patent 3415737, 1967.
9. Kluksdahl, H. E., US Patent 4012313, 1977.
10. Oyekan, S. O., Swan, G. A., US Patent 4436612, 1984.
11. Oyekan, S. O., US Patent 4539307, 1985.
12. Hayes, J. C., US Patent 3892656 A, 1975.
13. Carter, J., Sinfelt, J., US Patent 3729408A, 1973.
14. Schoennagel, H. J., US Patent 46524475A.
15. Carter, J. L., McVicker, G. B., Weissman, W., Kmak, M. S., Sinfelt, J. H., Bimetallic Catalysts: Application in Catalytic Reforming. *Appl. Catal.*, 4, 327–346, 1982.
16. Biloen, P., Helle, J. N., Verbeek, H., Dautzenberg, F. M., Sachtler, W. M. H., The role of Rhenium and Sulfur in Platinum-Based Hydrocarbon Conversion Catalysts. *J. Catal.*, 63, 112–118, 1980.
17. McClung, R. G., Oyekan, S. O., Sulfur Sensitivity of Pt/Re Catalysts in Naphtha Reforming. *1988 AIChE Spring National Meeting*, 1988.
18. Burch, P., Platinum-Tin Reforming Catalysts: I: The Oxidation State of Tin and the Interaction between Platinum and Tin. *J. Catal.*, 71, 346–359, 1981.
19. Burch, P., Garla, L. C., Platinum-Tin Reforming Catalysts: II: Activity and Selectivity in Hydrocarbon Reactions. *J. Catal.*, 71, 360–372, 1981.
20. Serban, M., Negiz, A., VandenBussche, K. M., US Patent 8926828, 2015.
21. Serban, M., Costello, C. K., Lapinski, M. P., US 92666091 B2, 2016.
22. Wooten, J. T., Dense and Sock Loading Compared. *Oil and Gas J.*, 96(41), 66–70, October 12, 1998.
23. Honeywell UOP, UOP Dense Loading Service. Accessed February 15, 2017. http://unidense.com/downloads/6brochure-uop.pdf.
24. Catalyst Handling, Catalyst Handling services. www.contractresources.com.
25. Vinci Technologies, *Catalyst Crushing Strength Testers, ASTM D4179, ASTM D6175, ASTM D7084, SMS-1471.* Vinci Technologies, France. http://www.vinci-technologies.com/products-pilots.aspx?IDR=82296&IDM=754024.
26. Kumar, M., Lal, B., Singh, A., Saxena, A. K., Gangwal, V. S., Sharma, L. D., Dhar, G. M., Control of Mesoporosity in Alumina. *Indian J. Chem. Technol.*, 8, 159–161, May 2001.
27. Davies, B. H., Antos, G. H., Characterization of Naphtha-Reforming Catalysts, in *Catalytic Naphtha Reforming*, 2nd edition, by Antos, G. J., Aitani, A. M., Marcel Dekker, 2004.
28. Russell, A. S., Cochran, C. N., Alumina Surface Area Measurements. *Ind. Eng. Chem.*, 42(7), 1332–1338, 1950.
29. Oxford University, 1.06 Determination of the Surface Area by Nitrogen Adsorption at 77 K, Physical Chemistry Practical Course, January 2008.
30. Waters, R. F., Peri, J. B., John, G. S., Seelig, H. S., High temperature Changes in Alumina Structure and Activity of Platinum- Alumina Catalysts, 1960. *Ind Eng. Chem*, 52(5), 415–416, 1960.
31. Unmuth, E. E., Fleming, B. A., US Patent 4703031 A.
32. Sinopec Catalyst Company, SCC Catalysts. Accessed February 15, 2017. http://scc.sinopec.com/scc/en/Resource/Pdf/products/SCCCatalysts.pdf.
33. Doolin, P. K., Zalewski, D. J., Oyekan, S. O., Continuous Regeneration and Continuous Reforming Issues, in *Catalytic Naphtha Reforming* by Antos, G. J. and Aitani, A. M., Marcel Dekker, 2004.
34. Weeks, S. A., Dumbill, P., Method Speeds FCC Catalyst Attrition Resistance Determinations. *Oil and Gas J.*, 88(16), 38–40, April 16, 1990.

35. Forsythe, Jr., W. L., Hertwig, W. R., Attrition Characteristics of Fluid Cracking Catalysts. *Ind. Eng. Chem.*, 41(6), 1200–1206, 1949.
36. Amblard, B., Bertholin, S., Bobin, C., Gauthier, T., Development of Attrition Evaluation Method Using Je-Cup Rig. Accessed February 15, 2017. https://hal.archives-ouvertes.fr/hal-01149043.
37. Oyekan, S. O., Catalytic Naphtha Reforming Lecture Notes in Catalytic Processes, in *Petroleum Refining* by Chuang, K. C., Kokayeff, P., Oyekan, S. O., Stuart, S. H., AIChE Today Series, 1992.
38. McVicker, G. B., Collins, P. J., Ziemiak, J. J., Model Compound Reforming Studies: A Comparison of Alumina-Supported Platinum and Iridium Catalysts. *J. Catal.*, 74(1), 156–172, 1982.
39. Elfghi, F., *Catalytic Reforming of Model Compound and Real Feedstock*, Lambert Academic Publishing, Saarbruken, Germany.
40. Clem, K. R., Catalytic Reforming of Heptane, *LSU Historical Dissertations and Theses*, 3057. http://digitalcommons.lsu.edu/gradschool_disstheses/3057?utm_source=digitalcommons.lsu.edu%2Fgradschool_disstheses%2F3057&utm_medium=PDF&utm_campaign=PDFCoverPages.
41. Oyekan, S. O., McClung, R. G., Moorehead, E. L., Optimized Pretreatment for Reforming Catalysts. *AIChE Spring National Meeting*, Houston, 1991.
42. McClung, R. G., Oyekan, S. O., Sulfur Sensitivity of Catalysts in Naphtha Reforming. *AIChE Spring National Meeting*, New Orleans, 1988.
43. Gregg, S. J., Singh, S. K. L., *Adsorption, Surface Area and Porosity*, Academic Press, New York.
44. Delgass, W. N., Wolf, E. E., Catalytic Surfaces and Catalysts Characterization Methods, in *Chemical Reaction and Reactor Engineering* by Carberry, J. J., Varma, A., Marcel Dekker, New York, 1987.
45. Doraiswamy, L. K., Tajbl, D. K., Laboratory Catalytic Reactors. *Catal. Rev.*, 10(1), 177–219. Editors Heinemann, H., and Carberry, J. J., Marcel Dekker, New York.
46. Biloen, P., Helle, J. N., Verbeek, H., Dautzenberg, F. M., Sachtler, W. M. H., Conversion of n-Hexane over Mono-functional Supported and Unsupported PtSn Catalysts. *J. Catal*, 63(1), 112–128, 1980.
47. Sachtler, W. M. H., Selectivity and Rate of Activity Decline of bimetallic Catalysts. *J. Mol. Catal.*, 25, 1, 1980.
48. Dautzenberg, F. M., Selected Topics from Applied Industrial Catalysis Eindhoven: Technische Universiteit Eindhoven D, 1994. Accessed February 10, 2017. https://pure.tue.nl/ws/files/2431567/426284.pdf.
49. Wagstaff, N., Prins, R., Alloy Formation and Metal Oxide Segregation in PtRe/γ-Al2O3 Catalysts as Investigated by Temperature-programmed Reduction. *J. Catal.*, 59, 434, 1979.
50. Bertolacini., R. J., Pellet, R. J., The Function of Rhenium in Bimetallic Reforming Catalysis, in *Catalyst Deactivation* by Delmon, B., Froment, G. F., Elsevier Publishing Company.
51. Barbier, J., Corro, G., Zhang, Y., Bourneville, J. P., Franck, J. P., Coke Formation on Bimetallic Platinum/Rhenium and Platinum/iridium Catalysts. *Appl. Catal.*, 16 (2), 169, 1985.
52. Davies, B. H., Westfall, G. A., Watkins, J., Pezzanite, J. J., Paraffin Dehydrocyclization: VI. The Influence of Metal and Gaseous Promoters on the Aromatic Selectivity. *J. Catal.*, 42, 247–256, 1976.
53. Johnson, M. F. L., LeRoy, V. M., The State of Rhenium in Pt/Re/alumina Catalysts. *J. Catal.*, 35, 434, 1974.
54. McNicol, B. D., The Reducibility of Rhenium on Gamma Alumina and Pt/Re on Gamma Alumina. *J. Catal.*, 46, 434, 1977.
55. Webb, A. N., Reducibility of Supported Rhenium. *J. Catal.*, 39, 484, 1975.

56. Bolivar, C., Charcosset, H., Froty, R., Primet, M., Tournayan, L., Betizeau, C., Leclercq, G., Maurel, R., Platinum-rhenium/alumina catalysts: I. Investigation of reduction by hydrogen. *J. Catal.*, 39(2), 249–259. 1975.
57. Scelza O. A., De Miguel, S. R., Baronetti, G. T., Castro, A. A., Performance of Pt–Re/A12O3 Catalysts with Different Radial Distribution Profiles React. *Kinetic Catal. Lett.*, 33(1), 143, 1987.
58. McClung, R. G., Oyekan, S. O., Kramer, R., Reformer Feedstock Pre-treatment: The Liquid vs Vapor Phase Sulfur Removal Processes. *NPRA (now AFPM) Annual Meeting*, San Francisco, 1989.
59. Oyekan, S. O., Laboratory Evaluation Procedures for Reforming Catalysis. *AIChE Spring National Meeting*, 1990, Orlando, Florida.
60. Trueba, M., Trasatti, S. P., Gamma Alumina as a Support for Catalyst: A Review of Fundamental Aspects. *Eur. J. Inorg. Chem.*, 3393–3430, 2005, © Wiley-VCH Verlag GmbH & Co. KGaA, 69451 Weinheim, Germany, 2005.
61. About-gheit, A. K., Al-hajjaji, M. A., Differential Scanning calorimetry Evaluates Relative Proportions and Strength of acid sites in catalytic materials and Adsorbents Alumina and Amorphous Silica-Alumina, *Anal. Lett.*, 20(4), 553–559, 1987.
62. Chang, C. H., Gopalan, R., Lin, Y. S., A Comparative Study on Thermal and Hydrothermal Stability of Alumina, Titania and Zirconia Membranes. *J. Mol. Science*, 91 (1–2), 27–45, 1994.
63. Kavarelle, R. M., Hahn, M. W., Copeland, J. R., Sievers, C., *AIChE Annual Meeting*, 2011.
64. Le-Goff, P. Y., Lopez, J., Ross, J., Redefining Reforming Catalyst Performance: High Selectivity and Stability, *Hydrocarbon Process.*, 47–52, September 2012.
65. Stiles, A. B., *Catalyst Supports and Supported Catalysts*, Butterworth Publishers, Boston, USA, 1987.
66. Bradley, S. A., Gattuso, M. J., Bertolacini, R. J., Characterization and Catalyst Development. *ACS Symposium Series*, 1989.

6 Catalytic Reforming Technologies

6.0 INTRODUCTION TO REFORMING TECHNOLOGIES

The focus of the preceding chapters was to emphasize the key position and importance of catalytic reformers in oil refineries. Naphtha sources and feed types, the processing of naphtha to generate hydrotreated naphtha that meets catalytic reformer feed qualities, reforming reactions, catalysts, and catalytic properties were covered. The importance of an effective comprehensive catalyst evaluation program for selecting high-performance catalysts was reviewed. In this chapter, the focus is more on the three major distinctive catalytic reforming technologies and how they impact planning in oil refineries with respect to operations, fuels, and other product production and routine and turnaround maintenance. Key objectives of the catalytic reforming process are to produce high-octane reformates; chemical compounds such as benzene, toluene, and xylenes; and hydrogen. Semiregenerative, cyclic regenerative, and continuous catalyst regenerative reformers are the main catalytic reforming technologies. The hybrid catalytic reforming process unit is the combination of fixed-bed catalyst semiregenerative and continuous catalyst regenerative technologies. This is accomplished by adding a reactor and its dedicated catalyst circulation and regeneration system to a fixed-bed semiregenerative reformer. The catalyst regeneration system is designed to be used solely for the regeneration of the catalyst from the added, last reactor. The other reactors in the catalytic reformer are operated as before in a strictly semiregenerative mode.[1,2]

Honeywell UOP reported that data from a PIRA Energy Group study indicated that the total global reforming capacity was 13.9 million barrel per stream day (BPSD) in 2010 and that the capacity for continuous catalyst regenerative reforming units was growing relative to fixed-bed catalytic reforming units, which was declining, as shown in Figure 6.1.[1]

In the same paper, Honeywell UOP reported that over the period of 2001 and 2010, China, India, and the CIS (Commonwealth of Independent States or Russian Federation) ranked in that order as the leaders in regional growth rates for CCR reforming capacity, as shown in Figure 6.2.[1,3]

Despite the declining rate for fixed-bed semiregenerative reforming processes, that reforming technology provides a good starting point in the review of developments in catalyst and process technologies for the catalytic reforming process. The steady progression of advancements in catalyst and process technology led to the invention and establishment of the CCR reforming technology as the premier processing unit for the reforming of heavy naphtha in oil refineries.

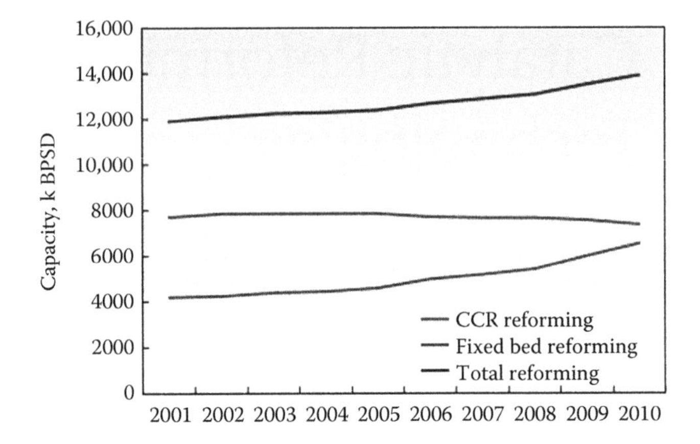

FIGURE 6.1 Reforming capacity in the global refining sector.
From Poparad, A. et al., Reforming Solutions for Improved Profits in an Up-Down World, Paper AM-59, *National Petrochemicals & Refiners Association (NPRA), Annual Meeting*, 2011; PIRA Energy Group. World Refining database, Q4 2010 Capacity Data. Accessed July 26, 2017. https://www.pira.com/data-tools/world-refinery-database.[1,3] (Illustration Courtesy of Honeywell UOP.)

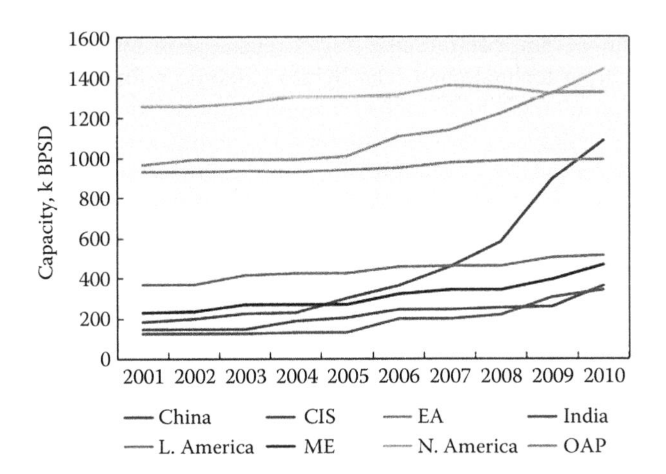

FIGURE 6.2 CCR reforming growth by region.
From Poparad, A. et al., Reforming Solutions for Improved Profits in an Up-Down World, Paper AM-59, *National Petrochemicals & Refiners Association (NPRA), Annual Meeting*, 2011; PIRA Energy Group. World Refining database, Q4 2010 Capacity Data. Accessed July 26, 2017. https://www.pira.com/data-tools/world-refinery-database.[1,3] (Illustration Courtesy of Honeywell UOP.)

6.1 SEMIREGENERATIVE CATALYTIC REFORMING

Fixed-bed catalytic reforming units replaced the earlier thermal cracking naphtha units that required that alternate processing and coke removal be conducted between reactors, as was the common practice in the pre-1940 era of processing of refining oil fractions. Molybdenum oxide catalysts were later used in fixed-bed reformers before the historic, landmark inventions of Vladimir Haensel and Honeywell UOP in the late 1940s. Haensel and Honeywell UOP inventions led to a quantum shift in the performance of catalytic reformers, as reformers were now operated with platinum-containing catalysts. Platinum catalysts exhibited higher activity, selectivity, and stability relative to molybdenum oxide catalysts. Honeywell UOP named their licensed process Platforming™. Platforming is now used synonymously with catalytic reforming in oil refining. Despite performance and productivity improvements with the use of platinum monometallic catalysts, fixed-bed reformer catalysts still deactivated within a short time and required reformer shutdowns to conduct periodic catalyst regenerations. These fixed-bed catalytic reforming units are now referred to as semiregenerative reformers. To extend the cycle lengths of the semiregenerative reformers, platinum/rhenium and other bimetallic catalysts such as platinum/tin and platinum/iridium are used and Pt/Re is now universally accepted as the predominant catalyst for fixed-bed semiregenerative and cyclic regenerative reformers.

A sketch of a semiregenerative reformer is shown in Figure 6.3.[1] The reformer has three, four, or five reactors in series and uses interreactor heating due to the overall endothermic heat requirement of the process. Key equipment is as shown and included the heaters or heater, combined feed/effluent exchangers, separator to separate gas from liquid product and recycle some of the gas to the reactors, and net hydrogen gas for use in hydrotreating and hydrocracking units in the refinery. Reformer liquid product and some residual gas are sent to the debutanizer (DEB). Reformate, hydrogen, LPG, and fuel gas are produced via fractionation in the debutanizer.

A number of licensed semiregenerative technologies are offered by Honeywell UOP (Platformer™) and Chevron (Rheniforming), and Magnaforming technology was offered for decades by ARCO and Engelhard (Magnaforming). Simplified drawings

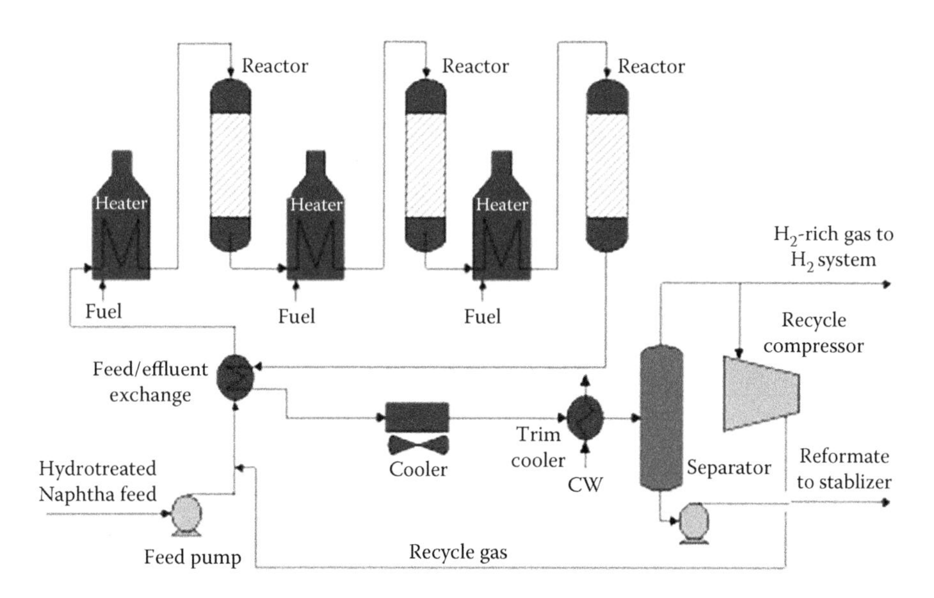

FIGURE 6.3 Semiregenerative catalyst reformer.

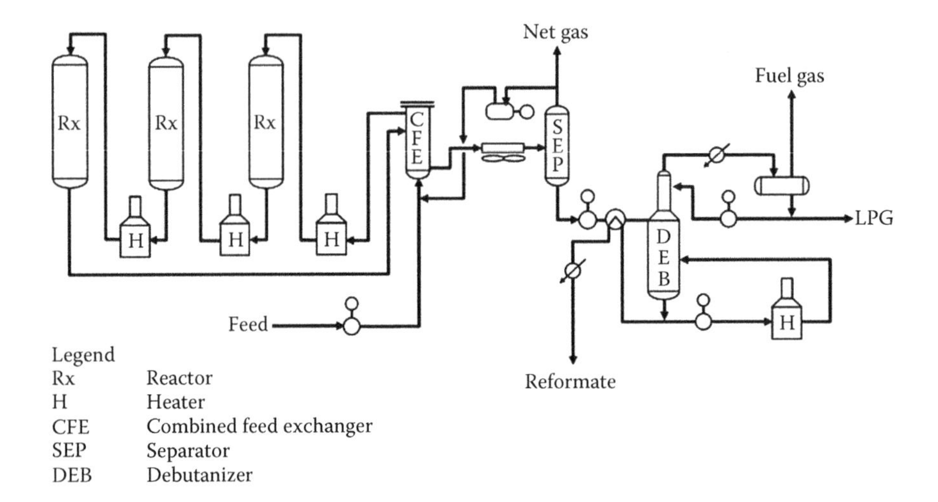

FIGURE 6.4 Honeywell UOP platforming unit.

of the Honeywell UOP Platformer, Chevron Rheniforming, and Magnaforming process units are shown in Figures 6.4, 6.5, and 6.6, respectively.[5] Later, Honeywell UOP incorporated a number of technology improvements that have led to significant reductions in capital and process operating costs for catalytic reformers. One of Honeywell UOP's technology contributory features is the use of a single large furnace box incorporating the charge and interreactor heaters. The ARCO/Engelhard Magnaforming reformer incorporates a split hydrogen recycle gas system to aid in maximizing reformate and hydrogen yields and extend reformer cycles by reducing the rate of coke make in the last reactor.[5]

Several key feed and process operating factors can negatively impact the activity, selectivity, and performance stability of the catalyst during naphtha reforming operations. Some of the same factors also negatively impact the reliability of semiregenerative catalytic reforming units. The factors that affect the activity and

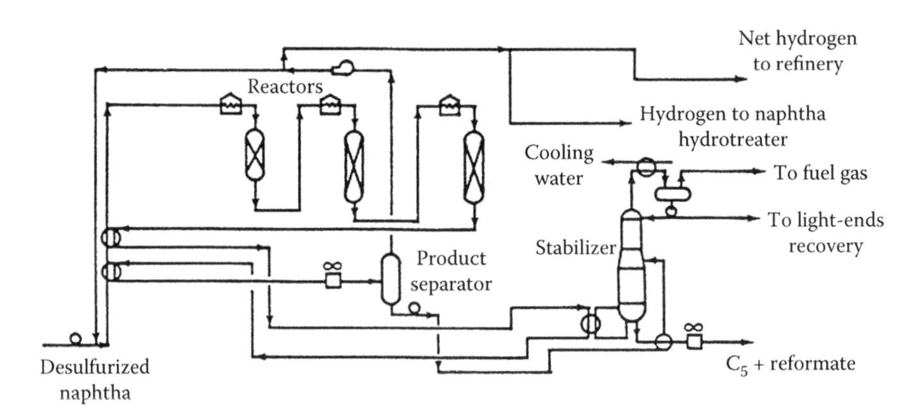

FIGURE 6.5 Chevron rheniforming unit.

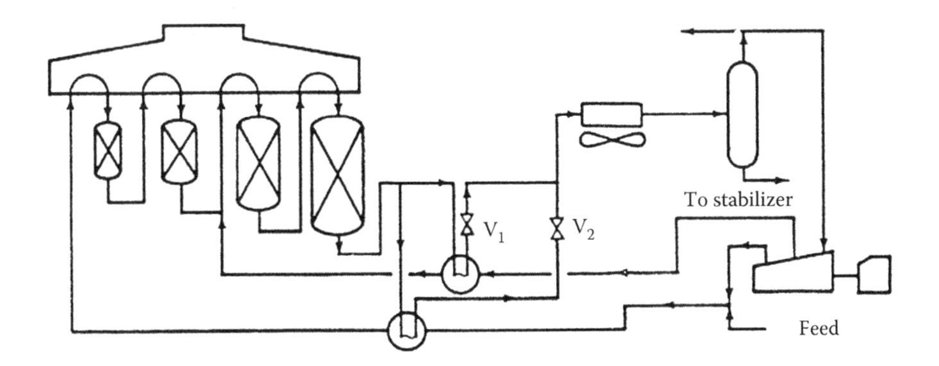

FIGURE 6.6 ARCO/Engelhard magnaforming unit.

selectivity of the catalysts are consistent with what are observed for fixed-bed and continuous catalyst regenerative units, and they are reviewed later in this chapter. The quality of the hydrotreated heavy naphtha as indicated by the concentrations of sulfur, nitrogen, silicon, arsenic, nickel, molybdenum, and chloride is crucial. Silicon, arsenic, and chloride concentrations in hydrotreated naphthas are indicative of the quality of the naphtha from the crude slate that the refinery is operating on. It can also be indicative of the extent of deactivation of the hydrotreater catalyst and the amount of contaminants loading on it. Reformer feed sulfur and nitrogen concentrations that are greater than those specified for the reformer process are likely to be due to refinery crude slate type operation, hydrotreater stripper upsets, and hydrotreater catalyst deactivation. High reformer feed sulfur could also be due to products of mercaptan reversion reactions, which tend to occur at the end of catalyst life in naphtha hydrotreaters.

In addition to catalyst contaminations caused by feed and equipment impurities for catalytic reformers, gradual coking of the catalysts occur over time. These contaminants negatively impact catalytic performances and reformer productivity. Continuous chloride addition is required to compensate for chloride losses and to maintain the acidic functionality of the catalyst. Due to the need for the oil refiner to maintain adequate catalyst chloride on the reforming catalyst, excess chlorides could accumulate due to operational challenges and possible surface area retention deficiency of the catalyst. The excess chlorides deposited on the catalyst in reactors and not retained are converted to hydrogen chloride and transported into the product separation systems and as hydrogen chloride in the net hydrogen gas. Ammonium salt depositions may also occur in the combined feed/effluent exchanger (CFE), debutanizer, and recycle gas compressors. These salt deposits can lead to reduced unit productivity and down times for periodic cleaning of affected areas and equipment. As a result of the accumulation of contaminant metals such as iron, silicon, nickel, and others, and coke deposition on the catalysts, the catalyst loses significant naphtha processing performance. To return the process unit to satisfactory productivity, naphtha reforming operations in semiregenerative units have to be discontinued. Scheduled, periodic turnaround maintenance and catalyst regenerations are conducted. After catalyst regeneration, the process unit is returned for another naphtha reforming production cycle. The cycle lengths of semiregenerative catalytic reforming unit operations could be in the range of 6 months to 2 years, and the actual cycle length achieved is dependent on naphtha feed qualities, process conditions, and reformate octane targets or severities. End-of-cycle criteria for semiregenerative catalytic reformers that are used by oil refiners vary and most are based on profitability measures. Some of the oil refiners do take their semiregenerative reformers down when their crude units are shut down for maintenance, while others may operate their reformers on purchased naphtha depending on the fuel market and opportunities that are available to them. Three main end-of-cycle criteria are used:

- 2 volume percent decline in the C5+ yield or reformate yield from start of cycle
- 30- or 40-degree rise in reactor WAIT
- A measure of unacceptable hydrogen gas loss as defined by negative impacts on the productivity of the other hydroprocessing units in the refinery[5,22]

6.1.1 Reformer Naphtha Feed

Effective hydrotreating of naphthas to meet feed sulfur, nitrogen, and metal contaminant specifications of catalytic reformer feeds was reviewed in Chapter 3 of this book. Most catalytic reformers that operate with Pt/Al_2O_3, $Pt/Re/Al_2O_3$, $Pt/Sn/Al_2O_3$, $Pt/Ge/Al_2O_3$, and trimetallic catalysts have dedicated naphtha hydrotreaters upstream to treat naphtha feeds. A few oil refiners operate catalytic reformers without the benefit of a naphtha hydrotreater, and platformer catalyst performances are usually subpar for those units. In those catalytic reformers, cycle lengths are usually of a short duration due to extensive catalyst deactivation. Catalytic performance is usually subpar despite the use of low-severity reforming process conditions that include low reformate octane targets, high recycle hydrogen to hydrocarbon ratios, low liquid hourly space velocity, and low reactor WAITs. Catalyst utilization is typically poor due to rapid deposition of sulfur, nitrogen, and contaminant metal impurities on the catalysts, which lead to rapid catalyst deactivation, mediocre productivity, and short cycles for fresh catalysts. Catalyst regenerations are usually not conducted, as successful catalyst reactivations are not achieved, and the oil refiner has to operate each cycle with a load of fresh platinum-containing catalysts. Such annual single-cycle catalyst reforming operations lead to high catalyst management costs due to frequent catalyst purchases, excessive platinum reclamation costs, platinum losses, and platinum lease costs.

Paraffins, naphthenes, and aromatics in heavy naphtha feeds are reformed in catalytic reforming units. Since there are several hydrocarbons of each type, the characterization of heavy naphthas as paraffinic or naphthenic is used in this review. It is more difficult to reform paraffinic naphthas and they are, therefore, preferentially processed in CCR reforming units by oil refiners who have a choice of naphtha reforming units. The heavy naphtha feed could also be a composite of a variety of naphthas, which could also include cracked and coker naphthas, referred to as unsaturated naphthas. A broad range of naphtha feed compositions and distillation ranges is given in Table 6.1.[4–6] The N+2A factor is used as a qualitative indicator of the ease of catalytically reforming a naphtha feed, and when coupled with process conditions, could provide an indication of reforming severity and projected coke make in the processing of a naphtha.

TABLE 6.1
General Range of Reformer Naphtha Feed Quality

Feed Properties	
Gravity, deg. API	56.8–59.2
Paraffins, vol. %	25–65
Naphthenes, vol. %	15–50
Aromatics, vol. %	10–40
N+2A, vol. %	30–80
ASTM D-86 Distillation, F	
IBP	150–180
FBP	Up to 420 F
Metal Contaminant Specs.	
Sulfur, wppm	<0.5 wppm
Nitrogen, wppm	<0.2 wppm
Arsenic, wppb	<1
Lead, wppb	<5
Copper	<5
Mercury	<5
Iron	<5
Silicon	<5
Nickel	<5
Chromium	<5

A broad range of initial and final boiling points of the naphtha is also provided in Table 6.1. Prior to recent decades of catalytic reforming, the initial boiling point of 160 F was recommended so as to eliminate hexanes and lighter hydrocarbons that are either not converted in low-pressure operations or undergo hydrocracking to lighter hydrocarbons in high-pressure naphtha reforming, as was often the case in fixed-bed semiregenerative reforming.

High pressures and high recycle gas hydrogen to hydrocarbon molar ratio process conditions are utilized to minimize the rate of catalyst deactivation and to extend operating cycles of semiregenerative catalytic reformers. The initial boiling points of naphtha feed for some catalytic reformers have been raised from 160 F and lower to the 180 F to 200 F range in order to reduce the concentrations of benzene precursor compounds in reformer feed.[5]

Final boiling points range of the reformer feeds have also been drastically reduced to 330–370 F from 380+ F for most catalytic reformers in order to enable oil refiners to produce more diesel products. Oil refiners have the option of operating the catalytic reformer at low octane severities and take advantage of the blending of ethanol and other oxygenates into gasolines. Operating with lower feed naphtha endpoints is beneficial for semiregenerative reformers, as that leads to lower coking rates and catalyst deactivation. Cycle lengths are extended and the cost of frequent downtimes and catalyst regenerations is greatly reduced. Additionally, fewer catalyst regenerations leads to extended life for a semiregenerative catalyst.

6.1.1.1 Heavy Naphtha Contaminants

Key naphtha feed contaminants and their negative impacts on catalytic performance and fouling and corrosion of materials of construction are reviewed below. A number of recommended solutions and procedures are offered for good and timely recovery of catalyst activity if the catalyst poisoning effect is temporary and reversible.

6.1.1.2 Sulfur in Naphtha Feeds

Sulfur is a common poison of hydrogenation/dehydrogenation functionality of catalytic reforming catalysts. Typically, sulfur in naphtha feeds at low concentrations of less than 1 wppm modifies and poisons some of the platinum promoter metals such as rhenium and iridium. Sulfur poisoning of catalysts can cause temporary or permanent deactivation of the catalytic reforming catalyst. The ultimate degree of catalyst poisoning is dependent on the duration of catalyst contact and concentration of sulfur in the naphtha feeds. In extreme cases of high sulfur poisoning, the sulfur deposited on platinum and promoter metals is converted after reactions that occur in catalyst regeneration to form sulfates of platinum and promoter or modifier metals.

Sulfur specifications for naphtha feeds to catalytic reformers are usually in the range of 0.2 to 0.5 wppm. The specific reformer feed sulfur specification used is dependent on the catalyst type and the requirements of the refiner's catalytic reformer and refinery reliability programs. Naphtha hydrotreaters treat naphtha feeds for catalytic reformers and remove organic nitrogen and organic sulfur compounds via hydrodenitrogenation and hydrodesulfurization reactions. Residual sulfur compounds in the hydrotreated naphtha are usually thiophenes and alkyl thiophenes, mercaptans from mercaptan reversion reactions, and hydrogen sulfide as a result of hydrotreater stripper operational inefficiency and upsets. Most of the organosulfur compounds entering the reformer react with hydrogen to produce hydrogen sulfide. Metallic contaminants, iron scales, and other equipment corrosion products entrained in naphtha feeds deposit on the catalysts. Some of the iron sulfide impurities react with hydrogen to produce hydrogen sulfide. Some of the sulfur from iron scales reacts with oxygen and hydrogen during catalyst regenerations and produces sulfur oxides and hydrogen sulfide during catalyst regenerations. Hydrocracker naphtha containing mercaptans could also be adding more sulfur to the reformer if the hydrocracker naphtha product is not treated in a guard bed of adsorbent to remove mercaptans.

The first significant effect of sulfur is instantaneous and that is the loss of catalytic activity, especially of the dehydrogenation/hydrogenation functionality contributions. This decline in catalytic performance is usually observed as loss of the delta temperatures in the reactors, especially in the first or lead reactor. Its negative poisoning impact is due to negative effects of the sulfur on the catalytic efficiencies of the platinum and promoter metal sites, especially for Pt/Re, Pt/Ir, and Pt/Sn catalysts. Depending on the duration of higher feed sulfur operation, the negative performance effects could progress from the lead reactor to the last reactor with time. Since sulfur is considered a "temporary" poison, its negative effects could be reversed with timely process unit operational responses.

Some of the first set of operational adjustments could be made, including the lowering of reforming severity and reactor WAIT and, if required, reductions in charge rates to minimize the rate of catalyst coking. The recycle gas rate should be increased to strip off excess sulfur as hydrogen sulfide. The source of the high sulfur in the reformer feed should be identified and corrected in parallel with the operational changes that are being effected.

Another negative impact of high feed sulfur in the reformer feed is that the hydrogen sulfide produced in the reactors reacts with ammonia to produce ammonium bisulfide (NH_4HS) in the cooler sections of product separation systems. The deposited ammonium salts cause fouling and corrosion and can lead to significant reformer reliability and productivity challenges.

The reader is referred to the review provided of the pilot plant studies in Chapter 5 on the sulfur sensitivity of Pt/Re catalysts. In that study, catalytic data from a test run of a balanced or equimolar Pt/Re catalyst from test runs on high 1.3-wppm sulfur naphtha feed were contrasted with data for a 0.3-ppm sulfur naphtha test run. The plot for the reactor temperatures versus time for the balanced catalyst is produced here to emphasize the effects of long-term operation with high-sulfur naphtha feeds. Catalytic activity and stability data for the 0.3- and 1.3-wppm sulfur test runs are shown in Figure 6.7.

Reformate yields and selectivity stabilities for the 0.3- and 1.3-wppm sulfur test runs are compared in Figure 6.8.[33] The impact of high feed sulfur operation is obvious, as the cycle length for 0.3-ppm sulfur naphtha feed based on a 2 volume percent C5+ yield decline is about five times that of the 1.3-wppm feed sulfur.

Based on the review and experiences of oil refiners, it is obvious that operating with even moderately high sulfur naphtha feed greater than 0.5 wppm would lead to losses in catalytic reformer productivity, reformer reliability and productivity, and refinery profitability declines.

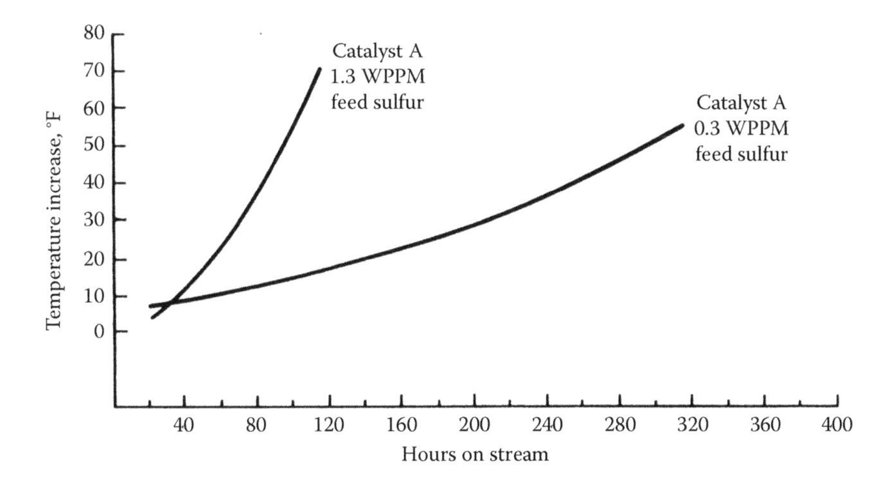

FIGURE 6.7 High naphtha sulfur effects on Pt/Re catalyst activity.
From Oyekan, S. O., McClung, R. G., Moorehead, E. L., Optimized Pretreatment Procedures for Reforming Catalysts, *AIChE Spring National Meeting*, Houston, Texas, April, 1991.[33]

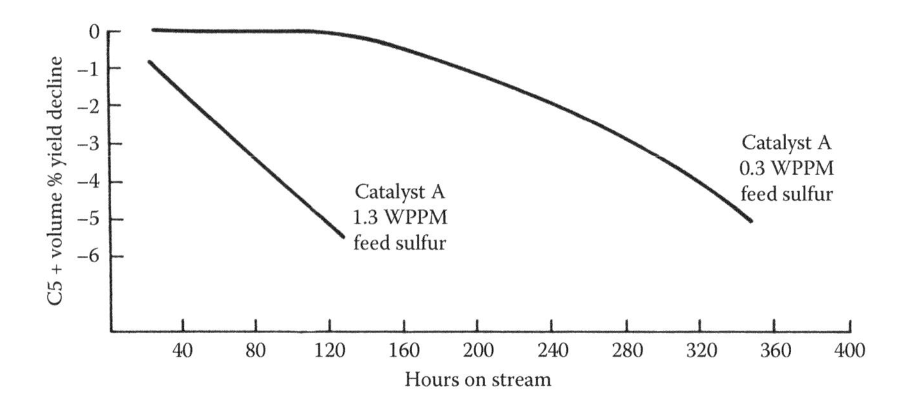

FIGURE 6.8 High naphtha sulfur effects on Pt/Re reformate yields and stability.

6.1.1.3 Nitrogen in Naphtha

Reformer feed specification for nitrogen is often set at 0.1 or 0.2 wppm depending on how focused the oil refiner is on managing feed nitrogen and driving reliability programs for its catalytic reformer and refinery. Due to the need for greater control of the rates of ammonium chloride (NH_4Cl) and ammonium bisulfide (NH_4HS) depositions and fouling in the product separation systems,[7,8] reformer feed nitrogen concentration specifications have now been reduced further to less than 0.1 wppm by some oil refiners.[5] Organo-nitrogen compounds entering the reformer reactors react with hydrogen to produce ammonia. Ammonia strips hydrogen chloride (HCl) from the catalyst and reacts with HCl to produce ammonium salts in cooler sections of the product separation systems and equipment. Stripping of catalyst chloride leads to acidic isomerization functionality losses with negative impacts on catalyst activity and selectivity. Acidic isomerization functionality can be recovered by adding more organic chloride and gradually recovering catalyst activity. Product separation systems and equipment that are fouled usually require cleaning if productivity losses become uneconomical for continuous catalytic reforming operations.

A few recommendations may aid the oil refiner in the management of high nitrogen feeds or upset conditions in the naphtha hydrotreater that lead to increases in reformer feed nitrogen. The first is to determine the factors that are leading to higher nitrogen relative to the reformer feed specification and address them as soon as feasible. Second, operational adjustments should be made to increase chloride injection rates to reformer reactors for some time to compensate for the expected high chloride losses from the catalysts. Another effort is to clean the stabilizer tower on-stream by washing out ammonium salts. A fourth suggestion is to increase naphtha hydrotreater temperature so as to increase dehydronitrogenation activity. In addition, the oil refiner should monitor the hydrotreated naphtha to ensure that there are no increases of sulfur compounds with increase in the naphtha hydrotreater temperature. Such increases in sulfur may be due to increases in the concentrations of mercaptans due to mercaptan reversion reactions that are typically evident at higher naphtha hydrotreater operating temperatures.

6.1.1.4 Naphtha Metal Contaminants

Three naphtha contaminants, arsenic, silicon, and mercury, are discussed briefly as worthy of attention. Programs that oil refiners use to manage them are reviewed. Arsenic, silicon, and mercury are usually not found in most crude oils. Contaminant metals such as nickel, copper, and vanadium concentrate in the heavier fractions of crude oils and are usually in the gas oil and residual oil fractions. Iron, chromium, and molybdenum found in naphthas are usually corrosion products from piping and equipment upstream of catalytic reformers. If nickel, molybdenum, and cobalt concentrations increase in catalytic reforming catalysts, that could be indicative of a low crush strength property of the naphtha hydrotreater catalysts used by the oil refiner. A brief review of arsenic, silicon, and mercury is provided below.

6.1.1.4.1 Arsenic

Arsenic occurs naturally in some crude oils from the United States, Russia, Venezuela, Canada, and Libya. The metal contaminant is also found in synthetic crude oils, as shown in Table 6.2.[9,10] Analyses of naphthas for arsenic have shown that there are significant concentrations of arsenic in the oils. It has been determined that up to 5 wppb of arsenic was present in a shale oil.

Arsenic should be removed in naphtha hydrotreaters, as hydrotreaters also function as a guard bed due to the adsorption and filtration attributes of catalyst beds. The metal is a permanent reforming catalyst poison and should be eliminated from the naphtha feed to the reformer. The need to better manage arsenic is greater for catalytic reformers that are operated without the benefit of a naphtha hydrotreater. If arsenic is present in the unhydrotreated naphtha feed of those reformers, catalyst poisoning by arsenic is initiated at the start of cycles and catalytic performances decrease drastically thereafter.

Metal contaminant poisoning is usually more severe for platinum-containing reforming catalysts in catalytic reforming relative to those of base metal oxide catalysts used in the naphtha hydrotreater. Rapid drops in hydrodenitrogenation and

TABLE 6.2
Arsenic in Crude Oils and Synthetic Oils

Crude Oil	Arsenic, wppb
Almein	2.4
Alberta	2.4–111
California	63–1112
Libya	77–343
Louisiana	46
Venezuela	20–284
Wyoming	111

Source: Olsen, C. AT734G: A Combined Silicon and Arsenic Guard Catalyst, Advanced Refining Technologies Catalagram. Accessed August 22, 2017. https://grace.com/catalysts-and-fuels/en-us/Documents/108SE-AT734G.pdf.[9]

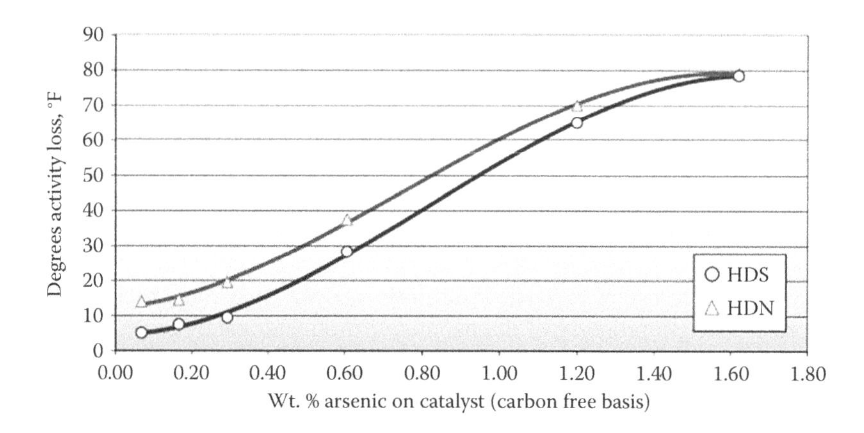

FIGURE 6.9 HDS and HDN activity loss as function of arsenic loading.
From Olsen, C. AT734G: A Combined Silicon and Arsenic Guard Catalyst, Advanced Refining Technologies Catalagram. Accessed August 22, 2017. https://grace.com/catalysts-and-fuels/en-us/Documents/108SE-AT734G.pdf.[9]

hydrodesulfurization activities as a function of the weight of arsenic on a naphtha hydrotreater catalyst are shown in Figure 6.9.[9]

The rapid losses in HDS and HDN activities for the catalyst reported in Figure 6.9 for a naphtha hydrotreater are indicative of the possible rate of activity and selectivity deactivation that could be expected for a catalytic reformer catalyst. Some of the catalyst and technology provider companies have excellent arsenic guard catalysts that oil refiners can use.[9,11,12,14,15] The impact of arsenic on the environment and the health of personnel is expected to become a key issue in the coming years, and the oil refining industry will need to focus on programs for effective management of arsenic.

6.1.1.4.2 Mercury

Mercury is found in crude oils in concentrations ranging from a few weight parts per billion (wppb) to weight parts per million.[61,62] The metal can be present as liquid metal, organometallic compounds, and inorganic salts. It was reported that under naphtha hydrotreater processing conditions, organically bound mercury is converted to elemental mercury, which is then entrained in naphtha and deposited on downstream equipment and catalysts.[13] Mercury is a known poison of platinum and other precious metal catalysts. In addition to catalyst poisoning, Honeywell UOP indicated that platinum and palladium can form stable amalgams with mercury at temperatures below 570 F and make it difficult to recover catalyst performance after catalyst regenerations. Honeywell UOP has advised of the need for awareness by oil refining and petrochemical industry operators of equipment attack by mercury and the negative impact of mercury on naphtha quality sold to the petrochemical industry. In addition, Honeywell UOP cautioned about the impacts of mercury on health, safety, and the environment.[12] Honeywell UOP indicated that the threshold limit value (TLV) for mercury is 0.025 mg/m^3 averaged over an 8-hour work shift.[12]

A number of catalyst and adsorbent technology companies offer suitable adsorbent guard bed adsorbents and catalysts to oil refiners for effective management of mercury, and the list includes Axens, Criterion Catalysts & Technologies, Honeywell UOP, Johnson Matthey, and Advanced Refining Technologies (ART).[9,11–15]

6.1.1.4.3 Silicon

Oil refiners who operate catalytic reformers in which coker naphtha is processed as a component of their naphtha feeds probably experience silicon poisoning of their reforming catalysts at silicon deposits on catalysts in the range of 1000–2000 wppm.[5] If an oil refiner is not running a top-notch catalyst management program, silicon poisoning–induced catalytic performance debits are not likely to be properly ascribed to silicon. Studies by Haldor Topsoe reported poisoning of naphtha hydrotreater catalysts due to high deposition of silicon.[19] Coker naphtha silicon is known to originate from the use of silicone antifoaming agents such as polydimethylsiloxane (PDMS). During the hold period in the delayed coker drum, foaming of hydrocarbon liquid and gases occurs, and PDMS is added to control and eliminate foaming. Use of PDMS antifoaming agents aids in enhancing the reliability and productivity of the delayed coker.

Polydimethylsiloxane is also applied as an aid in crude oil production and is present in significant levels of 2–5 wppm in some crude oils. The structure of the polymerized siloxane, PDMS, with {SiO $(CH_3)_2$} as the repeating monomer unit is shown in Figure 6.10.

PDMS undergoes thermal decomposition at temperatures greater than 650 F to yield D3 through D7 cyclic oligomeric siloxanes that are present in naphtha through distillate fractions from crude oils containing silicon. The numerical number in the Dx designation for cyclic oligomeric siloxanes represents the number of silicon atoms in the molecule of that oligomeric compound. D3 through D6 cyclic oligomeric siloxanes and their boiling points are listed in Table 6.3.[16] Sixty to 70% of the total silicon determined for a crude oil would partition into the naphtha fraction on decomposition of PDMS.

During the hydrotreating of silicon-laden naphthas, the entrained silicon deposits on the catalysts, leading to pore pluggage and catalyst deactivations.[17–19] After silicon breakthrough from naphtha hydrotreaters, silicon entrained in the reformer feed will then accumulate on reformer catalysts and begin to deactivate the catalyst. Significant reformer catalyst deactivations would occur with silicon deposition on the catalyst in the range of 1000–2000 wppm. Due to the likelihood of processing naphthas containing silicon and other metal contaminants from cokers, crude oils, synthetic oils, and opportunity oils, it is recommended that oil refiners establish

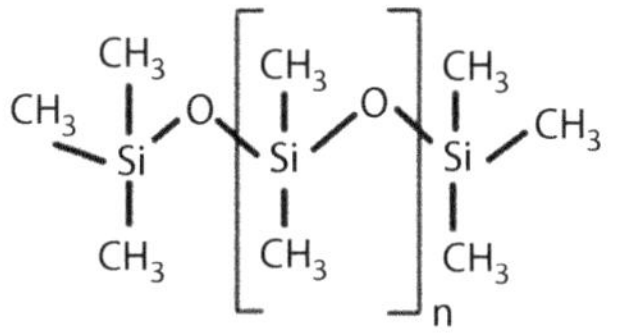

FIGURE 6.10 PDMS compound structure.
Courtesy Haldor Topsoe A/S.

TABLE 6.3
Boiling Points of Some PDMS Decomposition Products

Product	Boiling Point, C	Boiling Point, F
Cyclic D3	134.0	273.2
Cyclic D4	175.8	348.4
Cyclic D5	210.0	410.0
Cyclic D6	245.0	473.0

effective silicon management programs in order to improve naphtha hydrotreater and catalytic reformer operational reliability and profitability. Oil refiners could establish a comprehensive analytical inspection program for crude oils, purchased opportunity oils, synthetic crude oils, coker naphtha, catalytic reformer feeds, and other relevant refining oil fractions. In addition, despite possible concerns about "too many bells and whistles" in the plants, oil refiners could explore the use of silicon and contaminant metal guard bed technologies when reliable contaminant metal management options are being considered. With respect to profitable and reliable operations of the catalytic reformer, limiting silicon accumulation on catalytic reforming catalysts should be a key objective. Adherence to the reformer feed silicon concentration specifications in Table 6.1 is highly recommended.[9,11–15,17]

6.1.1.4.4 Other Metal Contaminants and Impurities

Due to age and the extent of corrosion of piping and equipment upstream of catalytic reforming units, corrosion products such as iron, chromium, molybdenum, and nickel are transported by the naphtha feed and deposited on catalytic reforming catalysts. Metal corrosion impurities could increase rapidly with time on catalytic reformer catalysts if the naphtha hydrotreater catalysts become saturated with metal contaminants and metal contaminant breakthrough occurs. The amounts of metal impurities from the naphtha hydrotreaters that are determined via metal analysis for CCR reforming catalysts could be lower than what could have been calculated from the rates of depositions due to the addition of fresh makeup catalyst for catalyst fines make and other reformer catalyst upsets. An example of iron loadings on a variety of catalysts is as shown for several CCR units in Figure 6.11. A similar set of catalyst silicon and sodium with age of catalysts is shown in Figure 6.12.

It is prudent to monitor the loadings of corrosion metal impurities on catalysts, as the information could assist in troubleshooting naphtha hydrotreater and catalytic reforming catalyst performance challenges and for making timely decisions for catalyst replacements. In addition, the rate of loadings of corrosion metals on catalysts could aid in monitoring piping and equipment areas that could fail such that timely routine and turnaround maintenance services could be applied to those areas to minimize the chances of catastrophic failures of materials of construction.

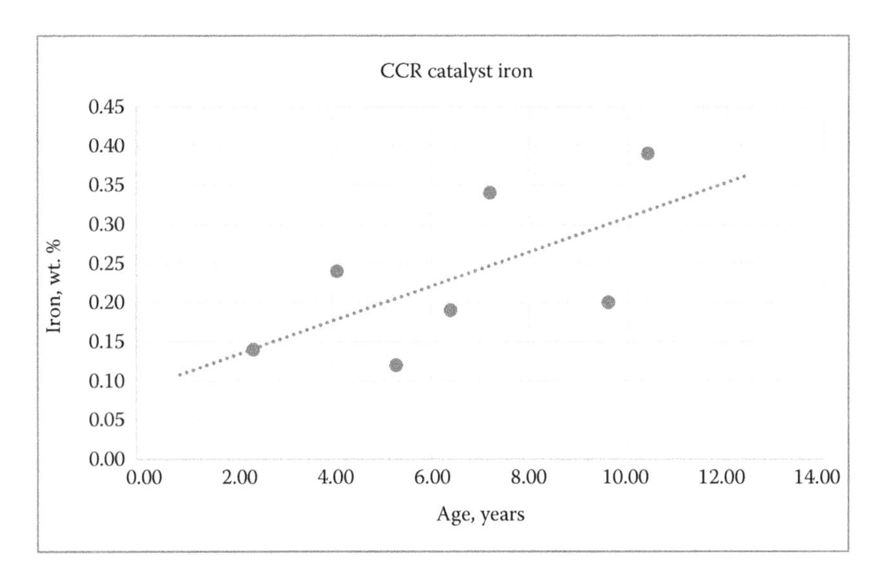

FIGURE 6.11 Iron deposition on CCR reformer catalysts.

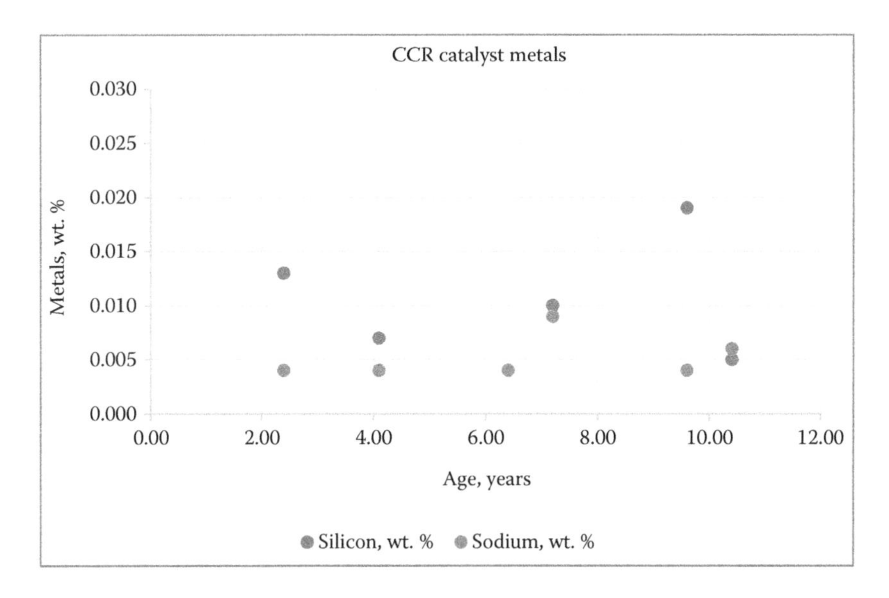

FIGURE 6.12 Silicon and sodium loadings on CCR catalysts with time.

6.1.2 Process Variables

The broad ranges of process variables for the three different catalytic reforming process units are provided in Table 6.4. Subsets of the listed variables are usually selected from within the ranges to meet specific production goals of CCR and fixed-bed reforming units that the oil refiner is operating.

TABLE 6.4
Range of Catalytic Reforming Unit Process Variables

Process Variable	Range
Pressure, psig	50–700
Reactor WAIT, F	850–1020
Recycle Rate, SCF/B	3000–10,000
H2/HC molar ratio	1.5–10
LHSV, 1/Hr	1.2–4.0
Reforming Severity (RON)	80–107
Naphtha Type	Paraffinic or naphthenic

For most semiregenerative reformers, process conditions and low reforming severities are typically selected to sustain stable catalyst performances for long cycles of many months and minimize the cost of frequent downtimes for turnaround maintenance. Sustained high performance of reforming catalysts and reformers is highly desired, especially for producing hydrogen for meeting the demands of the other hydroprocessing units in the refineries. Based on naphtha feed quality, process conditions, and reformate octane targets selected, semiregenerative cycles could be as short as 6 months and as long as 18–24 months. At the end of a cycle of catalytic operations, oil-reforming operations are discontinued and the reformer is then taken out of operation for much-needed turnaround work followed by catalyst regeneration for full recovery of catalyst performance and productivity.

6.1.3 Water/Chloride Management

Catalytic reforming catalysts are bifunctional, with the metal and acid sites assumed to be in some "balance" that enables total and effective utilization of the metal and acid sites. Metal-acid balance refers to the useful participation and appropriate contributions of metal and acid functionalities in the promotion of desired reactions such as the dehydrogenation, dehydrocyclization, and isomerization reactions of naphtha hydrocarbons relative to hydrocracking and demethylation reactions of hydrocarbons, as shown in Figure 6.13.

Gamma alumina support used in catalytic reforming combines with chloride to provide acid functionality via interaction with hydrogen chloride. Fresh catalytic reforming catalysts typically contain 0.9–1.0 wt. % of chloride. Chloride is stated by some researchers to interact with alumina hydroxyl groups to form Bronsted acid sites, and others have indicated that Lewis acidic sites are also present on gamma alumina.[20] In another study, it was shown that chlorination of alumina is in an equilibrium, as shown in Figure 6.14.[21]

Bronsted and Lewis acid sites most likely exist on gamma alumina, and this mix is what is modified in current reforming catalysts, as the gamma alumina is calcined to lower surface areas in the range of 175 to 200 m^2/gm relative to previous surface area ranges of 200 to 275 m^2/gm. Based on gamma alumina and chloride studies, surface chloride and hydroxyl groups are required and it is, therefore, essential during naphtha processing to add adequate organic chloride and operate with good moisture control

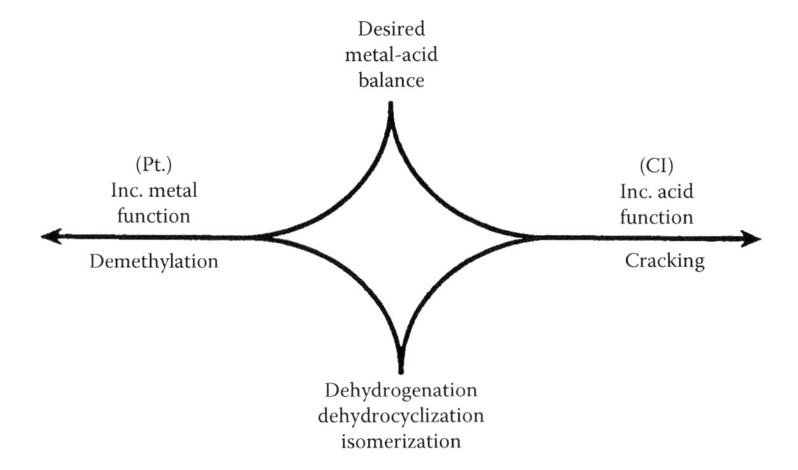

FIGURE 6.13 Metal-acid balance in catalytic reforming catalysts.

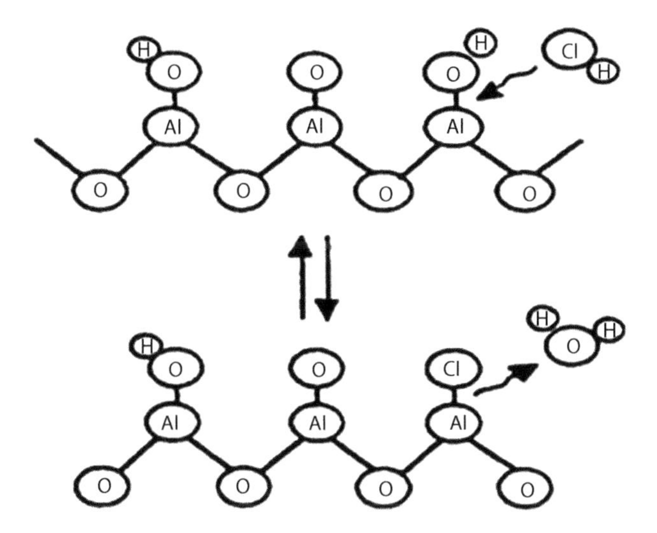

FIGURE 6.14 Acidic sites of alumina surface.

in reformers so as to maintain catalyst chloride at about 1.0 +/– 0.1 wt. %. A number of key factors impact oil refiners' capabilities to maintain catalyst chlorides within the specified ranges for catalysts. Surface area and chloride retention properties of catalysts, reformate quality, and reactor temperatures are known to impact water/chloride molar ratios and hence the concentrations of chlorides on catalytic reforming catalysts.

Catalyst and technology provider companies can provide recommendations for installing appropriate sample points for moisture and hydrogen chloride in the recycle gas system and online analytical equipment for measuring moisture and hydrogen chloride in catalytic reformers. At start of cycle (SOC), reactors should be operated with recycle gas water to chloride molar ratios of between 15/1 to 25/1 in order to maintain catalyst chloride in the 1.0 +/– 0.1 wt. % range in fixed-bed type catalytic

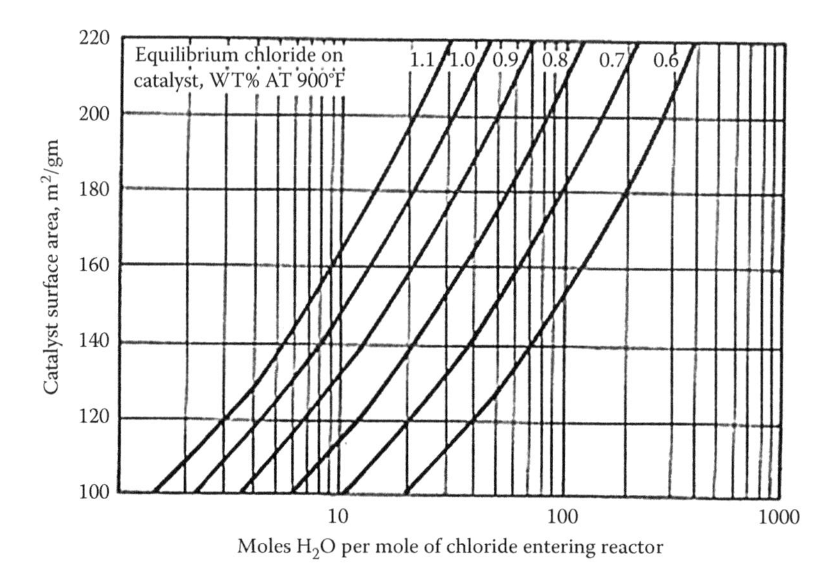

FIGURE 6.15 Curves for equilibrium chloride on catalysts at 900 F.
From Oyekan, S. O., Catalytic Naphtha Reforming Lecture Notes from AIChE Course, in *Catalytic Processes in Petroleum Refining* by Chuang, K. C., Kokayeff, P., Oyekan, S. O., Stuart, S., H., AIChE Today Series, 1992.[5]

reformers. Daily monitoring of the concentrations of recycle gas moisture and HCl should be conducted and key performance indicators compared periodically to targets. Reforming unit data could then be used for making adjustments to the water to chloride ratio in order to stay within the chloride specifications for the catalysts. An example of a graph with correlation curves for extracting information to manage fixed-bed reformer catalyst chloride at 900 F reactor temperature is given in Figure 6.15.[5]

6.1.4 Catalyst Regeneration

During a naphtha processing cycle in a fixed-bed reforming unit, activity and selectivity performances of the catalyst decline gradually due to catalyst coking and possibly deactivation by sulfur and metal contaminants deposited on the catalyst. In order to maintain catalyst activity and selectivity, reactor WAIT is increased and other process conditions are sometimes adjusted to compensate for the rate of catalyst deactivation. Process condition adjustments are required for extending the cycle for the production of desired reformate quality and hydrogen. When it is no longer profitable to operate, reforming operations are discontinued, shutdown of the unit is effected, and the unit is safely secured for turnaround maintenance work. A concise review of turnaround preparation and recommended work scope is provided in another chapter of this book. It is recommended that plans for a turnaround project be completed months in advance with respect to the proposed work by the maintenance service companies. Sufficient time should be allocated to ensure turnaround and catalyst regeneration programs are completed safely and satisfactorily. Three to 4 weeks may be required for the catalyst regeneration program. Required chemicals and equipment such as electrolytic hydrogen,

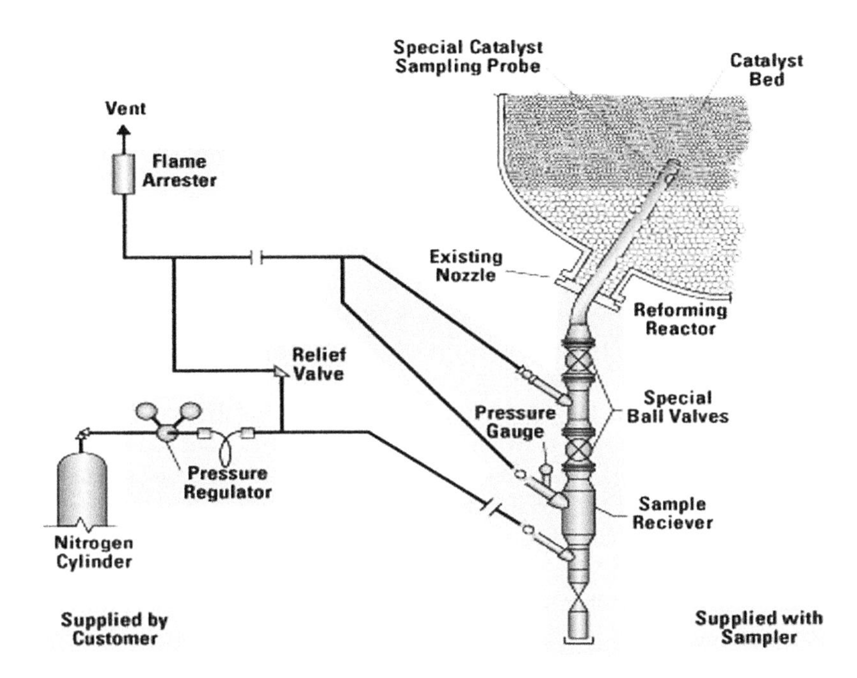

FIGURE 6.16 Installed Honeywell UOP catalyst sampler on a reactor.
From Honeywell UOP, Catalyst Samplers. Accessed July 26, 2017. www.uop.com/equipment/ccr-regeneration/catalyst-sampler.[23] (Illustration Courtesy of Honeywell UOP).

caustic soda, organic chloride, organosulfur compounds, and an air compressor should be available on site, and refinery personnel and services should be trained and available to conduct the turnaround maintenance work in a timely and safe manner. During turnaround and catalyst regeneration planning meetings, catalyst regeneration and startup procedures should be reviewed by all the personnel expected to participate in the catalyst regeneration and reformer startup. If catalyst samplers are installed on the reformer, it is recommended that servicing of the samplers be completed and that they be available and functional for use during catalyst regeneration. An example of an installed catalyst sampler on a reactor by Honeywell UOP is provided in Figure 6.16.[23] Axens also provides a catalyst sampler for similar applications. In addition, gas and liquid sampling units and organo-chloride and organo-sulfur compound injection systems should be serviced and prepared for trouble-free usage when required.

The overall goal is to conduct safe, satisfactory catalyst regeneration steps that result in "full" fresh catalyst activity and selectivity recovery and stable operations to meet the oil refiner's planned cycle length for the catalytic reformer. In addition, during catalyst regeneration stages when effluent hydrogen chloride concentrations are high, basic neutralizing solution, mostly water, should be circulated to minimize corrosion of piping and the separator after the combined feed exchanger. The catalyst regeneration process is usually initiated at an appropriate time after completion of turnaround maintenance work, which also includes establishing reliable sampling points after the reactors and separator. Refinery operations and laboratory personnel would then be in the position to conduct the catalyst regeneration and reformer startup. The primary stages of catalyst regeneration are the coke burn, rejuvenation or metal redispersion and rechloriding, platinum and promoter metal reduction, and sulfiding in the case of platinum and platinum/rhenium catalysts.

6.1.4.1 Catalyst Coke Burn Principles

Coke removal from the catalyst is the first necessary activity after a catalytic reformer shutdown. The coke burn is only initiated and conducted after ensuring that the reactor and relevant reformer sections are free of combustible hydrocarbons and gases. The coke burn activity must be completed before initiating the activation of the metal and acidic sites of reforming catalysts. A major objective of the coke (C_xH_y) burn stage is to burn off the coke and achieve residual catalyst coke of less than 0.1 wt. %. A coke combustion reaction is shown in Figure 6.17. An additional benefit of the coke burn process is the combustion of some of the adsorbed sulfur on the catalyst and iron scale and removal of sulfur oxide products through the regeneration vent gas. Some undesirable changes to the state of the catalyst also occur during the coke burning process. The first and most drastic is that platinum and promoter metals are agglomerated due to the oxidative conditions of coke combustion. The oxidative environment of water, air, and high temperatures in excess of 1100 F lead to the formation of metal oxides and agglomeration of platinum and promoter metals.

$$C_xH_y + (x + y/4)\,O_2 \longrightarrow xCO_2 + (y/2)\,H_2O$$

FIGURE 6.17 Complete coke combustion reaction.

Another key undesirable development is that a significant amount of catalyst chloride loss occurs as hydrogen chloride is released. Hydrogen chloride is produced and stripped by the by steam generated in the coke (C_xH_y) combustion reaction. Regeneration system effluent hydrogen chloride is removed by the regeneration vent gas. Volatile organic compounds that are formed during the combustion of the coke are also removed by the regeneration vent gas. Therefore, key objectives of the coke burning process are to achieve residual coke of less than 0.1 wt. % and minimize platinum agglomeration and chloride losses from catalysts. In addition, the rate of coke burn and reactor bed temperatures in fixed-bed catalyst regenerations should be controlled so as to minimize damages to the catalyst and reactor internals. Combustion bed temperatures should be controlled so as to minimize catalyst damage via alumina phase transitions and minimize damage to the reactor internals.

6.1.4.1.1 Primary Coke Burn

A catalytic reformer is prepared for the coke burn by discontinuing oil feed at 850 F and purging out oil and hydrogen at 750 F with nitrogen. The reformer is filled with nitrogen and cooled in nitrogen. For more details on safe shutdown procedures, the oil refiner should use established procedures or seek technical assistance from the oil refiner's catalyst and technology provider.

Heat-up of the reactors is controlled at about 50 to 75 degrees per hour to 750 F. Coke burn is initiated at 750 F with low oxygen concentration of about 0.2–1.0

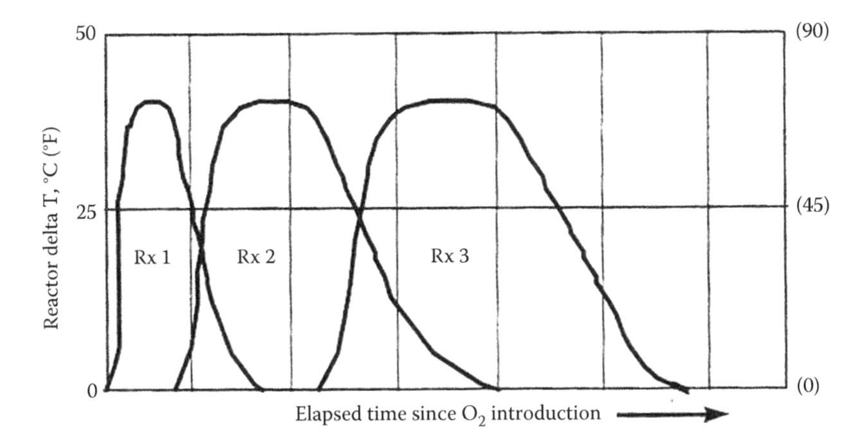

FIGURE 6.18 Coke burn profile for a three-reactor reformer.
From Oyekan, S. O., Personal Notes from a Honeywell UOP Platforming Process Technology Course, Chicago, 1998.[22]

volume percent, preferably at 0.5 volume percent, in nitrogen. Reactor pressure and recycle gas rates are maintained as high as feasible consistent with equipment and limitations. In an optimized coke burn procedure popularized by Honeywell UOP,[6] organic chloride is added at a rate of 20–1 molar ratio of water to chloride to minimize chloride losses from the catalysts and moderate the extent of agglomeration of platinum and promoter metals. Basic neutralizing water is circulated after the combined feed exchanger and through the separator to neutralize hydrogen chloride and minimize corrosion of piping and vessels. The circulating basic water should be greater than 50% of the design feed rate of the catalytic reformer. Basic solution pH is maintained between 7 and 8 and other properties such as alkalinity and total solids of the solution are monitored. Fresh caustic is added as required to meet basic water pH, alkalinity, and total solids guidelines. Reactor temperature deltas are maintained at less than 100 degrees. Regeneration gas oxygen, chloride addition rates, and basic water circulation are maintained as per specifications. Reactor temperature delta profiles are monitored and plotted as shown in Figure 6.18 for a three-reactor reformer.[22] Coke burn in a reactor is completed when the reactor temperature delta (delta *T*) drops to zero for that reactor. Coke burn temperature profiles and reactor exit oxygen concentrations can provide information on the amount of coke and flow characteristics of the catalyst bed. In addition, if the oil refiner has catalyst samplers, samples can be taken for analytical determination of residual coke on catalysts.

6.1.4.1.2 Secondary Coke Burn

Primary coke burn is considered completed when there are no observable changes in coke burn profiles and the reactor DTs are essentially zero. Secondary or proof burn is then initiated to ensure that nearly all the coke has been burnt off the catalyst. This is a major requirement to ensure that irreversible catalyst damage does not occur as a result of excessive temperature excursions in the reactors during the high-temperature oxidative catalyst rejuvenation process. For the secondary or proof burn process, reactor temperatures are raised to 850 F at 50 degrees per hour and the oxygen concentration in nitrogen is raised in stages to 2 volume percent. In addition to reactor temperature and oxygen concentration, reactor pressure, chloride addition, and recycle gas rates are controlled as was done during the primary coke burn. Circulating basic water is monitored and adjusted with respect to pH, alkalinity, and total solids, as was practiced during the primary burn process. Chloride addition during the coke burn process aids in minimizing catalyst chloride loss and achieving better chloride distributions in the reactors. The secondary burn is completed when there are no observable temperature deltas for the reactors and there is essentially no observable burn in any of the sections of the reactors for 4–6 hours. In addition, there is no change in the oxygen concentration of the gas through the reactors. Meeting the listed end of coke burn criteria and no evidence of coke burning are indicative that the catalyst is essentially "coke free."

6.1.4.2 Catalyst Rejuvenation

A major benefit of the coke burn process is the elimination of coke from catalysts and, as indicated, two undesirable changes also occur in the catalyst. One major negative consequence of the coke combustion process is the agglomeration of platinum and promoter metals and formation of metal oxides. The other is loss of some of the catalyst chloride, and as a result, loss of some of the effectiveness of the acid isomerization functionality. As a result of the undesirable changes, the rejuvenation procedure must be applied to recover catalyst activity. It is necessary to redisperse the platinum and promoter metal oxides and increase catalyst chloride to meet specified targets. The specified targets of catalyst chloride are usually in the range of 1 +/– 0.1 weight percent. The conditions for successful oxy-chlorination and redispersion of agglomerated metals are sufficient air soak time, high reactor temperature of 900–980 F, partial pressure of oxygen of about 15 psia, dry air environment, and addition of a chloriding agent. It is generally accepted that an oxy-chlorinated complex of platinum, such as PtO_xCl_y, and similar complexes for the promoter metals are active in the rearrangement and dispersion of agglomerated metals. The set of reactions that involves the use of ethylene dichloride, oxygen, and water leads to the production of chlorine and hydrogen chloride, as shown in Figure 6.19. Chlorine is essential for the redispersion of platinum and promoter metals and hydrogen chloride for increasing catalyst chloride to the specified targets for the catalyst.

Chlorine produced from the organic chloride compound in conjunction with oxygen is responsible for forming the oxy-chlorinated complex required for effecting

$$C_2Cl_4 + 2O_2 \longrightarrow 2Cl_2 + 2CO_2$$

$$2Cl_2 + 2H_2O \longrightarrow O_2 + 4HCl$$

$$\text{Metals} + O_2 \longrightarrow \text{Oxidized \& Dispersed}$$

Metal Oxides

Oxides of Pt, Re, Sn, Ir, etc.

FIGURE 6.19 Chemical reactions of ethylene chloride to chlorine.

redispersion of the metals. At the end of a successful catalyst rejuvenation step, platinum and promoter metals are dispersed as the oxides of the metals and anchored to the catalyst surface. The catalyst is still not considered active, as the metals now exist as oxides. It is necessary to reduce the metal oxides that are now dispersed on the catalyst to their catalytically active states.

There is a potential risk of negating the metal oxide redispersion that has been accomplished if the oil refiner does not proceed as soon as feasible to the metal oxide reduction stage and complete catalyst reactivation and oil in. If there is significant delay on the order of days, the rejuvenation process may have to be repeated, which could lead to loss of production time of some days. Since the reformer is also a hydrogen producer, the delay could negatively impact refinery planning and profitability. It is highly recommended that if time is required for maintenance work on the reformer that work should be done after the coke burn stages and definitely not after the rejuvenation stage. The practice in oil refining has been to use the time after the coke burn as safe hold times for maintenance and other related work for the catalytic reformer.

6.1.4.3 Catalyst Metal Oxide Reduction

The rejuvenation or redispersion stage is one of the activities in the course of activating the metal sites or the dehydrogenation/hydrogenation functionality and acid isomerization sites of a catalyst. Though the acid isomerization functionality is restored after the metal redispersion step due to the fact that chloride has been increased to target levels, the dispersed platinum and promoter metals are still inactive. Full activity recovery after regeneration for the catalyst can only be achieved after reducing the redispersed metal oxides to their respective active states in reactions with hydrogen.

The factors that are conducive for achieving good reduction of platinum oxide and promoter metal oxides are use of high purity hydrogen and preferably electrolytic hydrogen, appropriate reduction temperatures for achieving required catalytically active states of the metals, dry reducing gas environment, and optimal time for completing the reduction process. Numerous studies have been conducted on fresh and activated platinum and platinum/rhenium catalysts that focused on the states of platinum and promoter metals after the rejuvenation stage and on the reducibility of the catalytic metals.[24–32] Researchers have shown that platinum and rhenium can exist as bimetallic clusters on the catalysts and that the reductions of platinum and rhenium to their zero valent states are dependent on the moisture content during reduction. Some researchers have concluded that the reduction of platinum occurs between 500 F and 600 F, whereas that of rhenium occurs in the higher temperature range of 750 F to 1150 F. Results from temperature programmed reduction studies by Scelza are shown in Figure 6.20 and show the effects of small changes in catalyst chloride on the reduction temperature ranges for Pt/Al_2O_3, Re/Al_2O_3, and $Pt\text{-}Re/Al_2O_3$ catalysts.[31]

Past studies by some of the cited researchers concluded that oxidized platinum mostly likely existed predominantly in the plus-4 oxidation state and that of rhenium was in the plus-7 state.[25] Key reduction reactions of the oxides of platinum and rhenium are shown in Figure 6.21.[5]

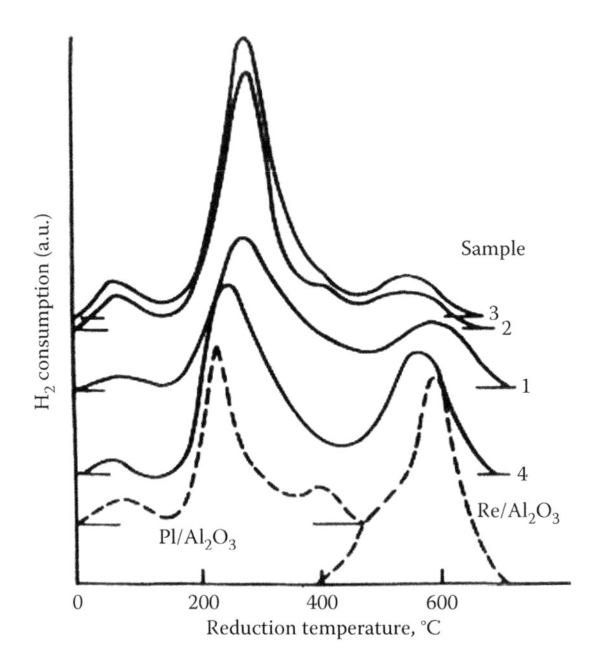

FIGURE 6.20 TPR of Pt/Al_2O_3, Re/Al_2O_3, and $Pt\text{-}Re/Al_2O_3$ catalysts.
From Scelza, O. A. et al., *React. Kinet. Catal. Lett.*, 33(1), 143, 1987.(31)

Similar reduction reactions can be written for platinum-tin alumina catalysts and, in that case, the tin oxide is assumed not to be reduced to its zero valent state. A significant amount of water is produced during the reduction of the metal oxides, and the water must be continuously removed to achieve optimal reduction of platinum oxides and oxides of the promoter metals, as shown in Figure 6.21. Since basic water neutralization is not used during the reduction stage, the pH of the effluent gas must be monitored, and the oil refiner is advised to discontinue the reduction process if the pH is dropping precipitously. A cutoff point of 3 or 4 pH could be used to terminate the reduction process so as not to corrode reformer piping and separator vessels.

Additionally, in order to facilitate the reduction of rhenium and other promoter metals, Oyekan introduced the two-stage reduction process for platinum-containing catalysts. The first step is the facile reduction of platinum between 600 F and 700 F, and the second step involves a higher temperature greater than 900 F. During the reduction process, the reduced active platinum of the catalyst catalyzes the reduction

$$PtO_2 + 2H_2 \longrightarrow Pt^0 + 2H_2O$$

$$Re_2O_7 + 7H_2 \longrightarrow Re^0 + 7H_2O$$

FIGURE 6.21 Reduction reactions of platinum and rhenium.
From Oyekan, S. O., Catalytic Naphtha Reforming Lecture Notes from AIChE Course, in *Catalytic Processes in Petroleum Refining* by Chuang, K. et al., AIChE Today Series, 1992.(5)

of the promoter metals, thus minimizing the time and temperature required to complete the reduction of promoter metal oxides.[32,33]

Catalytic reforming catalysts are most active after the metal reduction stage. Reformer startup should be initiated either immediately or as soon as feasible for the activated platinum/tin catalysts. For reduced platinum and platinum/rhenium catalysts, sulfiding of the metals and metal oxides should follow either as soon as feasible or immediately after the reduction step, and the sulfided catalyst can then be kept isolated and protected in that state in the reactors until reformer startup with no catalytic performance losses at startup of the reformer.

6.1.4.4 Catalyst Sulfiding

Platinum-containing catalysts, with the exception of platinum/tin catalysts used mostly in continuous catalyst regenerative reformers, require sulfiding of the active metal sites to moderate hyperactive sites on platinum and minimize excessive hydrocarbon hydrocracking activity by rhenium.[33–35] Catalyst sulfiding is usually conducted by injecting a predetermined amount of an organo-sulfur compound into hydrogen and into each reactor at between 750 F and 900 F. The organo-sulfur compound addition typically starts with the sulfiding of the catalyst in the last reactor. Sulfur is then added sequentially from the preceding reactor before the last reactor to the lead reactor. This sequence should provide enough time for proper sulfiding of the last two reactors, which usually contain over 70% of the total reformer catalyst. The weight of organo-sulfur compound added is based on a targeted concentration of 0.05–0.15 weight percent sulfur. Target sulfur is usually based on the rhenium content of the $Pt/Re/Al_2O_3$ catalyst. Higher sulfur is recommended for high-rhenium, skewed bimetallic Pt/Re catalysts.

Catalyst sulfiding was deemed complete when sulfur breakthrough, as hydrogen sulfide, was determined from the last reactor of fixed-bed catalytic reformers. This catalyst sulfiding procedure was utilized by some oil refiners for their fixed-bed catalytic reformers in the earlier years of fixed-bed catalytic reforming until the 1980s. The sulfur breakthrough procedure essentially deposited up to 0.25 to 0.30 wt. % sulfur on catalysts. Such high start-of-cycle catalyst sulfur led to lower reformate and hydrogen productivity for reforming cycles. Higher sulfur in the reformer also led to increased fouling and corrosion in product separation systems after the reactors. Organosulfur compounds that can be used for catalyst sulfiding include hydrogen sulfide, dimethyl sulfide (DMS), dimethyl disulfide (DMDS), methyl mercaptans (CH_3SH), and tertiary nonyl polysulfide (TNPS).[36]

Several studies that were conducted between 1978 and 1985 led to the development of a range of acceptable start-of-cycle concentrations of catalyst sulfur for optimizing the performance and productivity of catalytic reformers that operate with platinum/rhenium catalysts. Specifically, start of cycle (SOC) catalyst sulfur was determined for bimetallic Pt/Re catalysts with varying rhenium to platinum ratios. Extensive pilot plant studies by Oyekan provided a "working" hypothesis for a better understanding of the impact of rhenium in the optimization of bimetallic platinum/rhenium catalysts. One interesting set of pilot plant data indicated that platinum monometallic catalysts produced more methane relative to other lighter hydrocarbons in naphtha processing, whereas rhenium only–containing catalysts nonselectively cracked the naphtha feed to produce almost equivalent yields of methane through butane.[5,37]

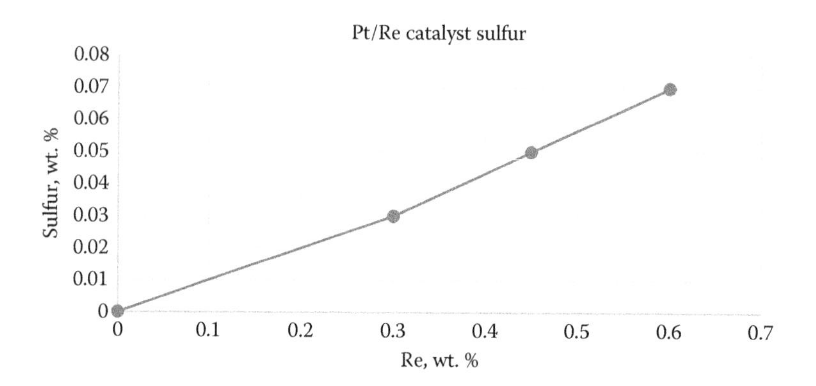

FIGURE 6.22 Catalyst sulfur correlates with rhenium for $Pt/Re/Al_2O_3$ catalysts. From Oyekan, S. O., Personal Notes – Pilot Plant Studies Conducted at Exxon Research and Engineering Lab, Baton Rouge, 1979 in Baton Rouge.[37]

Additionally, sulfur inspection data for spent bimetallic catalysts showed conclusively that residual sulfur on $Pt/Re/Al_2O_3$ catalysts was tightly bound to the rhenium sites, and there is a linear correlation of residual catalyst sulfur with the rhenium content of Pt/Re catalysts, as shown in Figure 6.22.

This brief Pt/Re bimetallic catalyst sulfur review can help guide the oil refiner with respect to the optimal amount of sulfur required during catalyst sulfiding, as SOC catalyst sulfur could negatively impact cycle average reformate and hydrogen productivity.

6.1.4.5 Sulfate Removal Procedure

The typical progression in the regeneration and reactivation of platinum and platinum rhenium reforming catalysts from coke burn to catalyst sulfiding is as described in the preceding sections of this chapter. The usual sequence of regeneration steps is depicted in Figure 6.23 for spent catalysts that have been deactivated mainly by the deposition of catalytic coke on the active sites. In some fixed-bed catalytic reformers, deactivation of catalysts is accelerated due to the processing of naphtha feeds with sulfurs that are much higher than the recommended sulfur specification of less than 0.5 wppm sulfur. It must also be added that many fixed-bed catalytic reformers operating with Pt/Re catalysts specify naphtha feeds containing less than 0.2 wppm sulfur for good catalytic performance.[63] Catalytic performances of platinum-containing catalysts operating with naphtha feed sulfurs that are higher than 0.5 wppm sulfur are usually not optimal due to significant poisoning of the hydrogenation/dehydrogenation functionality of the catalysts. Reformers with

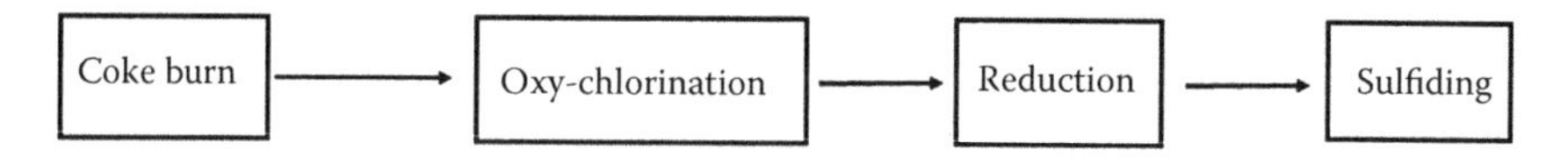

FIGURE 6.23 Regeneration sequence for Pt and Pt/Re catalysts.

$$PtS_2 + 4O_2 \longrightarrow Pt(SO4)_2 \text{ (Coke burn)}$$

$$Re_2S_7 + 14O_2 \longrightarrow Re_2(SO4)_7 \text{ (Coke burn)}$$

FIGURE 6.24 Platinum and rhenium sulfates formation.

sulfur-contaminated catalysts exhibit lower reformate and hydrogen productivities and poorer performance stabilities relative to catalytic reformers with the same catalysts that are operating with recommended feed sulfur and equivalent process conditions. Analytical inspections of spent catalyst samples that operate with high-sulfur naphtha feeds usually show high catalyst sulfur concentrations in excess of 0.05 weight percent for a spent equimolar Pt/Re catalyst.

Platinum-containing catalysts that are contaminated with sulfur have to be subjected to additional regeneration steps to activate them by removing sulfur from the catalyst in a "sulfate" reduction step before conducting platinum and promoter metal reductions.

During standard catalyst regenerations, platinum sulfide, rhenium sulfide, iron sulfides, and chemisorbed sulfur on alumina are usually converted to sulfates of platinum, rhenium, and iron, as shown in Figure 6.24 for platinum and rhenium. Platinum and rhenium oxidation states are assumed for platinum and rhenium to be +4 and +7, respectively, in the reactions. The sulfate species are not dispersed in the oxychlorination step, as the sulfates inhibit the formation of the platinum oxychloride complexes that are required for platinum and rhenium redispersions.[(64)]

In the reduction step of the catalyst regeneration process, platinum and rhenium sulfates are essentially reduced to platinum and rhenium sulfides, and these metal sulfide states hinder the reduction of platinum and rhenium to the desired active zero-valent states. The usual standard catalyst regeneration is, therefore, not adequate for achieving full reactivation of sulfur-contaminated platinum and platinum/rhenium catalysts.

To achieve good reactivation of sulfur-contaminated platinum containing reforming catalysts, a sulfate removal step is recommended. An appropriate sulfate removal procedure can be applied to recover catalytic performance. The procedure involves the use of high-temperature treatment of the catalyst with hydrogen and organic chloride.[(33)] Essentially, after a brief oxychlorination treatment of 10–12 hours, air is purged out of the reactors in preparation for introduction of hydrogen. Reactor temperatures are raised to 950 F in hydrogen and the pressure is set at 100 psia. Hydrogen and an organic chloride are added continuously. Organic chloride reacts with hydrogen to form hydrogen chloride. Hydrogen reacts with the sulfates to form hydrogen sulfides, which are then stripped out of the reactors. Since hydrogen chloride is continuously produced, the sulfate removal procedure is conducted with a continuously included caustic circulation solution to neutralize the hydrogen chloride in the effluent treat gas. The same basic neutralizing solution procedure that is used during the catalyst coke burn step is used to protect piping and separator in the product separation section. The hydrogen sulfide stripping process continues until less than 5 ppm hydrogen sulfide is detected. At the completion of the sulfate removal step, the catalyst regenerations process is then restarted with the coke burn, as shown in the scheme in Figure 6.23.

After the completion of catalyst regeneration and startup, the catalytic reformer catalytic performance should be accorded the usual diligent monitoring and assessments of efficacy of the sulfate removal procedure from the catalyst. Catalytic performance is compared to that expected for low naphtha sulfur feed operations. The catalytic reforming catalyst can be replaced if it is determined from catalytic performance and available catalyst inspection data that catalyst performance was not recovered due to high catalyst sulfur.

After catalyst regeneration, the catalytic reformer performance and sulfur contents of constituent naphtha feeds to the reformer should be monitored closely to determine if "reasonable" catalytic performance recovery has been achieved for the reforming catalyst and, if not, the next set of options to rectify catalytic reformer performance should be explored. Naphtha hydrotreater catalyst could be replaced if operating data from that unit indicate end of catalyst life hydrotreating performance. If hydrocracker and opportunity naphthas are contributing to the high sulfur in the reformer feed, options such as sulfur removal treatments can be explored. For the hydrocracker, a sulfur guard with an appropriate adsorbent could be used if the hydrocracker naphtha sulfur is high, especially at the end of cycle of some of the hydrocracker system catalysts. Opportunity naphthas are typically coprocessed with straight-run naphtha through the naphtha hydrotreater and these could be checked as well. Ultimately, if the sulfur is contributed intermittently by operational upsets in the naphtha hydrotreater, a sulfur guard unit could be used for the removal of sulfur from the naphtha feed to the catalytic reformer.[63–66]

A key point needs to be shared for operators of catalytic reformers that do not have the advantage of an upstream naphtha hydrotreater and that today appear to be relics of the era of sweet crude oil topping configuration refinery operations. Operators of those catalytic reformers typically practice catalyst replacements at the end of short cycles of moderate catalyst performance. Catalyst regenerations, including the sulfate removal step, for such catalytic reformers do not result in achieving the desired full catalytic performance recovery.

6.2 CYCLIC CATALYST REGENERATIVE REFORMERS

Fixed-bed cyclic regenerative process units such as ExxonMobil's Powerformer and BP's (previously Amoco) Ultraformer represent process technologies that permit longer on-stream operations relative to fixed-bed semiregenerative reforming units. The cyclic regenerative reforming technology enhances the cycle length of fixed-bed reformers from months to years by eliminating the need to periodically discontinue reforming operations in order to conduct in-situ catalyst regenerations. Cyclic fixed-bed regenerative technologies enable oil refiners to operate a number of reactors in naphtha reforming for years while simultaneously regenerating catalyst in a parallel system in the same catalytic reformer configuration. Basically, the reactors are operated to produce high-octane reformate and hydrogen while selected reactors are regenerated at set frequencies. The frequency of a selected reactor's regeneration is dependent on catalyst type and state, processing conditions, and oil refiners' production plans. A simplified flow diagram of a Powerformer is shown in Figure 6.25.[5]

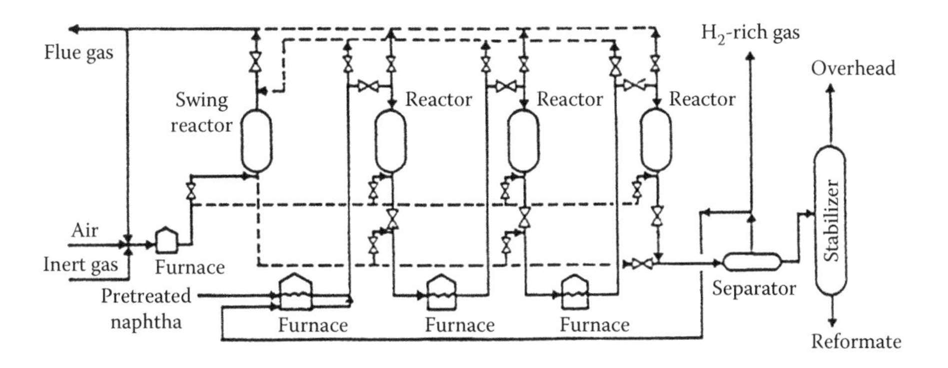

FIGURE 6.25 Cyclic powerformer.
From Oyekan, S. O., Catalytic Naphtha Reforming Lecture Notes from AIChE Course, in *Catalytic Processes in Petroleum Refining* by Chuang, K. et al., AIChE Today Series, 1992.[5]

The unique features of selected reactor regenerations and simultaneous processing of naphtha enable oil refiners to operate cyclic regenerative reformers at much more severe process conditions. Cyclic reformers are operated at low pressures between 125 and 300 psig, low recycle gas hydrogen to hydrocarbon molar ratios, high space velocities, and high reformate severities. These processing conditions contrast significantly with the higher pressures and mild processing conditions used in fixed-bed semiregenerative reformers. Platinum alumina, platinum/rhenium alumina, and bimetallic and trimetallic platinum-containing catalysts used in fixed-bed semiregenerative reformers are also used in cyclic regenerative reformers. There are several advantages associated with operating cyclic regenerative reformers. In a manner similar to fixed-bed regenerative reformers, oil refiners can load a different catalyst in each of the reactors in order to optimize naphtha reforming performance. If catalyst samplers are installed in some of the reactors, cyclic regenerative reformers have additional advantages with respect to more frequent catalyst sampling opportunities due to frequent isolation of selected reactors for regenerations. Cyclic regenerative reformers also provide oil refiners with the flexibility of conducting dump and screen of catalysts from selected reactors and replacing catalysts in selected specific reactors with no need to take a total reformer outage to work on all the reactors.

Due to the requirement of frequent switches of reactors between the hydrocarbon/hydrogen processing system and the air regeneration system, special motor operated valves (MOVs) are required to isolate the incompatible systems. This necessary isolation is required to ensure safe and reliable separations of the oil processing and catalyst regeneration systems within the same catalytic reformer. To facilitate reactor regenerations, large-diameter motor operated valves are used. MOVs are typically Class 600 NPS 8 to NPS 16 API 600 gate block valves. Double block-n-bleed arrangements of the MOVs are used to assure isolation of the 125–300 psig, high-temperature reforming process and catalyst regeneration system. Regeneration MOVs have to operate over a wide range of temperatures and be highly reliable.[38] It has been reported that each cyclic regenerative reformer could have as many as 50

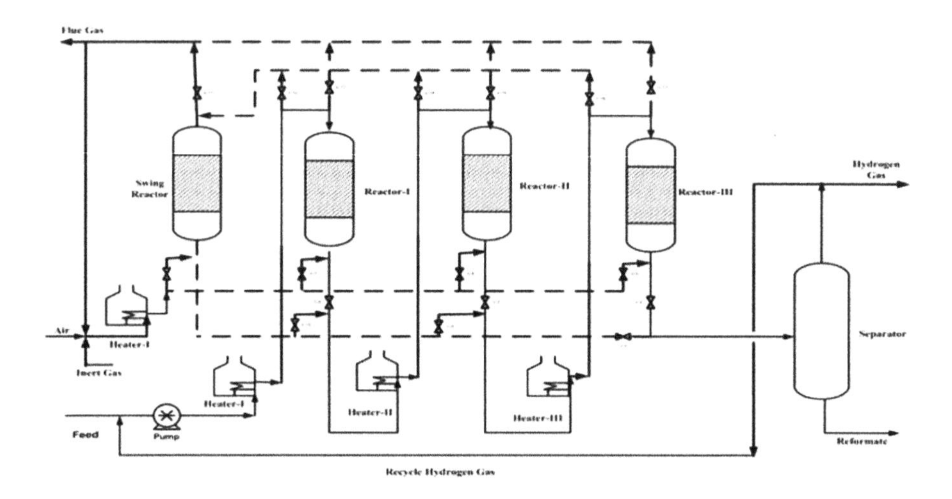

FIGURE 6.26 A simplified drawing of ultraformer.

motor operating valves. A number of MOVs are shown in the simplified drawing of an Ultraformer in Figure 6.26.

Another key differentiating structural feature of cyclic regenerative reformers relative to semiregenerative reformers is that spherical reactors are used for cyclic regenerative reformers, as shown for an Ultraformer in Figure 6.27.[39]

FIGURE 6.27 Five spherical reactors of an ultraformer.
From Univ. of Michigan, Elements of Chemical Reaction Engineering Class Notes. Accessed July 26, 2017. http://umich.edu/~elements/5e/01chap/prof-reactors.html#seriessphere.[39]

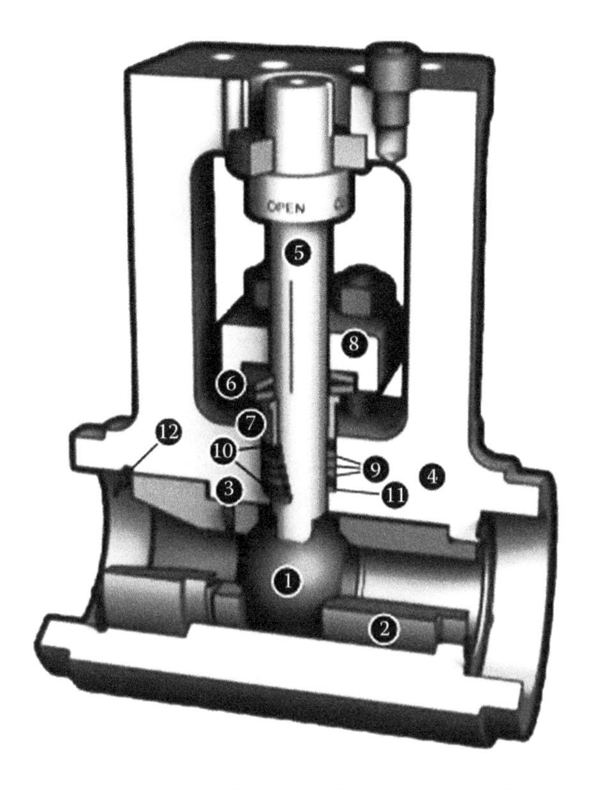

FIGURE 6.28 MOGAS Industries ASME 600/900/1500 Limited Class MOV. Courtesy of MOGAS Industries.

A sketch of a motor operated valve courtesy of MOGAS Industries is given in Figure 6.28 for operation in severe services.[(40)]

6.2.1 Special Features of Cyclic Reforming Catalysts and Regenerations

High-severity operations for the production of high reformate octanes at low pressure, low recycle gas hydrogen to hydrocarbon ratio, and high space velocities lead to rapid catalyst deactivation and frequent reactor regenerations. For many cyclic regenerative reformers, over 15 reactor regenerations are conducted per month. Such a high frequency of catalyst regenerations requires the use of rugged catalysts with high crush strengths, high surface area retention, and good chloride retention characteristics. The last two reactors are subjected to more regenerations, as catalyst in these reactors usually experiences higher rates of catalyst coking and deactivation. Earlier, it was stated that it is beneficial to operate with monometallic catalysts containing platinum concentrations in the range of 0.4 to 0.75 wt. %. Over the past three decades, platinum concentrations of monometallic catalysts have been lowered. Platinum, platinum/rhenium, and platinum tin bimetallic catalysts are currently used in cyclic reformers, and oil refiners are deriving huge savings in platinum costs. In order to operate with an optimal number of monthly regenerations and better catalytic performance, high-stability and high-performance platinum/rhenium bimetallic catalysts should be used by oil refiners for their cyclic regenerative reformer operations.

Chloride and sulfur use during reactivations should be highly optimized so as to minimize negative impacts on overall reformate and hydrogen yields associated with reactor swings. A key deviation in catalyst regeneration from that of the fixed-bed semiregenerative regeneration procedure is the limit on oxygen concentration at any time in the cyclic regeneration section due to National Fire Protection Agency (NFPA). Standard 69 recommendations are that an explosive mixture would result when the concentration of oxygen is 5 volume percent in hydrogen. The NFPA Standard 69 also recommends that it is permissible to operate with 60% of that oxygen concentration or 3% maximum only with continuous oxygen monitoring. Thus, even during the oxy-chlorination step, oxygen must be maintained at less than 3 volume percent and preferably between 2.5 to 2.8 volume percent. Low oxygen concentration limitations and low pressures could limit attainable platinum and promoter metal redispersions, as the partial pressure of oxygen is typically lower than 10 psia. The two-stage reduction process could be applied advantageously to reduce reduction time, improve catalytic performance, and minimize the rate of corrosion of piping and equipment.[32]

6.2.2 Reliability Enhancement Programs

Oil refiners should work closely with their licensors as they operate cyclic catalyst regenerative reformers because of the great requirements for process unit safety and reliability of naphtha reforming operations in reactors and during regeneration of selected reactors. For fixed-bed cyclic regenerative reformers, motor operated valves and low regeneration gas oxygen concentration monitoring are critical items of concern with respect to maintaining safe operations. The NFPA Standard 69 recommendations provide necessary guidance for the use of oxygen without creating an explosive mixture of oxygen in hydrogen due to gas leaks from the reactor section to the regenerator section and from the regenerator to the reactor section.

For MOVs, it is recommended that a maintenance reliability program be established. The MOV program should include spare MOV inventory, frequent inspections, routine MOV maintenance, and replacement schedules so as to ensure safe operations of the reactors and catalyst regeneration sections. In addition, the reactor and regeneration sections should be inspected as often as necessary and, as required, during reactor catalyst dump and screens due to continuous use of organo-chloride and organo-sulfur compounds.

6.3 CONTINUOUS CATALYST REGENERATIVE REFORMERS

Universal Oil Products and Institut Francais du (now Axens) Petrole ushered in an era of high performance, high reliability, and cost-effective catalytic naphtha reforming technologies with the introduction of their first versions of licensed continuous catalyst regeneration units between 1965 and 1975.[44–51] CCR process technologies enable oil refiners to operate at pressures as low as 35 psig, with low reactor WAITs and low hydrogen recycle molar ratios relative to fixed-bed catalytic reforming technologies. The severe process operations of CCRs enhance oil refiners' capability for the production

TABLE 6.5
Catalytic Reforming Unit Performance Data

	Semiregen	Cyclic Regen	CCR 1	CCR 2
Catalyst	Pt/Re	Pt/Re	Pt/Sn	Pt/Sn
Process Conditions				
Pressure, psig	375.0	190.0	116.0	65.0
H2/HC, molar ratio	6.2	3.50	2.0	1.4
Reactor WAIT, F	900.0	941.0	953.0	936.0
Charge Rate, MBPD	9.0	32.0	26.0	19.0
Feed Properties				
API Gravity, degrees	56.0	53.3	59.7	59.5
N+2A, vol. %	57.4	58.8	52.3	48.8
Unit Performance				
Reformate Octane, RON	89.0	87.0	95.8	95.1
C5+ Yield, vol. %	80.5	95.7	80.4	85.6
Hydrogen, SCF/B	720.0	1200.0	1190.0	1275.0

Source: Oyekan, S. O, Personal Notes.[50]
Note: Where CCR 1 and CCR 2 refer to two different CCR units, as shown by the process conditions above.

of high yields of liquid products and hydrogen. Currently, $Pt/Sn/Al_2O_3$ catalysts, with their superior catalytic performance, high attrition resistance, and good surface area and chloride retention characteristics are used in CCR reformers.[41,46,47,49–54]

Catalytic performance data for a set of semiregenerative cyclic regenerative and continuous catalyst regenerative units are provided in Table 6.5 for a wide range of process conditions.[50]

As indicated, Axens (IFP) and Honeywell UOP are the main licensors of continuous catalyst regeneration technologies. Honeywell UOP's CCR Platforming™ process features a stacked radial reactor section and catalyst circulation between the reactor and the catalyst regeneration section.[59] Naphtha and hydrogen flow radially across the annular catalyst bed, and catalyst moves vertically downward via gravity flow between the scallops or outer screen and centerpipe or inner screen of a reactor. Similarly, in the regeneration tower, catalyst moves vertically downward between the inner and outer screens, and regeneration or rejuvenation gas flows radially from the outer to inner screen. In CCR units, catalyst fines removal systems are provided to manage expected fines make from the attrition of catalyst particles as the catalyst circulates and in gravity flow in the annular zones between screens in the reactors and regenerators. Special vessels are provided for the management of catalyst circulation and inventory. Brief descriptions of the special vessels and their functions in catalyst circulation are provided in a later section. A simplified drawing of a Honeywell UOP CCR platforming process unit is as shown in Figure 6.29.[1]

A simplified flow diagram of a CCR process unit with its catalyst circulation and atmospheric catalyst regeneration sections is given in Figure 6.30. The simplified

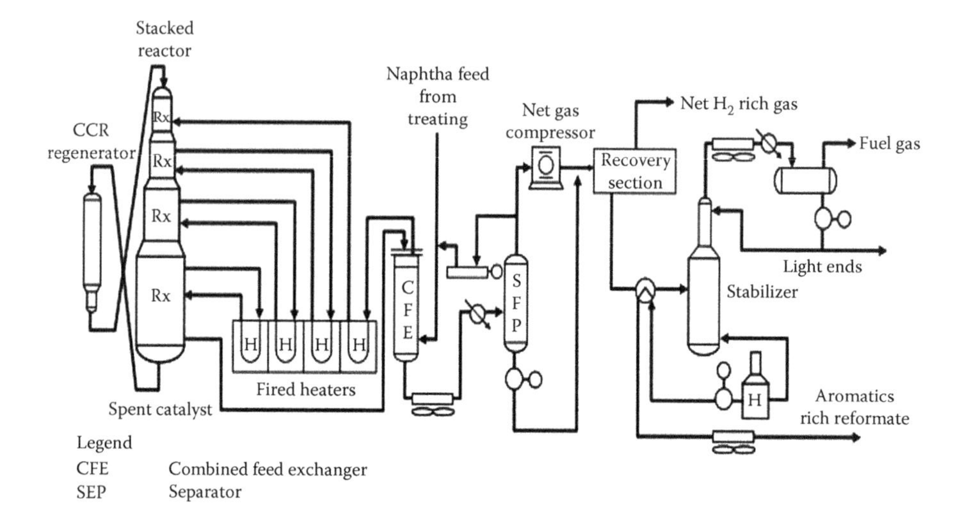

FIGURE 6.29 Honeywell UOP CCR platforming process unit.
From Poparad, A. et al., Reforming solutions for Improved Profits in an Up-Down World, Paper AM-59. *National Petrochemicals & Refiners Association (NPRA), Annual Meeting*, 2011.[1] (Illustration Courtesy of Honeywell UOP.)

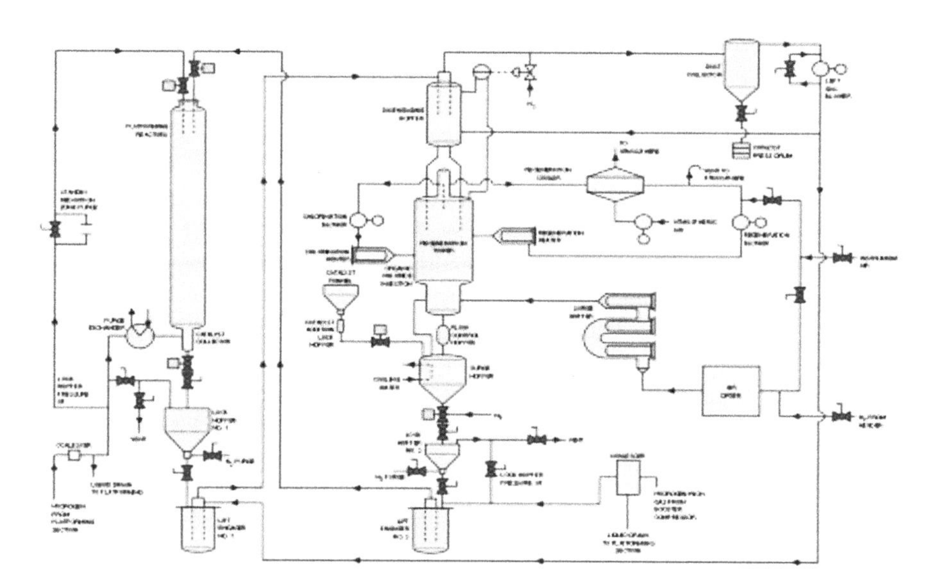

FIGURE 6.30 CCR Process units with catalyst circulation and regeneration sections.
From MOGAS Valves, Continued Catalyst Regeneration, Benefit of MOGAS Valves. www.mogas.com/en-us/resources/media-centre/public/documents/application-notes/application-note-continued-catalyst-regeneration.[60] (Courtesy Honeywell UOP.)

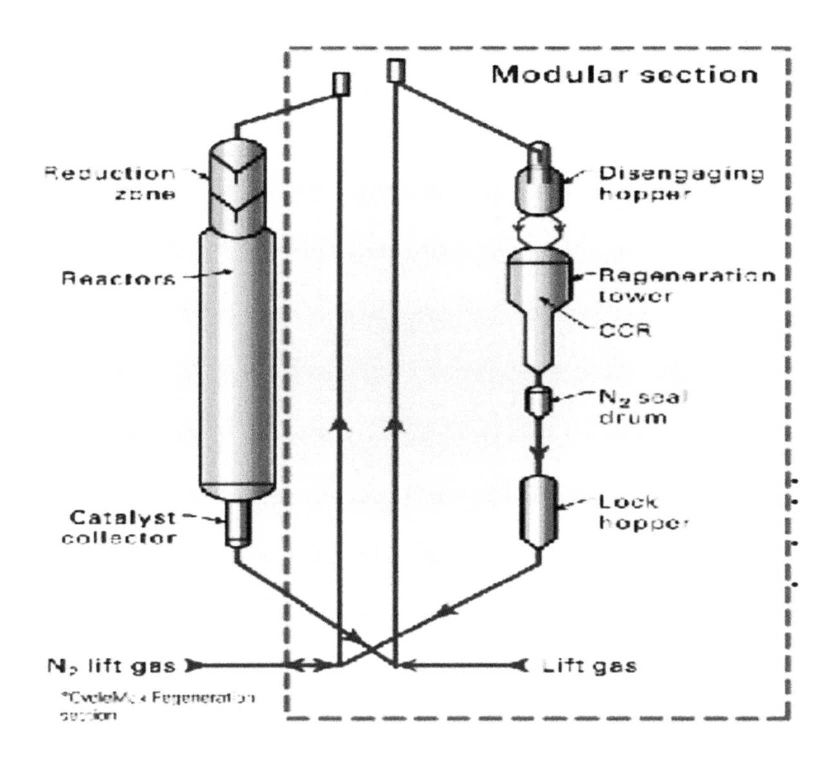

FIGURE 6.31 Honeywell UOP CycleMax platformer.
Illustration Courtesy of Honeywell UOP.

drawing shows the catalyst collector, lift engagers, lock hoppers, disengaging hopper, surge hopper, and regeneration tower.

Honeywell UOP has made several engineering improvements and upgrades over the last two decades on its initial atmospheric regenerators and CCR units. High-pressure regenerators, more reliable catalyst circulation, and improved process control systems are now being achieved with use of Honeywell UOP-licensed CCR process units. Honeywell UOP has incorporated the 35 psig catalyst regeneration section and reliable catalyst circulation systems into its licensed versions of the CCR process units that are now collectively referred to by Honeywell UOP as CycleMax CCR technologies.[58] A simplified diagram of a CycleMax process unit showing its vertical catalyst transfer lines and nitrogen and hydrogen lift gas systems is provided in Figure 6.31.

For more specific details, oil refiners can get more information via contractual agreements with Honeywell UOP. A picture of the Honeywell UOP CCR platformer in Kiwinana, Australia, showing the area of stacked reactors and regenerator sections on the left and the charge and interreactor heaters on the right is provided in Figure 6.32.

Axens Octanizer reactors are aligned in a side-by side arrangement with reactor internals that are similar to those of the Honeywell UOP CCR platforming reactors. Catalyst and naphtha flow are as described for the CCR reactors. The side-by-side

FIGURE 6.32 A CCR Catalytic reformer in Kwinana, Australia.

reactor configuration is touted by Axens to require low capital investments and easy-to-erect reactors with low maintenance costs due to easy accessibility of the internals of all the reactors.[48] Catalyst and regeneration gas flows in the Octanizer catalyst regeneration towers are similar to those of the Honeywell UOP CCR catalyst regeneration systems.

The latest Octanizer is advertised by Axens as having a more efficient catalyst regeneration section relative to that of an earlier licensed CCR type reformer by IFP. Per Axens, engineering improvements of the RegenC2 regenerator lead to increased reliability, ease of operation, and longer catalyst life.[43] Axens continuous catalyst regenerative reformer is described as having a catalyst circulation system that does not require the use of valves and provides flexibility in the control of the catalyst velocity to minimize catalyst attrition in the transfer pipes. The Axens Octanizer is shown in Figure 6.33.[43] A picture of an Octanizer showing the side-by-side reactor arrangement is provided in Figure 6.34.

Continuous catalyst regeneration process technologies have several advantages over the fixed-bed type semiregenerative process technology. The units have catalyst addition facilities that permit frequent makeup catalyst batches addition to replacing catalyst fines and maintaining total system catalyst inventory and process unit performance. In addition, the reactor section can be isolated for continued production of reformate and hydrogen while maintenance is conveniently conducted on the catalyst circulation and regenerator sections. On-stream total catalyst

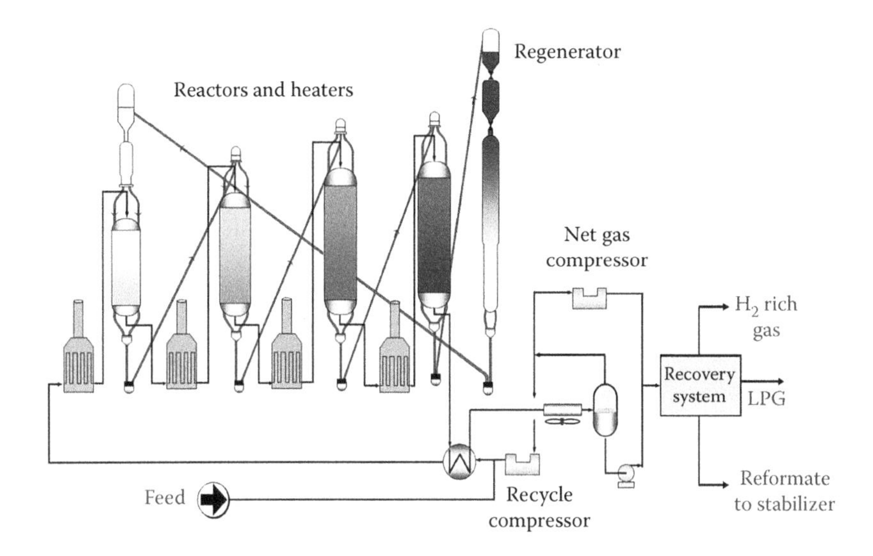

FIGURE 6.33 Axens octanizer reforming process.
Courtesy of Axens.

FIGURE 6.34 A picture of an Axens octanizer.
Courtesy of Axens.

replacements can also be conducted successfully without taking a shutdown for catalyst replacement.

Honeywell UOP has proven systems for total catalyst replacement while simultaneously operating the reactor section and regenerating spent catalyst for dumping. This mode of catalyst replacement while operating the reformer is referred to as "on the fly" catalyst replacement.[1] Successfully executed "on the fly" catalyst replacements in CCR process units save oil refiners millions of dollars, as CCR process unit outages and shutdowns are not required for catalyst replacements. Oil refiners also benefit immensely from uninterrupted production of reformate for gasoline blending and hydrogen to meet the demands of hydroprocessing units in the refineries when they can use "on the fly" catalyst replacement systems and procedures.

6.3.1 Catalyst Circulation Systems

The most significant features of the continuous catalyst regeneration reformer are the versatility and convenience that the technology provides for naphtha reforming in reactors while simultaneously circulating catalyst and conducting catalyst regeneration. High performance activated catalysts are continuously circulated to the lead reactor from the catalyst regeneration section, while spent coked catalysts are circulated to the catalyst regeneration sections. Honeywell UOP should be contacted for more details and specifics with respect to its licensed atmospheric CCR technology.

A brief overview of an atmospheric continuous catalyst regenerative unit is provided for basic understanding of the key elements of catalyst circulation in one of the licensed atmospheric CCR technologies by Honeywell UOP. Spent coked catalyst exits from the last reactor and enters the Catalyst Collector, as shown in the simplified Figure 6.35.[42] Liquid reformate product and gas are purged out and

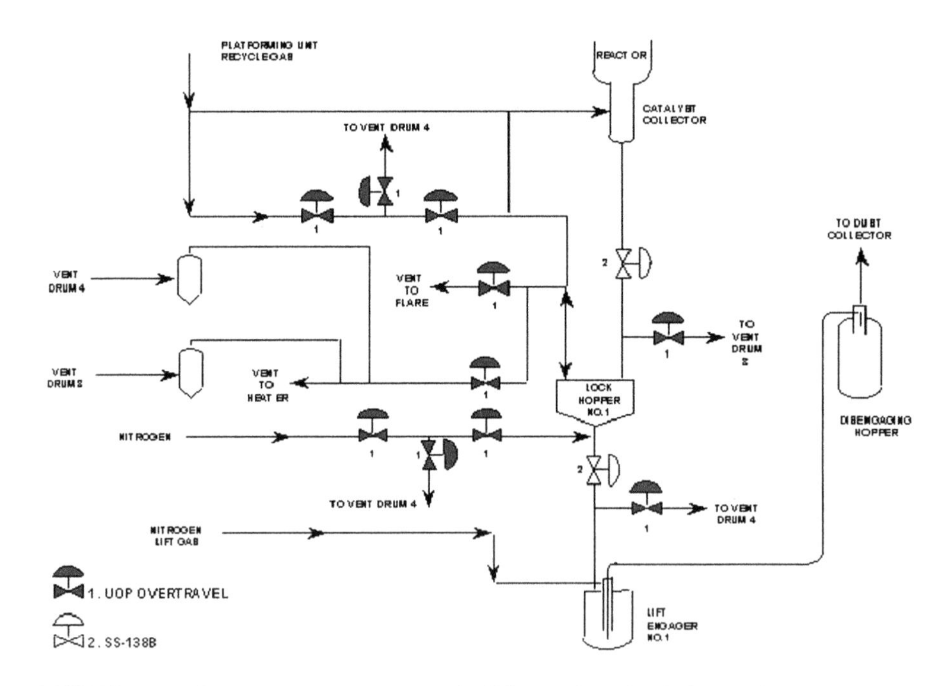

FIGURE 6.35 Spent catalyst in atmospheric CCR catalyst circulation section.

catalyst is transferred to Lock Hopper 1 as shown. Hydrogen gas is purged out and replaced by nitrogen in Lock Hopper 1, and nitrogen is used to move catalyst to the lift pot, which is referred to as Lift Engager 1, in Figure 6.35. Nitrogen is used as the lift gas to move the catalyst to the disengaging hopper, where the catalyst is elutriated and catalyst fines and dust are removed. The catalyst is transferred from the disengaging hopper to the catalyst regeneration section. After the regenerator section, cooled partially activated catalyst is moved into Lock Hopper 2, where the nitrogen and air atmosphere is replaced with hydrogen. Hydrogen is used to move the catalyst to Lift Engager 2 and from there to the catalyst reduction hopper on top of the lead reactor. Catalyst batches that are delivered to Lock Hopper 2 are partially activated, since at this stage, the redispersed platinum and promoter metals exist as their analogous metals oxides. Reduction of platinum oxide to the active zero-valent state and promoter metal oxides to their active states complete catalyst reduction and regeneration. After reduction, the fully activated catalyst enters the lead reactor, and the cycle of reactions, separations, spent catalyst transport, catalyst regeneration and activation, and transport of activated catalyst is repeated.

There are a number guidelines for successful, reliable catalyst circulation such as ensuring that the following conditions are maintained for steady-state operations of the catalyst circulation systems.

- Total adequate reformer catalyst inventory is maintained in reactors, hoppers, lift engagers, and other vessels.
- The catalyst exiting the catalyst collector is dry, as wet catalyst does not flow.
- There is uniform catalyst flow in the reactors, and catalyst pinning is not occurring. Pinning is the phenomena when some of the catalyst that should be moving vertically down via gravity flow within a reactor is held against the centerpipe of the reactor, "pinned" by the radial flow of hydrocarbons and hydrogen.
- There is uniform catalyst flow in the catalyst transfer pipes from the catalyst collector and disengaging hopper as determined from temperature determinations for the individual transfer pipes.
- There is good quality and adequate flow of nitrogen and hydrogen lift gases.
- Nitrogen and hydrogen lift gases are dry.
- A routine maintenance program for valves that also includes keeping several spare replacement valves is used.

Specific expertise of refinery mechanical and operational personnel are critical for operating reliable, profitable CCR process units. For oil refiners who do not have the necessary in-house expertise, it is recommended that they contract with relevant catalyst and CCR technology providers and available process and refinery consultants for technical support.

The catalyst circulation system for the Axens Octanizer technology is similar to that of the Honeywell UOP atmospheric only with respect to transporting spent catalyst from the last reactor section to the catalyst regenerator and returning regenerated catalyst to the lead reactor section. With the side-by-side reactors configuration for the Octanizer, catalyst is continuously moved between the reactors in hydrogen via the piping and lift pots. The gas fed to the lift pots is divided into a primary gas that controls lift operations and a secondary gas that controls catalyst flow. Axens states that the rate of catalyst flow is proportional to the secondary lift gas flowrate.[48] Spent catalyst is withdrawn from the last reactor and lifted in nitrogen to the regenerator section. After catalyst regeneration, thc catalyst is transported from a lift pot to a hopper on top of the lead reactor where catalyst reduction is conducted before moving the activated catalyst to the lead reactor. Axens indicates that there are no valves operating on catalyst in the catalyst circulation section of the Octanizer, thereby minimizing the crushing of catalyst into catalyst fines and frequent repairs of damaged valves.[48]

6.3.2 Catalyst Regeneration in Continuous Catalyst Regenerative Process Units

Platinum-tin alumina is the catalyst of choice for continuous catalyst regenerative processes for a variety of factors. First, platinum-tin catalysts produce higher reformate and hydrogen yields relative to platinum-rhenium alumina and other platinum-alumina containing catalysts in comparative naphtha reforming tests based the same naphtha feed and process conditions.[5,51]

C5+ yields of Pt-Sn and other reforming catalysts from comparative evaluation test runs are shown in Figure 6.36.

Second, as shown in Figure 6.36, the stability of the Pt/Sn catalyst is inferior to those of other platinum-containing bimetallic catalysts that arc uscd in fixed-bed regenerative reforming units. Third, Pt-Sn catalyst activation has been shown to be easy relative to those of platinum-rhenium and other platinum containing bimetallic catalysts. The activation of Pt-Sn reforming catalyst is essentially completed after the reduction of platinum to the zero-valent state and the bulk of the tin to the +2 oxidation state.[52–55] Catalyst sulfiding is not required for Pt-Sn catalysts, and that eliminates the need for incorporating a sulfiding system into the CCR process unit after the catalyst reduction hopper at the top of the lead reactor.

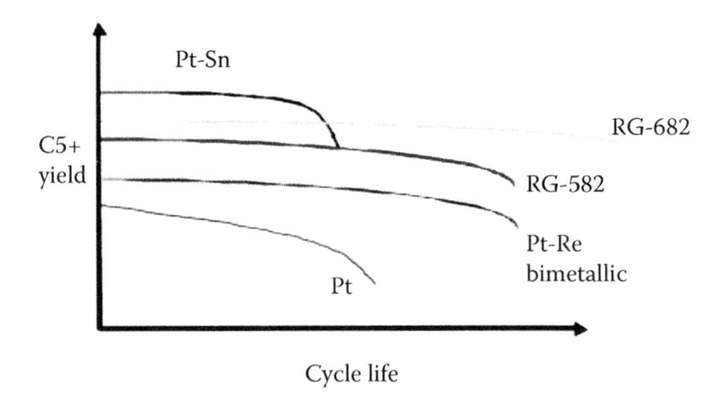

FIGURE 6.36 C5+ yields of reforming catalysts.
From Le Goff, P. et al., Catalytic Solutions for Improved Performance *Petrol. Technol. Q.*, 8(3), 27–29, 2003. www.axens.net.[51] (Courtesy of Axens.)

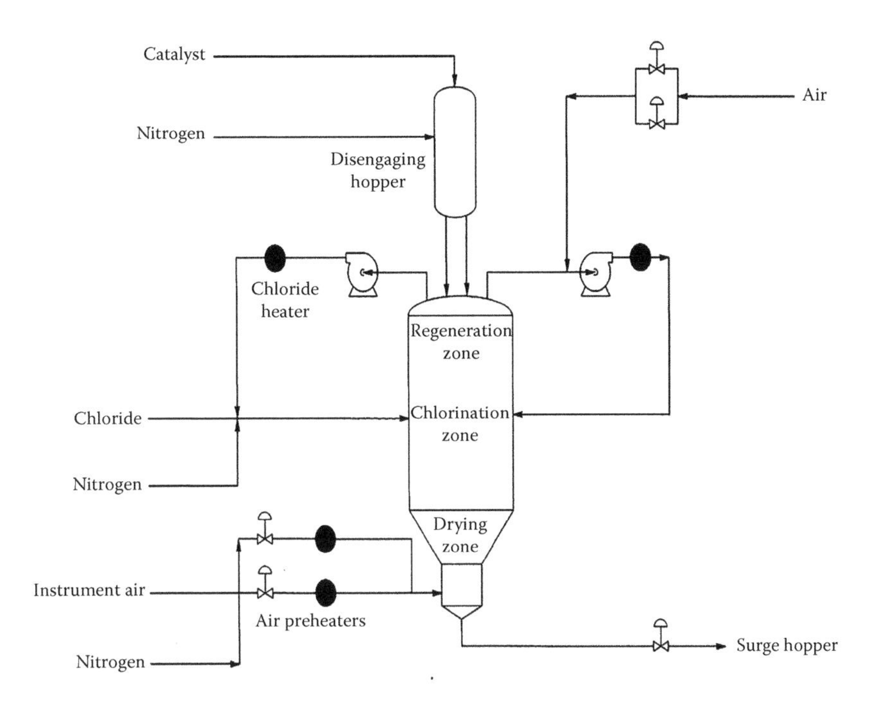

FIGURE 6.37 Regeneration section of an atmospheric CCR unit.
From Doolin, P. K., Zalewski, D. J., Oyekan, S. O., Catalyst Regeneration and Continuous Reforming Issues, in *Catalytic Naphtha Reforming* by Antos, G. J., Aitani, A. M., Marcel Dekker, New York, 2004.[56]

Within the Honeywell UOP CCR process unit, the catalyst regeneration section starts from the point of transfer of catalyst that is stored briefly in the disengaging hopper to the regeneration tower, as shown in Figure 6.37.

The regeneration tower consists of the regeneration, chlorination, and drying zones as shown for an atmospheric catalyst regeneration unit in Figure 6.37. The regeneration tower has a section of inner and outer screens that extends through the regeneration and chlorination zones. Within the annular section, catalyst moves vertically downward in gravity flow, and treat gases flow radially from the outer screen through the catalyst bed in the annular space between the screens. Air heaters and blowers are used for heating up regeneration gas and air to the regeneration tower. Two blowers are used in the atmospheric regeneration system; one is used for the regeneration gas and the other for the chlorination gas, as shown in Figure 6.37.

Catalyst regeneration in CCR process units consists of coke burn, oxy-chlorination, and the calcination or drying step. The listed steps are conducted in the catalyst regeneration section as shown in the simplified diagram in Figure 6.38. Regenerated catalyst is lifted to the reduction zone hopper at the top of the lead reactor where catalyst reduction is conducted in hydrogen.

The maximum hourly catalyst circulation rate is usually established during the design of the CCR process unit and regeneration system based on target spent catalyst

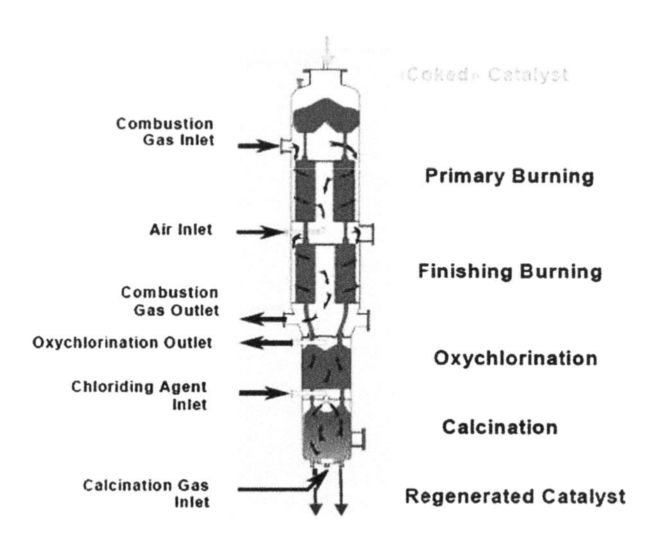

FIGURE 6.38 Key catalyst regeneration steps in the catalyst regeneration unit. Courtesy of Axens.

coke, naphtha feed, and process conditions. Catalyst circulation rate is a key variable in managing the catalyst flow in the regeneration section with respect to satisfactory completions of the primary burn, secondary or proof burn, oxy-chlorination, and catalyst drying or calcination.

6.3.2.1 Primary and Secondary Coke Burns

Some continuous catalyst regeneration units operate at atmospheric pressure and others at pressures in the range of 25–35 psig. To minimize catalyst and equipment damage, combustion or regeneration oxygen gas concentration is monitored continuously via use of an oxygen analyzer. Based on correlations that are usually provided by catalyst and technology providers, catalytic reformer operators can determine required oxygen concentrations and catalyst circulation rates that are sufficient to achieve "complete" combustion of the coke on spent catalysts in the coke combustion zone. Air rate is set such that oxygen concentration in the regeneration gas of nitrogen/oxygen is 0.7–1.4 vol. %, and it is typically maintained at about 1 volume percent. Regeneration zone temperatures are maintained in the range of 900 F to 1100 F for the primary burn. Catalyst circulation rate is adjusted to ensure that primary and secondary coke burns are conducted at steady state. During the primary coke burn, special monitoring of the burn zone temperatures is required to ensure that maximum burn zone temperature increases are kept below 200 degrees. Maximum regeneration zone temperatures should be less than 1200 F, as higher burn zone temperatures lead to higher rates of catalyst surface area losses and alumina phase transitions. If the metallurgy of the regeneration tower is stainless steel, the maximum peak temperature in the burn zone is limited to 1100 F. The burn profile is monitored as well as peak temperatures in a manner similar to those of spent catalyst coke burns in fixed-bed semiregenerative catalyst regenerations. Catalyst samples are usually taken after the regeneration tower to determine if catalyst carbon of less than 0.1 wt. % has been achieved as a critical step before proceeding to the "white burn" oxy-chlorination operation from the "black burn" operation. If the residual carbon on catalyst of less than 0.1 wt. % has not been achieved, the oxy-chlorination process is not initiated and adjustments are made in catalyst circulation rates to ensure maintaining steady state burning in "black burn" to achieve the less than the 0.1 wt. % target carbon on catalyst. The main goal during primary and secondary burns is to achieve less than 0.1 wt. % residual carbon on catalyst before proceeding to the oxy-chlorination step or white burn step. Other key objectives are to achieve less than 0.1 wt. % residual coke without operating at temperatures that can lead to rapid catalyst surface area losses and catalyst and regenerator equipment damages.

Catalyst regeneration systems for the Octanizer process have excellent features that should be highlighted. Specifically, the Axens RegenC-2 consists of the four independent zones: primary burn, finishing or secondary, oxy-chlorination, and calcination, which are similar to those that were previously listed for the atmospheric catalyst regenerator. The Axens RegenC-2 is different from standard CCR catalyst regeneration processes, as a dry burn loop system is provided, as shown in Figure 6.39. The dry burn loop permits washing and drying of the circulating combustion and oxy-chlorination gases.

The dry burn loop is added to reduce the amount of water in the circulating combustion and oxy-chlorination calcination gases. Catalyst surface area losses are significantly reduced by operating with the dry burn loop relative to cold and hot burn loop regenerators. Surface areas are shown in Figure 6.40 for catalysts in CCR reformers with different catalyst regeneration systems.[47]

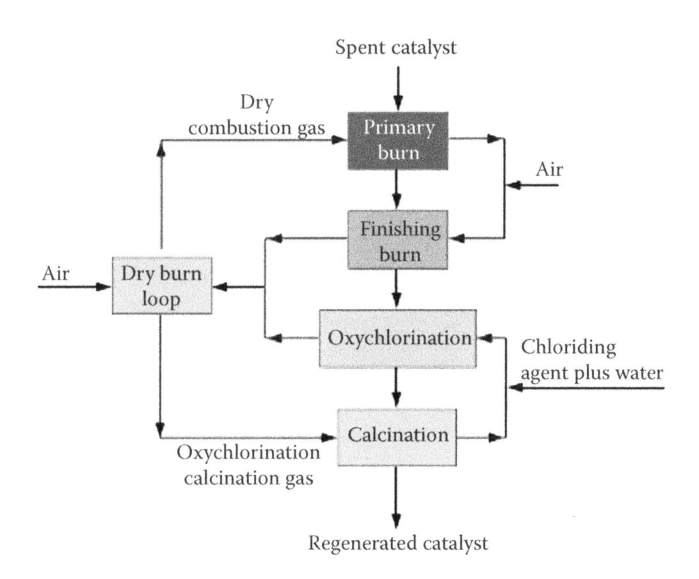

FIGURE 6.39 Axens CCR reforming process RegenC-2.
From Domergue, B., Le Goff, P., Ross, J., *Petrol. Technol. Q.*, 11(1), 67–73, 2006.[47] (Courtesy of Axens.)

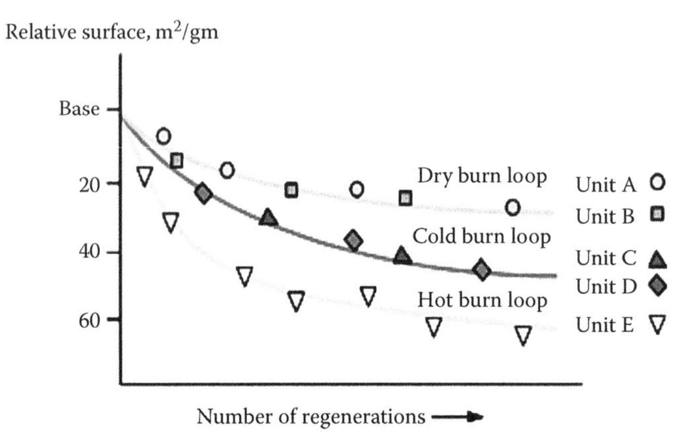

FIGURE 6.40 Effect of regeneration burn loop and regenerations on surface area. From Domergue, B., Le Goff, P., Ross, J., *Petrol. Technol. Q.*, 11(1), 67–73, 2006.[47] (Courtesy of Axens.)

In addition to better surface area retention, the dry burn loop catalyst regeneration system extends catalyst life, reduces downstream corrosion due to better chloride management, and discharges a clean regeneration vent gas that does not contain chloride and eliminates the need for a technology to reduce catalyst regeneration vent gas hydrogen chloride.[47,57] The special chloride benefit is covered in Chapter 8 in some detail with respect to meeting the environmental quality regulation in the United States via use of refinery maximum achievable control technology for reducing catalyst regeneration vent gas hydrogen chloride and volatile organic compounds.

6.3.2.2 Oxy-Chlorination Process

The relevant key points reviewed in Section 6.1.4.2, "Catalyst Rejuvenation," relate to similar factors that are required for successful redispersion of agglomerated oxides of platinum and promoter metals. They are factors for replacing catalyst chloride that was stripped from the catalyst in the primary and secondary burn stages of the regeneration process. Relevant equations for the oxy-chlorination process from Figure 6.20 are reproduced here as in Figure 6.41 to serve as reference.

$$C_2Cl_4 + 2O_2 \longrightarrow 2Cl_2 + 2CO_2$$

$$2Cl_2 + 2H_2O \longrightarrow O_2 + 4HCl$$

$$\text{Metals} + O_2 \longrightarrow \text{Oxidized \& Dispersed}$$

Metal Oxides

Oxides of Pt, Re, Sn, Ir, etc.

FIGURE 6.41 Some of the reactions in the chlorination zone.

Successful redispersion of oxides of platinum and promoter metals is dependent on time, temperature, partial pressure of oxygen, dryness of air, and chloride addition to the catalyst. Catalyst circulation rates range from 300 to 6000 pounds per hour, and depending on the size of the chlorination zone, catalyst residence times in CCR regenerators are much shorter relative to the long air/chloride soak time of over 10 hours typically used in fixed-bed semiregenerative reformers. Other than catalyst residence times in chlorination zones, other pertinent variables are optimized as much as feasible. The higher-pressure continuous catalyst regeneration systems lead to higher partial pressures of oxygen in the zone relative to those in atmospheric regenerators. Chlorination zone temperatures of 950 F to 980 F are recommended and organic chloride is usually heated to facilitate transfer in the oxy-chlorination gas and vaporization of the organic chloride. Air is introduced into the chlorination zone with the objective of operating at as high an oxygen partial pressure as feasible. Oxygen partial pressures in the 10–15 psia range are desirable and operating with 35 psig regeneration unit provides appropriate chlorination zone conditions for achieving high platinum and promoter metal dispersions. Organic chloride reacts to produce hydrogen chloride and chlorine as shown in Figure 6.41. Chlorine reacts to form oxy-chloride metal complexes, which lead to redispersion of desired relevant platinum and promoter metal oxides. Hydrogen chloride reacts with the catalyst alumina hydroxyl groups to increase chloride concentrations and maintain the required concentrations of acidic isomerization sites.

Air containing 21% oxygen is used in the chlorination zone, and while this higher oxygen concentration is good for and facilitates metal oxide redispersion, it could lead to major catalyst and regenerator screens damage if there were significant carbon on the catalyst entering the chlorination zone. If the residual carbon on catalyst from the secondary burn zone is greater than 0.1 wt. %, thermal damage of the catalyst can occur. In addition, depending on the amount of residual coke on the catalyst entering the chlorination zone and how high the chlorination zone temperature rises, damage could also occur to the inner and outer regenerator tower screens. Burning catalyst carbon in the chlorination zone could lead to burn zone temperatures in excess of 2500 F. Such high chlorination zone temperatures on some CCR regeneration units have led to the fusing of catalyst as a consequence of alumina phase transitions, screen damage, and loss of use of the regeneration section. Loss of the regeneration section forces operation of the CCR process unit as a semiregenerative reformer until necessary repairs are conducted satisfactorily for damaged screens and replacement of damaged catalyst in the regenerator section. Operating the reactor section as a semiregenerative reformer would lead to major reformer and refinery profitability losses.

6.3.2.3 Drying and Cooling

At the end of the oxy-chlorination process, catalyst is moved into the drying zone, where it is dried at about 980–1000 F to remove any more water that could be entrained in the catalyst. In order to successfully accomplish adequate drying, the drying air must be treated with an air dryer, and the moisture content of the treated air should be less than 5 ppm. Some catalyst circulation systems use several valves, and the valves have to be protected from thermal damage. Dried catalyst is cooled and exits from the bottom of the regeneration tower.

It is important to emphasize that circulation of wet catalyst must be avoided, as that can lead to water and chloride deposits on piping in the cold section, which could lead to corrosion, holes in piping, and ultimately fires. Additionally, circulating wet catalyst would lead to greater rates of catalyst fines generation in the catalyst circulation system and reactors.

6.3.2.4 Catalyst Reduction

The final step in the catalyst regeneration process is the reduction of platinum oxides and promoter metal oxides to their active states. Metal reduction was covered in detail in Section 6.1.4.3. A concern with respect to achieving successful catalyst metal oxide reduction in catalytic reforming is the quality of the hydrogen used for the reduction of metal oxides. For semiregenerative reformers, electrolytic hydrogen is recommended for platinum oxide and promoter metal oxide reductions due to the need to achieve optimum catalyst regeneration and activation to ensure meeting refinery's projected cycle lengths of 6–24 months. Net hydrogen gas is usually used for the CCR process units, and for some, a small hydrogen purification unit is used. The catalyst metal reduction process for the CCR units takes place in one or two hoppers on top of the lead reactor. Older generations of CCR reformers typically reduce the catalyst metal oxides in hydrogen at temperatures between 850 F and 930 F, and the reduced catalyst, reduction section effluent gas including the water evolved from the reduction reactions, and stripped hydrogen chloride are sent directly to the lead reactor. Platinum and tin oxide reduction reactions are shown in Figure 6.42. The reduction of platinum oxide to zero-valent platinum is well established, the oxidation states of reduced tin oxides are not fully established, and the bulk of tin oxides are assumed to end up in the +2 oxidation state on the reduced Pt-Sn catalyst.[52–54]

A two-stage reduction process for platinum-containing catalysts is now applied in some new CCR units for the optimization of the reduction process of regenerated Pt-Sn catalysts. It had been shown that a two-stage reduction process is beneficial due to the immediate catalyzing of the promoter metal oxide reduction by the activated platinum, which leads to improved catalyst activity and product selectivity.[32,33] First, reduction of platinum oxide on the catalyst is conducted with heated net hydrogen gas between 500 F and 750 F, and water from the reduction of the platinum oxide is purged out by the effluent hydrogen gas. The partially activated catalyst moves into another reduction zone where heated reduction gas between 900 F and 920 F reduces the tin oxide in the second stage. In the second reduction stage, reduced platinum catalyzes the reduction of the promoter metal oxide and, in this case, stannous oxide. Water generated from the second reduction stage and any hydrogen chloride that is stripped from the catalyst are purged out of the catalyst, the catalyst flows into the lead reactor, and the catalyst cycle is restarted with reactor operations.

$$PtO_2 + 2H_2 \longrightarrow Pt^0 + 2H_2O$$

$$Sn_pO_x + yH_2 \longrightarrow Sn_mO_n + vH_2O$$

FIGURE 6.42 Reduction reactions of Pt-Sn catalysts.

6.4 HYBRID CATALYST REGENERATIVE REFORMERS

The review of the three catalytic reforming process technologies aided in highlighting activity and product selectivity benefits associated with operating a continuous catalyst regenerative reformer relative to fixed-bed catalyst regenerative reformers. Continuous catalyst regenerative process units provide oil refiners with much-needed flexibility for processing highly paraffinic naphthas at the high reformate severities required while realizing high product revenues and low process unit maintenance costs. A hybrid catalyst regenerative reformer is a fixed-bed semiregenerative reformer that has been revamped to take advantage of continuous catalyst regeneration for a fraction of the total reformer catalyst inventory. Low maintenance costs are realized due to lengthening of the times between turnarounds for the fixed-bed semiregenerative section of hybrid catalytic reformers after revamping to include a moving bed reactor and a continuous catalyst regeneration unit for the catalyst from the moving bed reactor. A variety of catalysts can be used for fixed-bed reactors, though Pt-Re is the preferred catalyst and Pt-Sn catalyst is used in the moving bed reactor section. Honeywell UOP suggests that a number of factors should be considered before starting a project to convert a fixed-bed semiregenerative reformer to a hybrid reformer. Honeywell UOP indicated that the age of the fixed-bed semiregenerative unit and ultimate processing goals should be considered for such semiregenerative process unit revamps. For a relatively modern fixed-bed semiregenerative unit, Honeywell UOP suggests that a new stacked reactor system, one fired heater cell, and a regeneration section would be required. Honeywell UOP also indicates that a new combined feed exchanger and additional net gas compression would also be included in a fixed-bed semiregenerative revamp.(1)

Axens discussed an attractive staged investment strategy that was applied in the conversion of a three-reactor semiregenerative reformer of an oil refiner in Southern Europe operating at 98 RON reformate target and 300 psig pressure to a hybrid semiregenerative unit.(47)

In its paper, Axens discussed a first upgrade of a semiregenerative reformer to a four-reactor catalytic reformer, which was completed in 1991 and is referenced in Table 6.6. That catalytic reformer was originally designed with that expectation that it could be converted to a CCR process unit later. In the first upgrade, the reactor pressure was lowered from 300 psig to 240 psig. The benefits listed included a reformer throughput increase of 50% and a reformate yield gain of over 1 volume percent at constant 98 octane severity operation.

Later in 2000, Axens reported that the oil refiner successfully completed the next upgrade of the same reformer that had been operated as a semiregenerative unit to a CCR process unit. Reactor pressure was lowered from 160 to 120 psig, and a regeneration unit and other necessary equipment were added, as shown in Figure 6.43. This upgrade to a CCR enabled the oil refiner to produce 78 volume percent reformate at 103 octane severity.

In that same Axens paper, Axens provided a good depiction of the dualforming reformer, which features the addition of the large reactor, regenerator, air cooler, and a larger feed/effluent exchanger. Dualforming, the hybrid catalytic reformer of

TABLE 6.6
Staged Revamp of a Semiregen Reformer to CCR

	Phase 1	Phase 2	Phase 3
	1979	1991	2000
Capacity, MBPD	Base	1.5 × Base	1.6 × Base
Reformate RON	98	98	103
Operation	Semiregen	Semiregen	CCR
Number of Reactors	3	4	4
Reactor pressure, barg (psig)	20 (300)	16 (240)	8 (120)
Reformate Yield, vol. %	79.0	80.1	78.0

Source: Domergue, B., Le Goff, P, Ross, J., *Petrol. Technol. Q.*, 11(1), 67–73, 2006.[47]

Axens, is composed of a three-reactor fixed-bed semiregenerative section, a moving bed reactor, and the continuous catalyst regeneration unit, as shown in Figure 6.44.[47]

Axens estimated in 2006 that a product revenue benefit of about 14.1 million dollars was realized in operating a 20-MBPD dualforming unit relative to a 20-MBPD fixed-bed semiregenerative reformer. Similarly, for a 20-MBPD octanizer (CCR) relative to the fixed-bed semiregenerative reformer, the product revenue benefit was 28.1 million dollars. Reformate and hydrogen yields, revamp investments, catalysts, and utilities

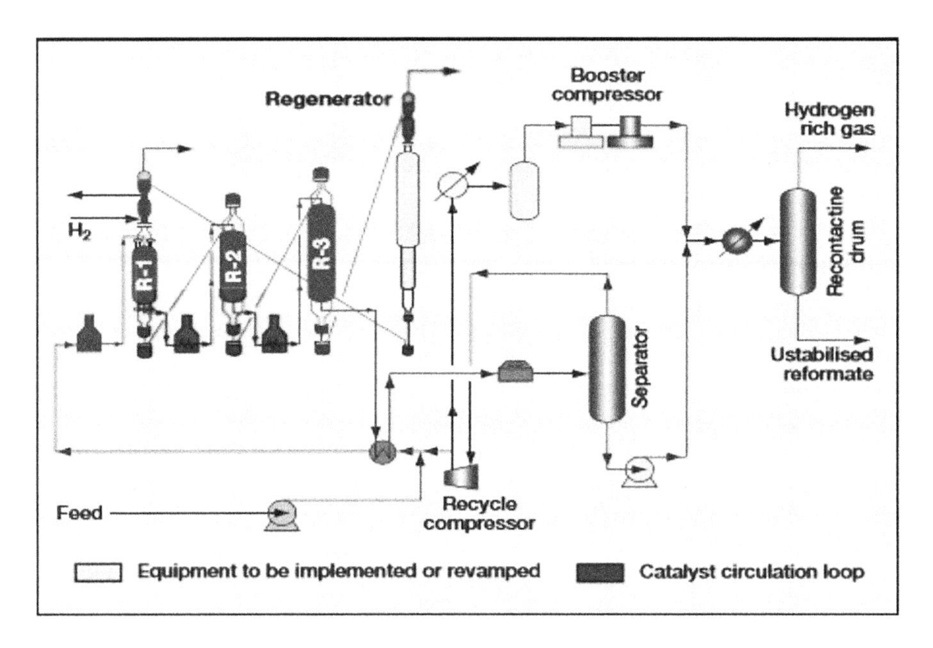

FIGURE 6.43 Revamped fixed-bed reformer to CCR.
From Domergue, B., Le Goff, P., Ross, J., *Petrol. Technol. Q.*, 11(1), 67–73, 2006.[47] (Courtesy of Axens.)

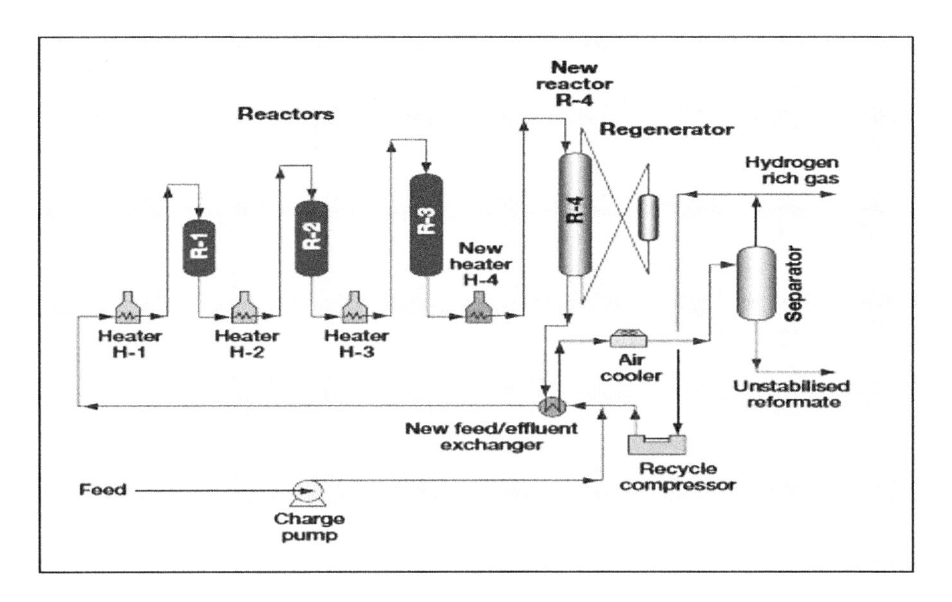

FIGURE 6.44 Axens dualforming reformer flow diagram.
From Domergue, B., Le Goff, P., Ross, J., *Petrol. Technol. Q.*, 11(1), 67–73, 2006.[47] (Courtesy of Axens.)

costs for revamping a 20-MBPD fixed-bed regenerative reformer to a dualforming unit or octanizer are as provided in Table 6.7.[47]

In this chapter, a review of the three major catalytic reforming technologies has been completed and the differences with respect to their range of operations, catalyst regeneration, and reliability challenges were highlighted. Foremost, and per the paper by Honeywell UOP, continuous catalyst regenerative reformers are becoming the favored grassroots reformers globally, especially in China, India,

TABLE 6.7
Semiregen Reformer Revamp to Dualforming or CCR Unit

	Existing Semiregen	Dualforming	Octanizing CCR
Capacity, MBPD	20	20	20
Reformate Octane, RON	100	100	100
Cycle Length, months	6	12 and continuous	Continuous
Product Yields, wt. %			
Hydrogen	1.7	2.4	3.1
C5+ Liquid	76.3	81.9	87.4
Revamp Investment, $MM	Base unit	+18	+40
Catalyst & Utilities, $MM/year	3.1	5.3	7.8
Products Rev. Benefit, $MM/year	Base	+14.1	+28.4

Source: Domergue, B., Le Goff, P, Ross, J., *Petrol. Technol. Q.*, 11(1), 67–73, 2006.[47]

and the CIS (or Russian Federation). It is expected that oil refiners will need to rely more heavily on their catalyst reformers, alkylation units, and other octane upgrading process units to produce projected regulated gasolines of 95 to 100 octane to meet US EPA CAFÉ of 54.5 miles per gallon for some automobiles by 2025. The ultimate blendstock gasoline produced in the United States and other global oil refineries is expected to be determined by the concentration of oxygenates and biofuel that can be blended into gasoline by 2025. CCR reformers and catalytic reformers in general will continue to play major pivotal roles in oil refineries as high-octane blendstock providers and refinery "gatekeeper" process units.

REFERENCES

1. Poparad, A., Ellis, B., Glover, B., Metro, S., Reforming Solutions for Improved Profits in an Up-Down World, Paper AM-59. *National Petrochemicals & Refiners Association (NPRA), Annual Meeting*, 2011.
2. Domergue, B., Le Goff, P., Ross, J., Octanizing Reformer Options. *Petrol. Technol. Q.*, 11(1), 67–73, 2006.
3. PIRA Energy Group, World Refinery Database Q4 2010 Capacity Data. Accessed July 26, 2017. https://www.pira.com/data-tools/world-refinery-database
4. Chemical Technology, Module 6, Lecture 6, Catalytic Reforming. Accessed July 26, 2017. http://nptel.ac.in/courses/103107082/module6/lecture6/lecture6.pdf
5. Oyekan, S. O., Catalytic Naphtha Reforming Lecture Notes from AIChE Course, in *Catalytic Processes in Petroleum Refining* by Chuang, K. C., Kokayeff, P., Oyekan, S. O., Stuart, S. H., AIChE Today Series, 1992.
6. Lapinski, M. L., Baird, L., James, Chapter 4.1, UOP Platforming Process, in *Handbook of Petroleum Refining Processes* by Meyers, R. A., McGraw Hill Companies, USA, 2004.
7. Diestelkamp, T., Kempen, H., Otzisk, B., (Kurita GmbH), Hugot, B., (Gelsenkirchen GmbH), Ammonium Salt Removal in Fuel Gas Burners. *Petrol. Technol. Q. (PTQ)*, Q4, 91–95, 2007. http://www.eptq.com/view_edition.aspx?intContentID=21&intEID=56
8. Buchheim, G. M., Danis, J., API RP 571 Damage Mechanisms in the Refining Industry Training Course, November 2005. Accessed August 22, 2017. https://inspectioneering.com/tag/api+rp+571
9. Olsen, C., AT734G: A Combined Silicon and Arsenic Guard Catalyst, Advanced Refining Technologies Catalagram. Accessed August 22, 2017. https://grace.com/catalysts-and-fuels/en-us/Documents/108SE-AT734G.pdf
10. Cassella, R. J., Alberto, B., Barbosa, R. S., Santelli, R. E., Rangel, A. T., Direct Determination of Arsenic and Antimony in naphtha by Electrothermal Atomic Absorption Spectrometry with Microemulsion Sample Introduction and Iridium Permanent Modifier. *Anal. Bioanal. Chem.*, 379, 66–71, 2004.
11. Axens, Mercury & Arsenic Removal. Accessed July 26, 2017. http://www.axens.net/our-offer/by-market/petrochemicals/64/mercury-a-arsenic-removal.html
12. UOP, Advanced Mercury Removal Technologies. UOP, Des Plaines, IL, USA.
13. Deady, J., Avoid Mercury Poisoning In Naphtha Hydrotreating Operations. Accessed August 22, 2017. http://refineryoperations.com/avoid-mercury-poisoning-in-naphtha-hydrotreating-operations/
14. Johnson Matthey Process Technologies, Mercury removal in Refineries. Accessed August 22, 2017. http://www.jmprotech.com/m/mercury-removal-absorbents-johnson-matthey

15. Criterion catalysts & Technologies, Grading & Poison Control. Accessed July 26, 2017. http://www.criterioncatalysts.com/products/product-applications/gradingpoison-control.html
16. Kremer, L., Silicon in Crude Oil, 2004 Baker Petrolite. Accessed July 26, 2017. http://www.coqa-inc.org/docs/default-source/meetingpresentations/20050127Silicon.pdf
17. Siegel, J., Olsen, C., *Feed Contaminants in Hydroprocessing Units*. Advanced Refining Technologies. Accessed July 26, 2017. https://www.scribd.com/document/290889530/104SE-Feed-Contaminants-in-Hydroprocessing-Units
18. XOS, Understanding the Silicon Issue. www.xos.com
19. Zeuthen, P., Andersen, K. V., Silicon Deactivation of Hydrotreating Catalysts in Coker Distillate Services.
20. Pines, H., Haag, W. O., Alumina: Catalyst and Support. I. Alumina, Its Intrinsic Acidity and Catalytic Activity. *J. Amer. Chem. Soc.*, 82, 2471, 1960.
21. Digne, M., Sautet, P., Raybaud, P., Euzen, P., Toulhoat, H., Hydroxyl Groups on Gamma Alumina Surfaces: A DFT Study. *J. Catal.*, 211, 1–5, 2002.
22. Oyekan, S. O., Personal Notes from a UOP Platforming Process Technology Course. Chicago, 1998.
23. Honeywell UOP, Catalyst Sampler. Accessed July 26, 2017. www.uop.com/equipment/ccr-regeneration/catalyst-sampler
24. Le Goff, P., de Bonnerville, J., Domergue, B., Pike, M., Increasing Semi-Regenerative Reformer Performance through Catalytic Solutions. *Axens*.
25. Johnson, M. F. L., LeRoy, V. M., The State of Rhenium in Pt/Re/alumina Catalysts. *J. Catal.*, 35, 434, 1974.
26. Bertolacini, R. J., Pellet, R. J., in *Catalyst Deactivation* by Delmon, B., Froment, G. F., Elsevier Publishing Company.
27. Wagstaff, N., Prins, R., Alloy Formation and Metal Oxide Segregation in PtRe/γ-Al2O3 Catalysts as Investigated by Temperature-Programmed Reduction. *J. Catal.*, 59, 434, 1979.
28. McNicol, B. D., The Reducibility of Rhenium on Gamma Alumina and Pt/Re on Gamma Alumina. *J. Catal.*, 46, 434, 1977.
29. Bolivar, C., Charcosset, H., Froty, R., Primet, M., Tournayan, L., Betizeau, C., Leclercq, G., Maurel, R., Platinum-Rhenium/Alumina Catalysts: I. Investigation of Reduction by Hydrogen. *J. Catal.*, 39(2), 249–259, 1975.
30. Nacheef, M. S., Kraus, L. S., Ichikawa, M., Hoffman, B. M., Butt, J. B., Sachtler, W. M. H., Characterization and Catalytic Function of Re^0 and Re^{4+} in $ReAl_2O_3$ and Pt/Re/Al_2O_3. *J. Catal.*, 106(1), 263–272, 1987.
31. Scelza, O. A., De Miguel, S. R., Baronetti, G. T., Castro, A. A., Performance of Pt–Re/Al_2O_3 Catalysts with Different Radial Distribution Profiles React. *Kinet. Catal. Lett.*, 33(1), 143, 1987.
32. Oyekan, S. O., US Patent 4539307, 1985.
33. Oyekan, S. O., McClung, R. G., Moorehead, E. L., Optimized Pretreatment Procedures for Reforming Catalysts. *AIChE Spring National Meeting*, Houston, Texas, April, 1991.
34. Cole, E., Coone, J., Kravits, S., US Patent 3850846A, 1974.
35. Viswanathan, B., Sivasanker, S., Ramaswamy, A. V., *Catalysis: Principles and Applications*. Narrosa Publishing House, New Delhi, India, 2007.
36. Gaylord Chemical Company, Catalyst Sulfiding Overview, Bulletin 205, 2005.
37. Oyekan, S. O., Personal Notes – Pilot Plant Studies Conducted at Exxon Research and Engineering Lab, Baton Rouge, 1979 in Baton Rouge.
38. Shipley, K., *Case History Fitness-for-Service Assessment of Cyclic Catalytic Reformer Motor Operated Valves in the Petrochemical Industry*. The Equity Engineering Group, 2011.
39. Univ. of Michigan, Elements of Chemical Reaction Engineering Class Notes. Accessed July 26, 2017. umich.edu/~elements/5e/01chap/prof-reactors.html#seriessphere

40. MOGAS Industries, RSVP ASME 600/900/1500 Limited Class Data Sheet. Accessed August 22, 2017. www.mogas.com/en-us
41. Olsen, T., An Oil Refinery Walk Through, AIChE Chemical Engineering Progress, (CEP), May 2014. Accessed July 26, 2017. http://www.aiche.org/cep
42. Metso, Continuous Catalyst Regeneration—Catalyst Valves, https://www.metso.com/solutions/solutions-for-continuous-catalyst-reforming1/
43. Nptel, Catalytic Reforming Course. Accessed July 26, 2017. http://nptel.ac.in/courses/103102022/20
44. Mansuy, C., Parillard, J. M., Andrews, J., Bonnifay, P., Octanizing, An Effective and Flexible Approach to Catalytic Reforming, Paper AM 92-53 NPRA, 1992.
45. Lapinski, M. P., Wier, M. J., Leonard, L. E., Lok, K., Chapter 12.1, UOP CCR Platforming, in *Handbook of Petroleum Refining*, 4th edition by Meyers, R. A., McGraw Hill, New York, USA.
46. Lapinski, M. P., Zmich, J., Metro, S., Chaiyasit, N., Woransinsiri, K., Increasing Catalytic Reforming Yields, Petroleum Technology Quarterly Catalysis, 23–25, 2008. Accessed July 26, 2017. https://www.uop.com/?document=uop-increasing-catalytic-reforming-yields-tech-paper&download=1
47. Domergue, B., Le Goff, P., Ross, J., Octanizing Reformer Options. *Petrol. Technol. Q.*, 11(1), 67–73, 2006.
48. Hennico, A., Mank, L., Mansuy, C., Smith, D. H., Texas Reformer Designed for Two-Step Expansion. *Oil & Gas J.*, 90(23), 54–59, June 1992.
49. Le-Goff, P., Lopez, J., Ross, J., Redefining Reforming catalyst Performance: High Selectivity and Stability. *Hydrocarbon Process.*, 47–52, September 2012.
50. Oyekan, S. O., Personal Notes.
51. Le Goff, P., Domergue, B., Le Peltier, F., Joly, J. F., Catalytic Solutions for Improved Performance. *Petrol. Technol. Q.*, 8(3), 27–29, 2003. www.axens.net
52. Burch, R., Garla, L. C., Platinum-tin Reforming Catalysts: II. Activity and Selectivity in Hydrocarbon Reactions. *J. Catal.*, 71(2), 360–372, 1981.
53. Davis, B. H., Srinivasan, R., The Structure of Platinum-tin Reforming Catalyst. *Platinum Met. Rev.*, 36(3), 151–163, 1992.
54. Adkins, S. R., Davis, B. H., The Chemical State of Tin in Platinum-Tin-Alumina Catalysts. *J. Catal.*, 89, 371, 1984.
55. Sexton, B. A., Hughes, A. E., Fogler, A. E., *J. Catal.*, 88, 466, 1984.
56. Doolin, P. K., Zalewski, D. J., Oyekan, S. O., Catalyst Regeneration and Continuous Reforming Issues, in *Catalytic Naphtha Reforming* by Antos, G. J., Aitani, A. M., Marcel Dekker, New York, 2004.
57. Axens, Catalytic Reforming Continuous. https://www.axens.net/our-offer/by-market/petrochemicals/aromatics-complex/22/catalytic-reforming-continuous-aro.html
58. Hammel, E. J., CCR Platforming Turnaround, CUOP LLC 2001.pdf. https://dokumen.tips/documents/ccr-platforming-turnarounds-cuop-llc-2001pdf.html
59. Galvin, T., Paper by Ganapati, M., Ding, R., Mooley, P. D., Modular Construction of Catalyst-Regen Unit Saves Time. *Costs, Oil and Gas J.*, 98(25), June 2000.
60. MOGAS Valves, Continued Catalyst Regeneration, Benefit of MOGAS Valves. www.mogas.com/en-us/resources/media-centre/public/documents/application-notes/application-note-continued-catalyst-regeneration
61. Hase, B., Vickery, V., Radford, R., Mercury in Hydrocarbon Streams. Sampling/Analysis Methods, Exposure Monitoring, Equipment Decontamination & Waste Minimization. Paper AM-12-22. *APFM Annual Meeting*, San Diego, California, March 2012, 1–11.
62. Wilhelm, S. M., Liang, L., Cussen, D., Kirchgessner, D. A., Mercury in crude Oil Processed in the United States (2004). *Environ. Sci. Technol.*, 2007.
63. McClung, R. G., Oyekan, S. O., Sulfur Sensitivity of Pt/Re Catalyst in Naphtha Reforming. *AIChE Spring National Meeting*, New Orleans, March 1988.

64. McClung, R. G., Oyekan, S. O., Kramer, R., Reformer Feedstock Pre-treatment: The Liquid vs Vapor Phase Sulfur Removal Processes. *NPRA (Now AFPM) Annual Meeting*, San Francisco, 1989.
65. McClung, R. G., Reformer Operations Improved by Feed Sulfur Removal. *Oil and Gas J.*, 88(41), October 8, 1990.
66. Hawkins, G. B., Naphtha Sulfur Guards. https://www.slideshare.net/GerardBHawkins/naphtha-sulfur-guards

7 Catalyst and Process Management Challenges

7.0 INTRODUCTION TO CATALYST AND PROCESS CHALLENGES

The types of catalyst and process-related management challenges are highly dependent on the performance of the catalytic reformers, reliability of equipment, and operation of upstream hydrotreating units. Subpar performing catalysts could negatively impact the productivities of excellently operated catalytic reforming process units, as the catalysts can create catalytic and unit reliability challenges. For example, the performance of a poor catalyst in a continuous catalyst regenerative reformer could limit full realization of the superior advantages of that process unit over fixed-bed regenerative reformers. Inferior physical properties of a catalyst and performance could lead to increased rate of catalyst fines make, catalyst circulation challenges, and catalytic performance. Mediocre catalyst and catalyst reformer performance, if not addressed in a timely manner, could lead to a greater need to commit more resources to manage catalyst and reforming operations. Likewise, process management challenges due to inferior operational and mechanical capabilities and inexperience of oil refinery personnel could waste the time and resources that were expended to acquire the catalyst that was deemed "excellent." The objectives of this chapter are to cover the management of fresh catalysts from the plants where they are manufactured, through catalyst handling in the refineries, catalyst loading, catalytic reformer startup, and the performance of catalysts in catalytic reformers. Additionally, a review is provided of the specific case of low coke naphtha reforming challenges in continuous catalyst regenerative reformers it includes as an example and options for successful operations of low coke naphtha in continuous catalytic regenerative reformers.

7.1 PHYSICAL AND CHEMICAL CHARACTERISTICS OF CATALYSTS

Fresh catalysts are typically delivered to oil refiners as oxidized forms of platinum and promoter metals deposited on a gamma alumina support. The oxidized forms of platinum and promoter metals are formed during the final calcination of the catalyst in manufacturing process. Catalyst calcination is the step after the metal impregnation procedures to rid the catalyst of water and impart some additional strength to the catalyst pellets or extrudates. Metal impregnation is the process wherein appropriate precursor metal ions are deposited in a catalyst's alumina support and chloride.

Oil refining companies usually conduct catalyst activation after loading catalysts in reactors unless an oil refiner has elected to use "fully activated" catalyst that could have been provided by Honeywell UOP. Honeywell UOP provides these "pre-activated" platinum catalysts. In that case, Honeywell UOP will provide catalysts that can be used in naphtha reforming after catalyst loading with no further activation.

Otherwise, most catalytic reforming catalysts are delivered with the active metals as inactive oxides on gamma alumina support. Oxy-chlorination and reduction of the metal oxides on the catalysts are usually completed for Pt-Sn catalysts before starting naphtha reforming operations in CCR units. For platinum, platinum/rhenium, and other platinum bimetallic catalysts, metal sulfiding is an additional treatment in the activation process and is conducted after the oxy-chlorination and reduction steps within the reactors before initiating oil-in and naphtha processing.

Catalytic reforming catalysts that are selected are typically "good" catalysts as determined and ranked after catalyst evaluation programs that are conducted by some of the oil refiners who have the necessary analytical laboratory capabilities and technical expertise. A number of major global and national oil companies have catalyst evaluation programs that are capable of ranking oil refining hydroprocessing catalysts, as was indicated in Chapter 5. Oil refiners who do not have technical centers that can conduct such comprehensive catalyst evaluation programs usually rely to some extent on catalyst performance projection data in bid packages that are supplied by catalyst provider companies for making their catalyst selections.

For those with testing capabilities and manpower assets, physical characterizations are usually conducted to determine surface area, pore volume and average pore sizes, particle size, crush strength for all catalysts, and attrition resistance specifically for catalytic reforming catalysts for CCR process unit applications. Attrition resistance has become a critical characteristic of Pt-Sn catalysts used in CCR process units because some catalysts in the past exhibited high rates of catalyst fines, with consequent detrimental impacts in catalyst circulation systems and platinum management. Recent Pt-Sn CCR catalysts are exhibiting better attrition resistance characteristics relative to those of previous reforming catalysts. The surface area retention characteristic is important and is tracked, as surface area is one of the end-of-life criteria for catalyst replacements. Catalysts with subpar surface area retention characteristics usually exhibit poor chloride retention characteristics. Current CCR catalysts exhibit good surface area and chloride retention characteristics. Surface areas for CCR catalysts over time from process units are shown in Figure 7.1.[1] The three catalysts exhibited excellent surface area retention characteristics over time, as shown in Figure 7.1.

Similarly, the other critical physical properties of catalysts are tracked. The attrition resistance characteristic is an invaluable key physical property that can be correlated with the rate of catalyst fines make in CCR reformer operations. Care must be exercised to monitor makeup catalyst and maintain catalyst inventory for the CCR reactors and for the catalyst circulation and regenerator sections. Good analysis of catalyst attrition resistance data can provide information on the equilibrium state of the catalyst and indications of mechanical damage within the reactors. Fines data monitoring and assessments could also provide information on undesirable changes that are occurring in the catalyst state during the regeneration process.

Chemical characterization data for the fresh catalyst should include platinum, promoter metals, and chloride concentrations for the fresh catalyst and equilibrium catalyst samples over time. During the catalyst evaluation program, and before the selection of a catalyst, platinum, promoter metal, and chloride distributions should be assessed. Platinum, promoter metal, and chloride distributions are key catalyst

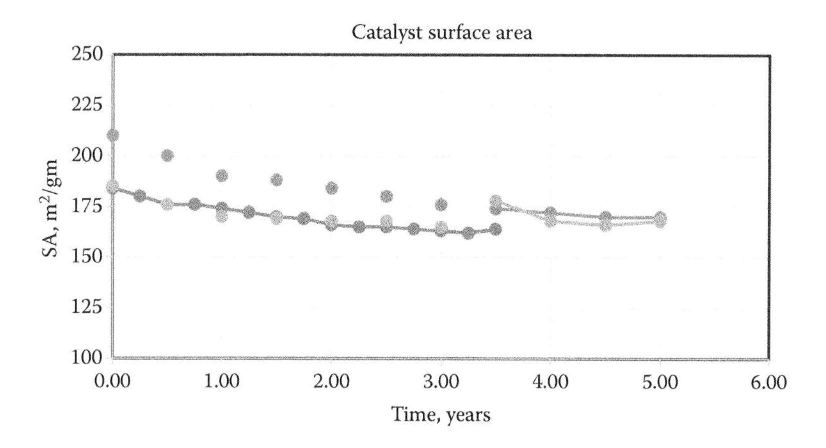

FIGURE 7.1 Surface area retention of CCR catalysts.

characteristics, as good distributions will lead to more efficient utilization of the metals and acid sites. If the distributions are subpar, especially for platinum, that could lead to some platinum losses if the platinum is deposited preferentially on the outer surface of the catalyst. Electron microscopy is usually the analytical methodology used to determine metal and chloride distributions. For clarification, it should be emphasized that metal distributions are indicative of concentration profiles for platinum, promoter metals, and chloride across a catalyst extrudate and they are different from what is usually referred to as metal dispersions, as reviewed previously. With the exception of metal and chloride distributions, several other key relevant chemical properties of catalysts should be determined and monitored regularly. The database of catalyst inspections can then be used to optimize activity and selectivity performances of catalysts in reforming operations.

In the same program, trace metal concentrations of the fresh catalyst will enable the oil refiner to establish base levels of potential metal contaminants that could accumulate over time during naphtha reforming operations. Such a database of metal contaminants would permit applying appropriate, timely responsive actions to mitigate negative impacts of contaminant metal accumulations on catalyst performance and profitability of reformers. Contaminant or trace metals that are usually monitored include iron, silicon, sulfur, sodium, nickel, cobalt, chromium, and molybdenum.

Based on the use of such a catalyst inspections database and continuous reforming process monitoring, rapid silicon deposition and loss of catalyst performance was observed in a catalytic reformer and that led to timely responses and eventual catalyst replacement for a 30-MBPD CCR.(1) Silicon on catalysts for two CCR process units over a period of 6 years is shown in Figure 7.2. In CCR 2, silicon was essentially at base or reference levels for the catalyst for many years and only increased briefly in the third year. For CCR 1, there was a rapid increase of silicon in the first year until the source of the reformer naphtha feed silicon was eliminated.

Iron is another metal contaminant that is recommended for monitoring in reforming catalysts, especially for Pt-Sn catalysts used in CCR reformers. Iron is

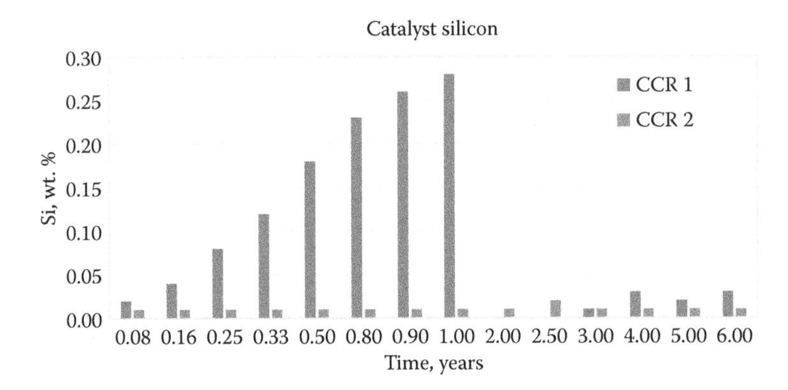

FIGURE 7.2 Catalyst silicon in CCR units.

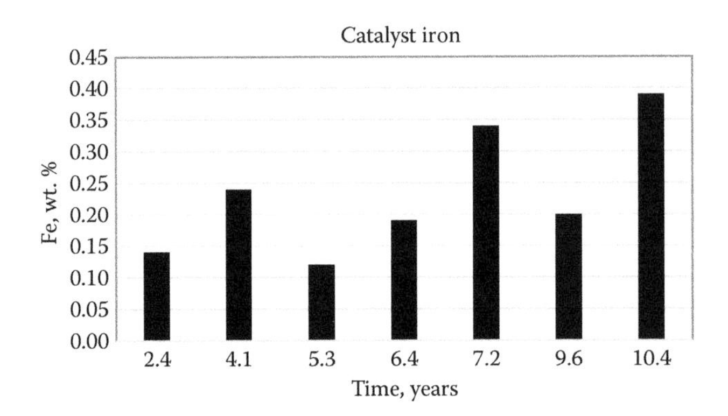

FIGURE 7.3 Iron concentrations on CCR unit catalysts.

essentially one of the elemental products of corrosion of materials of construction in tanks, piping, and equipment and typically accumulates on catalysts in catalytic processes in oil refineries. Iron concentrations for a number of catalysts in CCR units are shown in Figure 7.3.

Iron present in the range of concentrations that is close to the platinum concentrations of 0.25–0.30 wt. % on the catalysts could negatively impact the oxy-chlorination process and possibly lead to unsatisfactorily low percentages of platinum oxide and promoter oxide redispersions in the oxy-chlorination step in catalyst regenerations.

For fixed-bed catalytic reformers that have catalyst samplers, chemical characterizations should be conducted whenever opportunities arise, such as during catalyst regenerations, catalyst dump and screens, and catalyst replacements. In the case of CCR process units, a high frequency of scheduled physical and chemical characterizations of Pt-Sn catalysts is possible and recommended based on the oil refiner's analytical laboratories and manpower resources. Good chemical inspections data of catalysts in catalytic reformers usually enable oil refining engineers and operators to troubleshoot and make timely corrections to manage catalytic performance challenges in reforming process units.[1,2]

7.2 CATALYST HANDLING AND MANAGEMENT

In the reviews of catalytic reformer operations and catalyst management conducted in this book, it is evident that there should be good and timely coordination of the various functional groups or departments in an oil refining company. Catalyst handling requires planning and coordination of safe storage facilities and management of reforming catalysts and inert materials such as ceramic beads. It is also necessary to conduct safe and secure management of spent catalysts to ensure maximization of the value of platinum during the reclamation process. As a result, plant management, technical service, procurement services, the mechanical department, and warehouse staff should be fully involved in the management process. A number of factors concerning catalyst management should be reviewed in joint meetings of the relevant listed groups. Appropriate amounts of catalysts and inert materials should be ordered, and upon delivery, the drums should be inspected, properly labeled, and stored in a warehouse to ensure maintaining the integrity of the drums and contents. If feasible, catalyst could be transferred from drums into catalyst bins or containers that can hold up to 12 drums of catalyst or 2800–3000 pounds of catalyst. Use of large catalyst bins facilitates loading and unloading of catalysts and aids in achieving significant reductions in catalyst and platinum losses, and hence reduces process unit downtime and ultimate reformate and hydrogen production costs.[3]

Due to a number of factors and issues to consider with respect to reforming catalysts and the processes, best practice programs could be established to ensure flawless execution of key phases of planning, storage, and conduction of catalyst loading and unloading into reactors and platinum reclamation from dumped spent catalysts. It is vital in a rapidly changing workforce that experienced oil refining process technologists and corporate process consultants are available to develop and coordinate required best practice programs for oil refining companies. Companies that do not have such corporate or oil refining best practice programs for their process units may seek the services of proven global process oil refining consultants and companies.

7.2.1 Catalyst Unloading Principles

Effective planning and coordination of catalyst and catalytic reforming process unit programs for catalyst selection, catalyst quality assessments, and management have been emphasized through several sections of this book. Working collaboratively, mechanical engineering personnel, a turnaround team, procurement groups, and catalyst handling companies should develop and complete turnaround plans several months before the shutdown of the reformer for maintenance. Typically, such reviews by the turnaround team would lead to ensuring that necessary equipment such as cranes, temporary elevators, and scaffolding equipment are installed to permit sustained, trouble-free unloading of spent catalyst and loading of the fresh replacement catalyst. For detailed inspections of reactors and process units, experienced field service inspectors should be contracted and used. Honeywell UOP provides excellent field service inspectors who conduct detailed inspections of reactors and other critical equipment for catalytic reformer units, especially CCR process units.[4]

Oil refiners utilize different catalyst management programs that meet their specific oil processing objectives, which are highly dependent on the availability of resources and oil refining service companies. During catalyst unloading from fixed-bed regenerative reactors, some oil refiners unload their spent catalyst without conducting the coke burn process. It is recommended that regeneration of the catalyst be conducted to burn off catalyst coke, benzene, and sulfur before starting the catalyst dump due to the pyrophoric nature of coked catalysts. In some cases, and if catalyst replacement is the goal, the regenerated catalyst and inert ceramic materials used in the last naphtha reforming cycle are unloaded and sent later for platinum recovery and reuse or disposal of the inert materials. Spent unregenerated coked catalysts can also be dumped from reactors, and special care should be exercised to ensure that the spent catalyst and reactor contents have been cooled to ambient temperatures in nitrogen. The dumped catalyst and inert materials can be screened onsite or sent to a precious metals reclamation company for separation, screening, and platinum and promoter metal recovery. In the case that reuse of the catalyst is required, the regenerated catalyst and inert materials can be screened either onsite or offsite to recover catalyst, catalyst fines, and inert material. Recovered catalyst and makeup fresh catalyst can then be reloaded later with new inert ceramic materials. In all cases, it is recommended that catalyst bins be used, if feasible, to ensure good containment of catalysts and efficient handling of platinum and promoter metals. If catalyst bins are used, catalyst bins and contents should be weighed, labeled ,and securely and safely stored. A record should be kept of the weights of catalyst bins, and contents should be retained for reuse during platinum reclamation from the spent catalyst. The catalyst bins must be kept in a secure, safe, and dry environment.

For continuous catalyst regenerative process units, catalyst unloading should be handled with care and oil refiners should follow guidelines provided by process technology licensors. Honeywell UOP and Axens recommend unloading the spent catalyst in three stages. Free-flowing spent catalyst from the reactors should be the first batches unloaded, followed by the unloading of highly coked catalyst (heel catalyst) particles from the dead zones of the reactors, and finally the unloading of the catalyst contained in the regenerator and associated vessels.[(25)] Hammel reported results of a survey of oil refiners who indicated that for downflow reactors in UOP CCR units, heel catalysts averaged about 9 wt. % of the total reactor catalysts unloaded and heel catalyst percentages ranged from 2 to 27 wt. %. Heel catalysts are catalyst and inert catalyst particles that are in the "dead zone" sections of radial reactors. For upflow reactors, heel catalysts averaged 7 wt. % and varied from 1 to 18 wt. % from the survey of oil refiners reported in 1998.[(25)] As discussed in the section on catalyst unloading for fixed-bed regenerative reformers, good catalyst unloading procedures should be used and great care and safe practices should be utilized in the unloading of spent coked catalysts.

Heel catalyst particles in CCR reformer reactors usually consist of a mix of catalyst particles with coke in the range of 4–20 wt. %, and catalyst particles with higher coke are denser. Segregation of the dumped catalyst particles can be accomplished via length and density grading to recover equilibrium catalyst with less than 6 wt. % coke. Recovered catalyst can be reloaded into CCR reforming reactors and safely reused. Special precautions should be taken at startup of the catalyst regeneration section during the CCR catalytic process unit startup. Porocel Corporation provides a density grading segregation service that has saved oil refiners millions of dollars in catalyst replacement and management costs.[(25–27)]

7.2.2 Catalyst Loading Principles

Catalyst loading specifics vary considerably depending on whether the catalytic reforming reactors are cylindrical, radial, or spherical. Oil refining operators are advised to review and follow the details of the pertinent catalyst loading instructions. It is appropriate and beneficial to follow a good turnaround plan established by the turnaround team. When the reactors are opened, preliminary inspections should be conducted to get a good assessment of the undisturbed state of catalyst beds. At the completion of catalyst unloading, reactors and internals are inspected in greater detail to determine the scope of mechanical work that may be required. After completion of recommended mechanical work and repairs, final inspections of the reactors and other vessels are conducted in great detail to ensure that mechanical work and repairs have been completed as per the recommendations of the process technology licensors and inspection team. The final inspection should be thorough to ensure that debris and other external objects are not left in the reactors and other opened vessels, as such items would cause flow distribution problems. Catalyst loading is then conducted by using catalyst loading diagrams for the reactors. The goal is to load fresh catalyst properly and safely with full expectation of reliable and profitable reforming operations after startup of the units. A most important key condition is that catalysts should not be loaded during wet environmental conditions, as the hygroscopic catalysts would absorb water, leading to potential damage of catalyst during startup. For fixed-bed reactor systems, layers of bed support material consisting of different-sized inert ceramic balls are usually loaded as per catalyst loading diagrams. The layers of inert ceramic balls in the bottom section of the reactor serve as filter and containment barriers for catalyst within the reactors. Catalyst is loaded on top followed by layers of inert ceramic balls at the top to provide uniform flow distribution of hydrocarbon compounds and hydrogen through catalyst beds. A number of companies offer innovative reactor internals at the top that provide better flow distribution while minimizing the dead space at the top of the catalyst bed.(5,6) The Texicap reactor internal shroud is promoted by Axens for use in fixed-bed reactor systems, and the company states that oil refiners can realize cycle length improvements, more efficient coke burning, and lower platinum inventory in the reactors.(5)

CCR reactor systems such as UOP stacked reactors and Axens side-by-side reactors use inert catalyst spheres that are loaded in the dead zones of reactors, and inert catalyst spheres are ideal as they do not accumulate catalytic coke. If cokeless inert catalyst spheres are dislodged and carried out of the reactor with equilibrium catalyst, they would not cause catalyst and equipment damage in the catalyst regeneration sections.(1,4)

In the case of CCR process units, catalyst is loaded into reactors into vessels in the catalyst regeneration section and the catalyst regeneration section after the completion of a thorough inspection to ensure that debris is not left in the sections before loading catalyst. The use of catalyst bins obviates the need to use so many drums containing plastic sacks of catalyst during loading. One catalyst bin holds as much catalyst as 12 drums of catalyst. Potential problems such as plastic sacks left during catalyst loading that could block catalyst flow and cause a variety of catalyst circulation problems are eliminated by the use of catalyst bins.

The stacked reactors of the UOP CCR platformers can be loaded either via a reactor-by-reactor loading procedure through the manway of each of the reactors or the entire reactor stack-loading procedure after loading inert catalyst spheres in the dead zones of the reactors. Alternatively, catalyst is loaded via a large loading hopper and through the reduction zone at the top of first reactor. The catalyst flows through the first reactor to the second and to the last reactor through catalyst transfer pipes. Catalyst loading rates based on the use of catalyst drums and reactor stack procedures are usually faster than those of reactor-by-reactor loading and the rates are further increased by the use of catalyst bins that can hold up to 12 drums of catalyst.(25) Since catalyst particles are falling freely through the reactors, a higher rate of catalyst fines formation may occur and higher elutriation gas rates are recommended initially after startup of the platformer to remove catalyst fines. It is beneficial to maximize catalyst fines removal at startup so as to eliminate catalyst circulation challenges.

Two other catalyst loading operations in reactors are worth reviewing. The first involves the use of two parallel reactors in fixed-bed semiregenerative reformers operating with one common outlet for both reactors. This is not a common reactor configuration for fixed-bed semiregenerative catalytic reformers. Catalyst loading for such parallel reactors must be conducted with the great care to achieve similar catalyst loading and reactor pressure drops. Achieving this delicate balance in catalyst loading for the parallel reactors in fixed-bed catalytic reforming should lead to good, reliable reactor performances and catalyst regenerations.(1)

The second is loading of catalyst to achieve a higher catalyst loading density. Dense catalyst loading is the description given to the processes and technologies that enable the loading of a higher amount of catalyst in a fixed reactor volume. It is typically used by oil refiners for their fixed-bed reformer reactors and other hydroprocessing units. Basically, via a catalyst loading concept referred to broadly as catalyst-oriented packing, more catalyst particles can be loaded into reactors and this provides increased flexibility to run greater naphtha throughputs at constant reformate severity. Dense catalyst loading technologies could enable the oil refiner to load 15%–25% more catalyst into their reactors.(7–10) Crealyst provides a good diagram of its Calydens catalyst loading as a change is made from catalyst sock loading to dense loading as shown in Figure 7.4.(10)

7.3 CATALYST MANAGEMENT IN CATALYTIC REFORMING OPERATIONS

It is important to manage catalysts in catalytic reforming operations in a proactive manner so as to address changes in catalyst performance and process unit reliability challenges in a timely manner. To successfully accomplish catalytic process management goals, the oil refiner should have a laboratory with excellent analytical instruments and possibly pilot plants to generate the required data for timely and effective troubleshooting of deviations from expected catalytic performance. The laboratory should also be staffed with experienced personnel for supporting the analytical requirements for crude oil acquisition, oil refining in crude and process units, product blending, and quality control for finished oil-refined products. In addition, the laboratory or research and development center should have analytical and staffing capabilities for meeting an assortment of sampling frequencies that could include daily, weekly, monthly, and quarterly catalyst and hydrocarbon stream

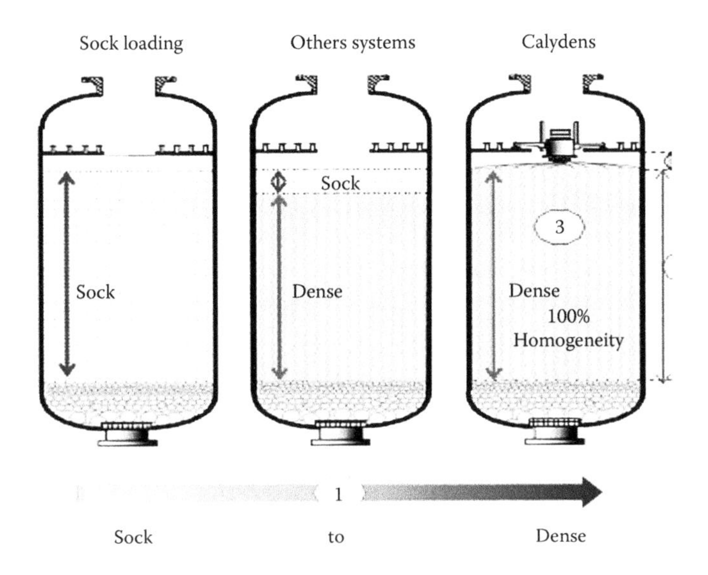

FIGURE 7.4 Catalyst sock and dense loadings.
From Crealyst, What's Dense Loading? http://www.crealyst.fr/whats-real-dense-loading/.[10]

samples for a variety of process units. Oil refining companies that do not have such assets need to partner with catalyst and process technology suppliers or external laboratories that can meet some of their analytical requirements. Many oil refining operators have clearly determined that when it comes to operating an oil refinery with a variety of processing units, it is beneficial to have an extensive array of data.

To manage literally thousands to millions of real-time data points on a frequent basis, data have to be organized, archived, and made available in useable formats for relevant personnel throughout the oil refining system and remotely to global corporate centers and offices. These capabilities are usually provided through plantwide information systems referred to as plant information management systems. One of the many key capabilities of a PIMS is that it can average, filter, and provide timestamped data; generate combinations of calculated values; and aggregate process data.[11,12] It can also provide timely visual depiction of real-time flow rates, temperatures, and compositions if appropriate analytical devices are connected to the PIMS.

The referenced assets such as laboratory and analytical support systems and staff and plant information management systems are required for effective and timely troubleshooting of process units. Data and information assembled from crude oil and oil fraction acquisitions, oil refining process units, oil refining operations, finished oil products, and laboratory analytical databases are usually readily available for use in the operations of top-tier oil refining companies. For fixed-bed catalytic reformers, the capabilities listed are of great use in operating the units and for making decisions for turnaround frequency and catalyst replacements. The demands for timely daily and weekly data required for operating CCR reforming units reliably and profitably are much greater than those for fixed-bed catalytic reformers. That is because three separate sections of the CCR process units, namely the reactors, catalyst circulation, and catalyst regeneration, must be monitored diligently such that timely changes can be made to correct for undesirable operations and performance deviations. In order to effectively use the amount of data collected for crude oil qualities, purchased naphtha, and naphtha from downstream units, that set of data is usually used with other specific key performance indicators (KPIs) of process units. For an example, and as in the case of a CCR platformer, the other key performance indicators are usually composed of those for the reactors, catalyst circulation, and regeneration sections. It is suggested that critical sets of key performance indicators should be established for crude distillation refinery process and CCR process units.

7.3.1 Key Performance Indicators and Unit Monitoring

Oil refiners' process unit monitoring programs operate at different levels of sophistication. The effectiveness of proactive monitoring of process units and refineries is highly dependent on an oil refiner's assets, corporate culture, and technical and business expertise. Reliability and profitability achieved in the operations of process units are dependent on the breadth and depth of expertise of various contributing personnel from the different departments of the oil refiner company. The successes of unit monitoring programs are dependent on the degree of use of automation in operating units, effective utilization of PIMS, effective utilization of predictive process models that provide guidance on expected process units' performances, and especially product selectivity and performance stability.

Profitability and reliability of an oil refinery are a function of its oil refining complexity as defined by the Nelson complexity index. As reviewed earlier, the complexity of the refinery is defined in terms of whether it is a topping, hydroskimming, catalytic cracking, or coking configuration. It is also useful if the oil refiner has access to external hydrogen supply for its hydroprocessing units and multiple similar process units for processing flexibility. In addition, it is beneficial if the oil refiner has capable, effective analytical and quality control laboratory support teams.

These capabilities and assets enable an oil refiner to run a variety of crude slates containing conventional crude oils, nonconventional bitumen-derived oils, and shale oils and to meet market changes and demands for fuels and environmental regulations.[14] In addition, oil refiners should have good established mechanical maintenance programs with metrics, for example, mean time between failure (MTBF), for key equipment such as heat exchangers, pumps, and heaters.[15,17,18]

Key performance indicators should be established for all key functional operating units of a refinery starting from the crude and vacuum distillation units to all the key downstream processing units. Oil refiners should identify useful key performance indicators that could be used in the monitoring of their process unit performances and apply appropriate data derived from key performance indicators for optimizing their refinery operations for greater reliability and profitability. In general, key performance indicators are used in the assessments of manpower utilization, equipment utilization, safety compliance within a refinery and within companies, process unit product yields and utilizations, and refinery reliability and profitability.

At this stage, it is necessary to digress in order to review and emphasize the flexibility and benefits of having several crude distillation and processing units in oil refineries of greater than 225 MBPD crude oil distillation capacity. Basically, the idea of operating multiple crude oil distillation and processing units is referred to as having multitrain operations of crude distillation and processing units. Multitrain refining operations provide much-needed flexibility and assurance for continuous oil refining, productivity, and profitability for oil refiners. The construction of a single-train oil refinery could provide modest initial capital costs savings during its design and construction; however, massive oil refining productivity and profitability losses may have to be sustained due to any outage of the crude distillation unit or one of the key process units for extensive routine maintenance of a major equipment. This loss of production could also be exacerbated by delays in even planned outages for turnaround maintenance of the single crude oil distillation unit or one of the key process units. Conventional wisdom developed from over 100 years of oil refining now lead oil refiners to add flexibility with any opportunity for either refinery capacity upgrades or construction of a grassroots refinery by exploiting multitrain oil refining strategies where applicable. Multitrain operations make it feasible and convenient for continuous oil production by oil refiners who have such process unit operational flexibilities. Such multitrain assets are necessary for enhancing the competitiveness and profitability of an oil refiner.

For catalytic reforming units and other process units, key performance indicators are selected as subsets from relevant data from a company's laboratory information management systems and plant information management systems. They are usually selected based on applying process technology expertise to identify the relevant key performance indicators for the various key sections and operations of catalytic reformers. The following key process variables, key performance indicators, and conditions are recommended for selection for periodic reviews at different frequencies for catalytic reformers:[(1,2)]

- Naphtha feed: API gravity; feed composition; naphtha N+2A; feed boiling range; key feed contaminants such as sulfur, nitrogen, silicon, iron, and other metals.
- Process conditions: Charge rate, LHSV, WHSV, pressure, reactor temperature, WAIT, WABT, H_2/HC molar ratio, recycle gas rate, recycle gas purity, recycle gas water, recycle gas chloride and recycle gas H_2S.
- Reactor characteristics: SOR WAIT, reactor temperature deltas, total delta temperature and reactor pressure drop, amount of catalyst in individual reactors, total amount of catalyst, number of catalyst cycles, percent catalyst of the total in each reactor.
- Catalytic performance: C5+ octane (RON), C5+ liquid (vol. %), H_2 production (SCF/B), H_2 purity, (%), total aromatics (vol. %), naphthenes (vol. %), benzene (vol. %), Reid vapor pressure of reformate, reformate research octane number, reformate motor octane number, catalyst life, activity measures, light end products, light end product ratios, process unit utilization factor, material balance, energy consumption rates, average number of days for a catalyst cycle.

In addition, for CCR catalytic reformers, the following KPIs can be added:

- Catalyst circulation system: Catalyst circulation rate, days per catalyst cycle, catalyst inventory turnover number per month and per year, other relevant circulation data as per process technology licensor, rate of catalyst fines make, catalyst fines distribution data, catalyst makeup rate, number of catalyst regenerations, total catalyst inventory in system, gas and hydrogen flow rates, properties of the gases.

- Spent catalyst: Carbon, chloride, contaminant metals, sulfur.
- Regenerated catalyst: Carbon, chloride, sulfur, silicon, iron and other metal contaminants, platinum dispersion, surface area, attrition resistance data, sodium, crush strength, porosity, chloride retention characteristics.
- Regenerator: Necessary data on the physical state of the screens.
- Regenerator combustion zone: Combustion zone gas inlet temperature, maximum burn zone temperature, burn zone gas oxygen concentration, pressure, catalyst circulation rate, intermediate zone (section between combustion and chlorination zones) temperature.
- Regenerator chlorination zone: Temperature, oxygen concentration, pressure, organic chloride addition rate, catalyst circulation rate.
- Drying zone: Gas flow rate, temperature, catalyst circulation rate.
- Reduction zone: Temperatures, gas flow rates, reduction zone temperature delta.
- Regenerator cent gas HCl and VOC reduction technology: Caustic scrubber key monitoring variables, treated regeneration vent gas HCl and VOC.
- Regenerator vent gas HCl and VOC reduction technology: Chlorsorb; contact Honeywell UOP if the technology is licensed from them.

In addition to the aforementioned list of relevant key performance indicators that could be selected for monitoring catalytic reformers, especially CCR reforming units, it could also be necessary to monitor key properties of crude oils, unconventional oils, and individual oils in the crude slate. In addition, the expected naphtha properties of the individual oils may be useful during troubleshooting of performance deviations of catalytic reformers. Chemical composition and properties of constituent naphthas in reformer feeds from coking, hydrocracking, and catalytic cracking units as well as those from opportunity naphtha fractions are usually excellent sources of useful data for making decisions for performance improvements of catalytic reformers.

Catalyst inspection data are critically important, as they can provide troubleshooting leads with respect to loss of catalytic performance that could be due to feed quality changes. Feed quality changes such as either an increase in high end point naphtha or more paraffinic feed could lead to lower yields of reformates and hydrogen and increased coking. Good catalyst inspection data could also provide leads with respect to either possible catalyst poisoning via entrained metal contaminants in the naphtha feed or an inactive catalyst as a result of ineffective catalyst regeneration process. Catalyst data are also vital for making daily and weekly decisions on how to optimize the operations of a CCR process unit.

After establishing specific refinery key performance indicators for process units, it is also beneficial to benchmark the oil refiner's process units' operations, performances, and utilization factors relative to those of other refiners. An environment and platform where these comparisons can be independently accomplished is through the services offered by the Solomon Associates company to oil refiners.[13] Oil refiners who have good proven capabilities for collecting performance data from their oil refining units usually benefit from their participation in Solomon Associates surveys, as performance and reliability data for participant process units are shared in summary reports.

7.4 TROUBLESHOOTING FOR PROCESS PROFITABILITY

The first requirement for effective troubleshooting is to have access to timely relevant data or a database with necessary information to assist in resolving operational and performance challenges of a process unit or manufacturing plant. Possessing necessary expertise with respect to the specific process unit, as in this case, of the catalytic reformer and other oil refinery process units and operations expertise is highly beneficial.

In order to meet and provide timely troubleshooting of catalytic reformer operational, performance, and reliability challenges, a roadmap with which to navigate to resolve and make recommendations is required. Elements of that roadmap could include the following steps listed below:

- Define the operating problem
- Determine possible causes
- Develop methods of resolution
- Evaluate and correct systems after making changes

There are a variety of catalytic performance challenges that can arise in the processing of heavy naphthas in catalytic reformers. Foremost are lower yields of reformates and hydrogen and higher light-end product yields. The other common key catalytic challenges include increased rate of catalyst deactivation due to increased rate of coke make, loss of catalyst activity, golden color of reformates, and so on. Additionally, there could be reliability challenges due to fouling and ammonium salt deposition in the product separation systems, especially in the debutanizer.

For continuous catalyst regenerative reformers, due to the incorporation of catalyst circulation and regenerations, additional challenges could include poor and non–steady state catalyst circulation, high catalyst fines make, catalyst losses via reactor, and regenerator screens. Additionally, reactor internal damage and high pressure drops in reactors can cause significant catalytic reformer productivity losses. The performance of the catalyst regeneration section is critical for CCR reformers. Challenges may occur due to poor catalyst regeneration operations, thermal damage to the catalyst and regeneration units, poor regeneration vent gas system operations, and many others factors.

The first step is to define the problem and review and assess performance deviations from those of normal operations. Secondly, it is suggested that use be made of available resources of other refinery personnel, PIMS, LIMS, and KPI data to conduct detailed analysis and reviews and identify potential factors causing performance deviations. When analysis and reviews are completed, solutions and recommendations can be offered for implementation.(1,2,4)

Knowledge accumulated from oil refining and catalytic reforming books, technical courses, and oil refining process experiences should enable engineers, operators, relevant staff, and process technologists in refineries to resolve many of the catalytic reforming challenges. Oil refiners whose personnel could be stymied in their troubleshooting efforts should seek assistance from their catalyst and technology provider and oil refining process consultants. Timely resolution of some of the problems listed could easily save an oil refiner between 10 to 30 million dollars per year. The negative impact on reformer profitability of subpar process unit performance and production losses could be quite high if catalytic performance debits and reliability challenges of a catalytic reformer are not addressed in a timely manner. The actual magnitude of the cost of lost opportunity is also highly dependent on the size or reforming capacity of the catalytic reformer.

Some general investigations of performance changes in the catalytic reformer could start with the following questions.

- What is the deviation from expected performance?
- What are the changes from a reference steady-state performance or reliability state?
- What unit upsets have occurred and when?
- Any naphtha feed changes? Any change in crude slate? Any change in naphtha constituents, if some are coming into the naphtha hydrotreater from other downstream process units such as coking or hydrocracking units?
- Has any equipment showed signs of failing?
- Do we have sufficient process unit and catalyst inspections data to suggest that the problem is due to the performance of the catalysts?
- Have we conducted reference performance test runs and do we have data from those test runs that can assist in our work?
- If a CCR process unit, are there catalyst circulation challenges?
- Any regeneration section problems? Chlorination zone challenges?
- Review all pertinent, timely unit and catalyst data that could provide good leads for resolution.

7.4.1 High Pressure Drop in Continuous Catalyst Regenerative Reformer Reactors

One key processing challenge from the list of numerous naphtha processing and catalyst regeneration challenges in Section 7.4 is the situation wherein some of the typical free-flowing catalyst particles within the annular section of the reactors are stagnated in CCR reactors during reforming operations. In radial reactors, as described previously, hydrocarbon and hydrogen flow in a radial direction from the outer screen to the inner screen and catalyst particles flow vertically downward under gravity, as shown in Figure 7.5. In such radial reactors, the scallops or special optimizer type scallops constitutes the outer screen and the surface of the center pipe is the inner screen. In some CCR reforming units, a catalyst flow problem can occur that can manifest via either gradual or rapid increase in the pressure drop in one of the reactors. Usually, stagnation of some of the catalyst particles occurs within the lead reactor.

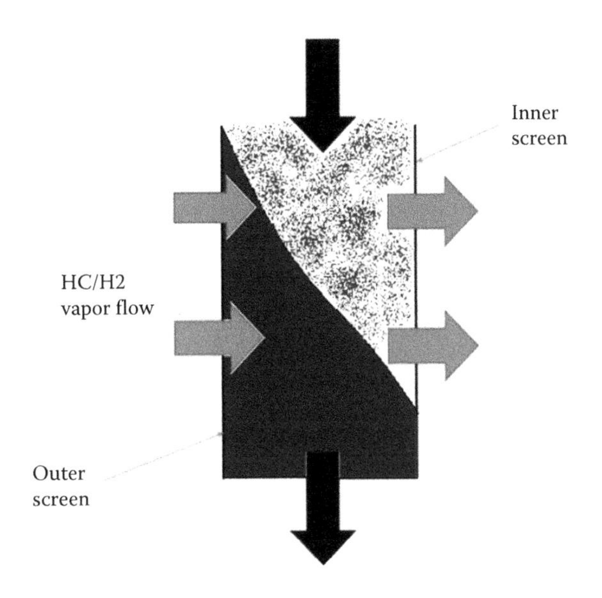

FIGURE 7.5 Catalyst pinning in radial reactors of CCR process units.

What is usually presumed to be occurring is that the radial flow of hydrocarbons and hydrogen "holds" or "pins" substantial amounts of catalyst particles against the inner screen. In radial reactors, pressure drops are usually in the range of 2–4 psig, and if there is restriction in the steady-state flow of catalyst, the reactor pressure can increase and exceed 15 psig. In the special case of localized stagnation of catalyst particles due to catalyst pinning, there are potentially several causative factors. The most probably cause is that the throughput of hydrocarbons and hydrogen is so much higher in the radial direction that it is holding or "pinning" catalyst particles against the inner screen. The second is that there is restricted flow of hydrocarbon and hydrogen through the inner screen that is caused by screen plugging with catalyst chips and fines.

Based on the nature of the catalyst pinning, some of the stopgap measures of changing processing and catalyst regeneration operations may alleviate the catalyst particle stagnation problem. For process unit upsets of this type in UOP-licensed CCR units, Honeywell UOP should be contacted and can provide a number of recommendations for managing and possibly alleviating the high pressure drop catalyst stagnation problem. Typically, preliminary suggestions are to maintain the reactor pressure drop at less than 15 psig for a short period of operation until the high pressure problem is rectified.[1,4] Operating below 15 psig pressure delta is usually maintained through a substantial reduction in charge rate and increased recycle gas rates. For detailed recommendations, oil refiners are advised to consult with their process licensors.

However, in addition to significant loss of productivity, there are other major operational risks and challenges. With the static catalyst particles against the inner screen, significant amount of coke deposition occurs due to the longer residence time, and coke on some of these catalyst particles could exceed 20 wt. %. A number of operational strategies have to be applied to the reactor section, catalyst circulation, catalyst fines management, and catalyst regenerator systems. Some critical concerns are that some of the highly coked catalyst particles could be dislodged, commingled with lower free-flowing catalyst with coke in the 4–6 wt. % range, and carried to the regenerator section.[16] Nonuniform coked particles could cause severe catalyst and equipment damage in the regeneration section. If the pinning problem is severe and prolonged, catalyst circulation would be negatively impacted. If the pinning problem persists unchecked, significant damage could also occur to the center pipe or inner screen, leading to loss of catalyst containment within the reactor.

Eventually, the CCR reforming unit would have to be taken out of service to dump catalyst from the reactors and possibly from the catalyst circulation and regenerator sections in order to conduct thorough inspections of the reactors and conduct necessary repairs. During the period before a process unit outage for turnaround and repairs, sustained and diligent management of the CCR process unit reactors, catalyst circulation, catalyst fines, and regeneration systems are required. Based on the inspection reports on findings on catalyst stagnation, an investigation is usually conducted to determine possible causes and ensure that the factors that led to the high pressure drop challenges are not repeated. In one severe case of such a catalyst pinning challenge, the inner and outer screens were found to be extensively covered with catalyst fines and particles due to cold hydrocarbon liquid that was entrained in the hydrogen gas used for metal reduction in the catalyst reduction hopper. The wet catalyst and entrained cold hydrocarbons entered the lead reactor. Cold hydrocarbon-laden catalyst shattered in the hot zone in the lead reactor. There was extensive coverage of the inner and outer screens (scallops) with catalyst chips and fines. Recommendations from the investigation team included eliminating hydrocarbon liquids from the reduction gas via greater attention to coalescer operations and possible eventual replacement of the coalescer.[1,4] It should be added that catalyst stagnation, pinning, or catalyst holding that leads to high reactor pressure drops could cause major loss of naphtha reforming production, profitability loss, and significant CCR unit turnaround maintenance costs.

7.4.2 Managing Low Coke Naphtha Reforming in Continuous Catalyst Regenerative Process Units

Continuous catalyst regenerative process units provide major advantages over fixed-bed reforming units due to CCR reforming units' capabilities of maintaining Pt-Sn catalysts within the regime of high catalyst activity and product selectivity for the production of reformates, hydrogen, and aromatics. These performance benefits are maintained by the use of excellent catalyst circulation systems that incorporate efficient catalyst fines removal sections to ensure that reactor and catalyst regenerator screens are not plugged and "steady state" catalyst circulation is maintained. As discussed previously, CCR process units operate on a variety of paraffinic and naphthenic naphtha feeds and at pressures as low as 35-psig, high reformate octane severities, low H2/HC molar recycle rates, and acceptable high endpoints for naphtha feeds. Since the rates of catalytic coke make are expected to be high at such high reforming severity operations, continuous catalyst regeneration is required. It is absolutely necessary to design for adequate catalyst regeneration capacity and catalyst circulation rates to permit operating in an optimum range of spent catalyst coke. For CCR naphtha reformer operations, some studies have shown that the range of catalyst coke should be 4–6 wt. %.[1,4,16] Though the burning of catalysts with higher catalytic coke is possible, great precautions have to be exercised to operate in a mode so as to preclude the use of high oxygen in the chlorination zone. Thus, for steady-state operations in the reactors and regeneration systems of CCR units, spent catalyst coke should be in the range of 4–6 wt. %.[1,4,5]

Gasoline/diesel pricing differentials are favorable for diesel production relative to gasoline, and that has led oil refiners to cut deeper into the naphtha fraction and transfer some of the heavier, high-boiling naphtha fraction to feed kerosene and diesel hydroprocessing units. US gasoline and highway diesel prices for the different production regions are as shown in Figures 7.6 and 7.7, respectively. The average on-highway diesel price was $2.472 per gallon for the United States on July 3, 2017, and that was about 5 cents more than the price a year before. The average gasoline price for the United States was $2.260 dollars on July 3, 2017, and that was 3 cents less than one year before.

Price differentials show an attractive average advantage of 21 cents a gallon for diesel relative to gasoline. Such price differential advantages for diesel have led to lower endpoint naphtha feeds of 330–350 F versus previous ranges of 370 F to 400 F. The second factor that has impacted CCR reforming operations for reformate and motor gas production is the significant lowering of reformate severity due to oxygenate blending in gasoline to meet environmental and fuel regulatory

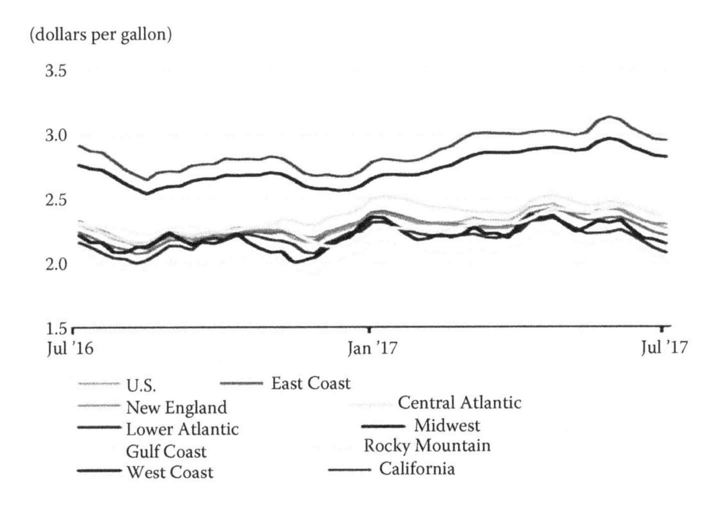

FIGURE 7.6 US gasoline prices for the different regions 2016–2017.
From US EIA, Petroleum Marketing Monthly, Petroleum and Other Liquids, July 3, 2017. www.eia.gov/petroleum/marketing/monthly/.[19]

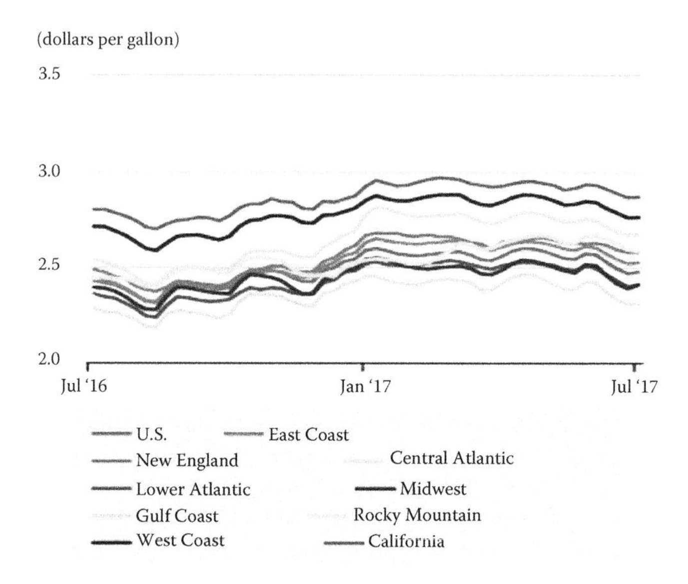

FIGURE 7.7 US diesel prices for regions 2016–2017.
From US EIA, Petroleum Marketing Monthly, Petroleum and Other Liquids, July 3, 2017. www.eia.gov/petroleum/marketing/monthly/.[19]

compliance. In the case of the United States, ethanol is now currently mandated at the 10 volume percent level in gasolines. Gasoline components produced in oil refineries are now referred to as blendstock for oxygenate blending (BOB) and of the blendstock, the United States produces mostly the conventional blendstock for oxygenate blending (CBOB) gasoline. This gasoline blendstock for regular gasoline has a road octane number or antiknocking index of 84. It should be noted that the road octane number is the average of the research octane number and motor octane number of the gasoline or blendstock component. The blending of the 10 volume percent ethanol boosts the CBOB from 84 to 87 RON for regular gasoline.[30]

While ethanol and other oxygenate blending is good for creating cleaner environments and less dependence on fossil fuels, especially crude oils, it can create special challenges for the operations of CCR process units. These challenges for CCR units negatively impact oil refiner planning due to the significantly reduced reformate octane severity and lower hydrogen production. Though higher reformate yields could be realized at lower reformate severities, the current combination of low reformate severity octane targets and low naphtha feed endpoints lead to much lower rates of catalytic coke formation in CCR reforming reactors. Due to concerns with gasoline octane giveaway, some oil refiners are also operating at low reformate severities. While some oil refiners employ strategies that require operating the CCR process units at high reformate severities, others are limited to low reformate severity operations. Low CCR reformer severity operations and low naphtha boiling feed ranges lead to very low rates of catalytic coke make in the reformers. Thus, spent catalyst coke for the CCR process units could be in the range of 2.0–3 wt. %. Since catalytic coke produced in the reactors is too low to permit safe and optimal operation of the catalyst regeneration process, catalyst regeneration systems are idled and sometimes operated intermittently with attendant risks for catalyst and regenerator hardware damages.

Frequent intermittent operations of the catalyst regeneration sections lead to poor CCR naphtha reforming operations with consequent significant losses in the production of reformates and hydrogen. Reformate yields (C5+ liquid) are compared for a period of continuous catalyst regeneration and to similar data for a period of intermittently (nonregenerating) operated catalyst regeneration system. Reformate yield loss at about 100 octane severity was 5 volume percent and hydrogen yield loss was 10–15 percent. C5+ liquid yield data for the two regenerating and nonregenerating operations are compared in Figure 7.8.

Catalytic performances of CCR process units that are not operating with steady-state catalyst circulation and continuous catalyst regenerations are in fact worse than those of fixed-bed regenerative reformers. Intermittent Pt-Sn catalyst regenerations lead to operations with catalyst particles that are not optimally active with respect to the contributions of the metal and acidic isomerization functionalities. It was estimated in 2011 that a refinery operating with a 20-MBPD CCR platformer lost about $10 million per year for a 2 volume percent loss in reformate yield and 10–15 volume percent loss of hydrogen production. Annual revenue losses for oil refineries operating with 30- to 85-MBPD CCR reforming process units are expected to be in the range of $15 million to $40 million per year. In addition, intermittent catalyst regenerator operations can lead to increased catalyst chips and fines that can plug reactor and regenerator screens. Furthermore, intermittent use of the catalyst regenerator could lead to major damage to the catalyst and regenerator equipment.

Honeywell UOP is addressing the low coke naphtha reforming challenges through its focus on engineering solutions that can be applied by modifying or revamping

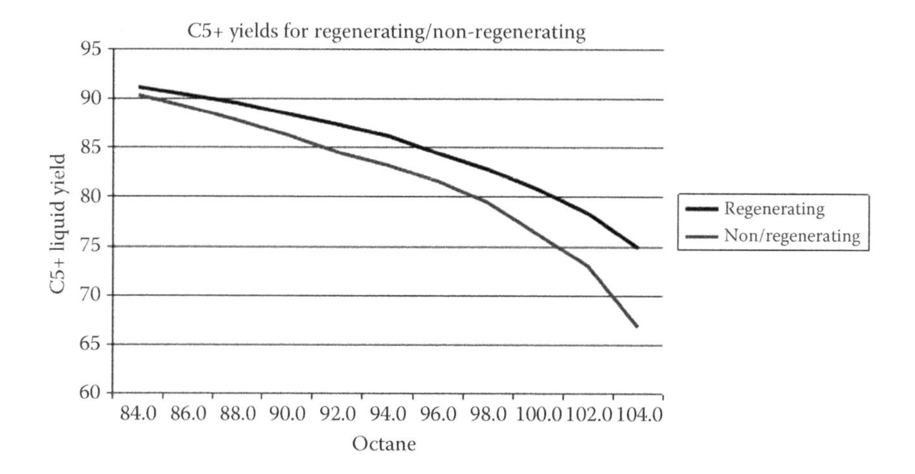

FIGURE 7.8 Continuous catalyst regenerating and nonregenerating systems.

TABLE 7.1
Comparative CCR Reformer Data

2012 CCR Data	Jan 8–Jan 12	Feb 2–Feb 17
Charge Rate, NBPD	19.9	18.5
Feed Sulfur, wppm	0.28	0.92
Feed N + 2A, vol. %	58.9	61.7
Reactor WAIT, F	929	927
Spent Catalyst Carbon, wt. %	2.7	5.7
Coke Rate, pounds/hour	37.6	43.5
C5+ Octane, RON	92.3	91.5
C5+ Yield, vol. %	85.1	85.8
H2 Production, SCF/B	1023	1082

Source: Oyekan, S. O., US9371493, Low Coke Reforming, June 2016.[20]

existing catalyst regeneration systems of their licensed CCR reforming process units. According to UOP's claims, the engineering solutions permit safe and efficient regenerations of spent catalysts with coke that is as low as 2.0–3 wt. %. Oyekan focused on the principles of the catalytic reforming process, specifically on catalytic reactions. His work has led to low-cost innovative solutions for increasing catalytic coke make for CCR reforming process units operating on low endpoint naphtha feeds and at low reformate severities.[20–23] Comparative performance data from a 20-MBPD CCR platformer operating with a design catalyst circulation rate of 1500 pounds per hour for the use of sulfur additive, taken from one of the US patents, is shown in Table 7.1.[20]

The comparative data for reforming operations for a reformate octane severity target of 97 RON show a significant carbon increase from a low 2.7 to 5.7 wt. % and higher reformate yield and hydrogen production due essentially to continuous use of the catalyst regeneration system. Spent catalyst carbon of 5.7 weight percent, which was much higher than the 3 weight percent lower limit of coke, permitted steady-state continuous use of the catalyst regeneration section.[20] In this type of CCR reforming process units, the regeneration sections are usually idled when spent catalyst carbon is less than 3 wt. % to eliminate damage to catalyst and catalyst regeneration systems as discussed.

7.4.3 Catalyst Regeneration Challenges

The three main catalytic reforming technologies are differentiated by a number of factors such as process conditions, catalyst circulation, and some specific modifications of the standard catalyst regeneration procedure to accommodate equipment, time, and critical process safety constraints.[24,29] Recommended catalyst regeneration procedures should be followed as closely as feasible and should only be modified by knowledgeable process technology experts who fully understand the possible impacts of the changes that they are making, even more so when the changes are temporary.

The changes should be carefully reviewed by the turnaround management team before implementation. Two examples are reviewed for catalyst challenges that can occur during fixed-bed catalyst regenerations and preventative actions that should have been taken to prevent the occurrence of the events. A more general challenge associated with intermittent catalyst regeneration operations in CCR process units is provided in Section 7.4.3.3.

7.4.3.1 Flow Distribution Challenges in a Fixed-Bed Regenerative Reformer

Fixed-bed semiregenerative catalytic process units have to be regenerated on a frequency of every 6–18 months or more depending on oil refiners' strategies and refinery plans. Since disturbances occur in the catalyst beds during naphtha processing and catalyst regenerations, a general recommendation is that catalysts in the reactors should be dumped and screened to remove catalyst fines and dust after three catalyst regenerations. The screened catalysts and makeup fresh catalyst can be reloaded and used in succeeding naphtha reforming cycles. Periodic dumping of catalysts from the reactors, screening, and reloading required amounts of catalysts into reactors will improve flow distributions of hydrocarbons and hydrogen during naphtha processing. The catalyst dump and screen also improve regeneration gas and chemical flow during catalyst regenerations. Flow distribution problems in reactors usually lead to poor catalytic performance during naphtha reforming and high catalyst coke burning rates. Despite the high burning coke rate, localized "dead zones" of catalyst may be impervious to gases during coke burns. Detailed analysis of naphtha reforming performance data could provide some indication of flow maldistribution problems. However, a better indication of flow distribution problems in catalytic reforming reactors is identifiable from coke burning characteristics data during catalyst regenerations. If there are indications of long, slow coke burns or "tailing," that may be indicative of poor distribution of gas flow in an affected section of the reactor catalyst bed.

One of the fixed-bed catalytic reformers in an oil refinery in Ohio in1996 exhibited evidence of excessive poor flow distribution and nonuniform coke burns, and that led later to major catalyst damage. After one coke burn and catalyst dump, it was discovered that catalyst particles were fused and the catalyst bed of the reactor was heavily damaged. The recommended catalyst dump and screen of reactors had not been done for some years due to competing refinery operations and turnaround priorities. Due to poor flow distribution of the regeneration gas and localized burning of highly coked catalyst particles within the bed, excessively high temperatures were reached in excess of 2300 F, which led to alumina phase transitions and the fusing of catalyst particles. It was determined that chiseling to remove the damaged catalyst was going to be time consuming, and a catalyst handling company later expertly removed the fused catalyst via the use of strategically located mini-dynamites. Despite our concerns, the catalyst handling company conducted an excellent operation of breaking up the dense, highly coked catalyst bed section and did not damage the reactor and its reactor internals. Catalyst replacement costs, extended turnaround time, and lost production costs due to maldistribution of hydrocarbons and gas flows in the small reformer were estimated to be greater than $2 million of losses for that oil refiner in 1996.[1,2]

7.4.3.2 Eliminate Delays after the Catalyst Oxy-Chlorination

In the catalyst regeneration process reviewed in some detail in Chapter 6, it was emphasized that catalysts were very active at certain stages, such as after the oxy-chlorination and metal reduction stages. It was also suggested that catalysts could be kept safely in the reactors after the coke burn and catalyst sulfiding stages to provide time for oil refiners to complete required maintenance work for fixed-bed semiregenerative reformers during catalyst regenerations. However, sometimes, other high-priority parallel events may compete for the time of key personnel and technical management in oil refineries and this may cause some delays after completion of a key critical stage of the catalyst regeneration process.

At the completion of oxy-chlorination for redispersing platinum oxide and promoter metal oxides of Pt-Re and other bimetallic reforming catalysts and after the coke burn stage, catalysts are usually reduced in hydrogen and sulfided. In 1996, at a refinery in Pennsylvania, personnel and management could not proceed readily after the oxy-chlorination stage for a high-rhenium Pt-Re catalyst and a decision was made to safely fill the reactors with nitrogen and maintain them at room temperature. Unfortunately, the planned delay of less than 12 hours turned into 3 days. Later, after completion of catalyst reduction, sulfiding, and startup, the activity and selectivity performance of the regenerated catalyst were poor. A catalyst supplier at the time offered the grim assessment that the catalyst metals had suffered "irreversible nitrogen reduction" and replacement catalyst should be ordered immediately. Umansky and Oyekan collaborated on a short troubleshooting effort that included conducting reactivation of catalysts and platinum redispersion studies in the laboratory.[1,2] Realizing from the data from our studies that we had only reagglomerated the Pt-Re catalyst, a second regeneration and activation of the catalyst was conducted and the regenerated catalyst exhibited full activity and selectivity performance recovery. The brilliant, timely troubleshooting efforts of Umansky and Oyekan led to revenue loss savings of over $3 million. Over $1 million was saved by not purchasing fresh catalyst and incurring the associated platinum management costs. The second catalyst regeneration was based on a modified catalyst regeneration procedure that incorporated a longer oxy-chlorination time and more conducive metal redispersion conditions. Modifications to the catalyst oxy-chlorination and reduction stages led to significant improvements in the naphtha reforming performance of that Pt-Re catalyst.

7.4.3.3 Transition Smoothly between Black and White Burns in Continuous Catalyst Regenerative Regenerators

Fixed-bed reforming catalyst regeneration procedures, reviewed in Chapter 6, contain distinct primary burn, secondary or proof burn, and oxy-chlorination stages. These stages are used for conducting desired catalyst reactivation activities in an environment of progressively higher oxygen and chloride. In the case of CCR reforming process units, the first step is to purge the spent catalyst from the reactors of entrained hydrocarbon liquid to facilitate circulation through pipes, equipment, and hoppers and finally to the regeneration tower. The regeneration tower consists of two major zones, the carbon burn and chlorination zones. A simple diagram of a CCR process unit with atmospheric regeneration section is provided in Figure 7.9.[28]

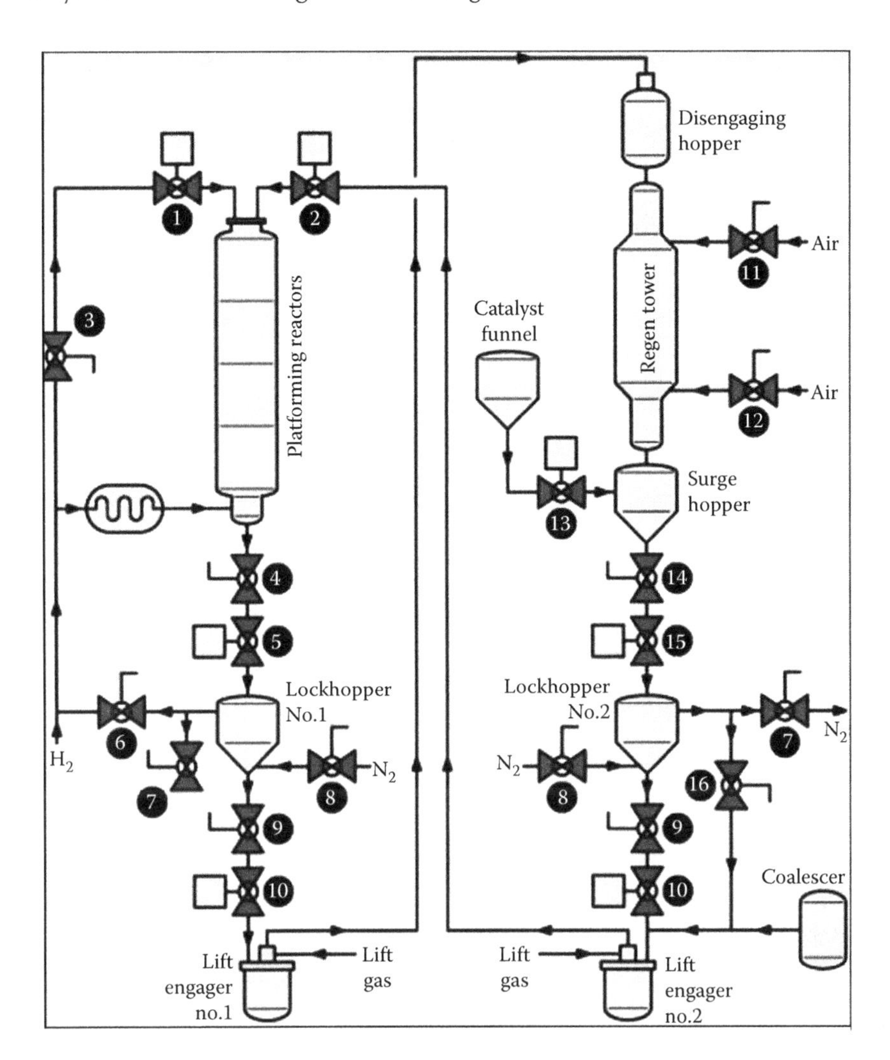

FIGURE 7.9 CCR process units with atmospheric regeneration section. From MOGAS Valves, Continued Catalyst Regeneration, Benefit of MOGAS Valves, www.mogas.com/en-us/resources/media-centre/public/documents/application-notes/application-note-continued-catalyst-regeneration.[28] (Courtesy of MOGAS Industries, Inc.)

Catalyst regeneration systems offered by a CCR process technology licensor are designed either for operations at atmospheric pressure or, for the more recent ones, for 35-psig operation. In steady-state operations, catalyst flows from the combustion zone operating in low oxygen concentrations of 0.8–1.3 volume percent to the chlorination or oxy-chlorination zone, which is operated in air. Temperatures in the chlorination zone are usually stipulated to be in the range of 950–980 F to maximize platinum redispersion. Combustion, chlorination, and drying zones are located within the regeneration tower, as shown in Figure 7.9. The combustion and chlorination zones are part of a structure within the regeneration tower consisting of inner and outer

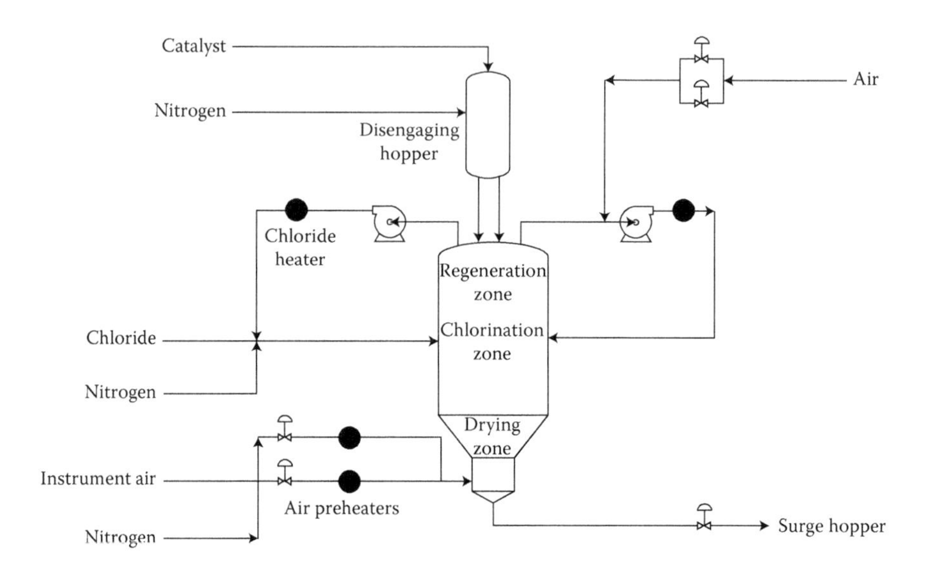

FIGURE 7.10 Diagram of a CCR process unit regeneration section.
Courtesy of Doolin, P. K., Zalewski, D. J., Oyekan, S. O., Catalyst Regeneration and Continuous Reforming Issues, in *Catalytic Naphtha Reforming* by Antos, G. J., Aitani, A. M., Marcel Dekker, 1995.(24)

screens and an annular space between the screens. Catalyst flows vertically downward under gravity. A simplified diagram of the regeneration tower and regeneration and chlorination gas systems is shown in Figure 7.10.(24)

After the coke burn in the combustion zone, catalyst moves from that zone and enters the chlorination zone. Oxygen in nitrogen at levels in the range of 0.8–1.3 volume percent is used in the regeneration or combustion zone. Nitrogen only or air is used as required in the chlorination zone. The mode of operating the regenerator for coke burn and nitrogen only in the chlorination zone is referred to as the "black burn." It is usually stipulated that coke on catalyst from the combustion zone should be less than 0.1 weight percent before operating the chlorination zone with air. When nitrogen is replaced with air, the chlorination zone is then being operated for redispersion of agglomerated platinum and other promoter metal oxides on the catalyst. The mode of operation to permit optimizing platinum dispersion is referred to as "white burn." For normal steady-state white burn operation of the catalyst regenerator, both the coke burn and chlorination zones would have oxygen in different concentrations in the zones, as stated previously.

Critically, if the catalyst coke is higher than 0.1 weight percent, air is not introduced into the chlorination zone. Honeywell UOP differentiates two critical phases of catalyst regeneration operations involving the use of different concentrations of oxygen. Coke burn operations in the regeneration tower with only nitrogen in the chlorination zone are referred to as black burn operations, as indicated earlier. The regeneration tower with operations with the use of oxygen concentrations of 0.8–1.3 volume percent in nitrogen in the regeneration zone and 15–21 volume percent oxygen in nitrogen in the chlorination zone are referred to as white burn operations.

If highly coked catalyst enters the oxy-chlorination zone from the combustion zone, and if oxygen is present in the chlorination zone, coke burn is reinitiated and can lead to high temperature excursions in the batch of catalyst. Temperatures higher than 2700 degrees can be reached. Such temperatures lead to major catalyst damage and possible melting of the Inconel alloy C-278 metallurgy used for some of the regeneration tower screens.

Refinery operators and engineers have to ensure that residual coke on catalyst after the combustion zone is less than 0.1 weight percent before operating in white burn steady state. The nomenclature is highly instructive and recommends staying at black burn conditions with essentially no oxygen in the chlorination zone until regenerated catalyst coke is less than 0.1 weight percent. That is achieved by ensuring that the key valve for introducing air into the chlorination zone, fondly referred to as either the "idiot" or "unemployment" valve, is not opened to introduce air to the chlorination zone until catalyst carbon analysis shows conclusively that catalyst carbon or coke is less than 0.1 weight percent. It is probably preferable to label the valve the "profitability," valve and other appellations can be offered. Thus, in Figure 7.9, the valve identified as 12 is the profitability valve, according to Oyekan. As long as the chlorination is operating in black burn mode, the regenerated catalyst coke is representative of the degree of decoking that can be achieved in the burn zone for the catalyst coke load and operating conditions in the carbon burn zone. The profitability valve should only be turned on and air introduced into the chlorination zone when several regenerated catalyst samples indicate definitively that the regenerated catalyst coke is less than 0.1 wt. %. Starting at that point and thereafter, after the introduction of air, the regeneration tower is then considered to be operating in white burn. Oxy-chlorination zone conditions can then be optimized to achieve good redispersion of the oxides of platinum and promoter metals for the catalyst.

The profitability valve is an appropriate label for the valve that needs to be opened to make the transition from black to white burn in CCR process units. Effecting this transition from black to white burn prematurely has led to damage to catalyst particles, high losses in catalyst surface area, and major damage to regeneration section screens. Severe damages to the screen have led to bent or "banana"-shaped screens and melting of Inconel alloy C-278 metallurgy. Banana curved screens and other screen damage usually require regeneration section downtimes for significant repairs and replacements of inner screens.

Reliable, safe operations of the regeneration tower can be achieved by diligently following catalyst regeneration operation procedures and ensuring that operational and technical personnel have the necessary expertise and experience to make appropriate timely decisions during catalyst regenerations. The use of a good analytical laboratory support group to provide timely inspection results for spent and regenerated catalyst carbon and chloride data is most crucial for making decisions with respect to safe and reliable operations of catalyst regeneration sections. Smooth transitions from black to white burns can be reliably completed via strict adherence to catalyst regeneration procedures and continuous monitoring of the performance of the catalysts in the reactors and the regeneration section.

REFERENCES

1. Oyekan, S. O., Personal Notes.
2. Oyekan, S. O., Catalytic Naphtha Reforming Lecture Notes in AIChE Course on Catalytic Processes, in *Petroleum Refining* by Chuang, K. C., Kokayeff, P., Oyekan, S. O., Stuart, S. H., AIChE Today Series, 1992.
3. Hoover Ferguson, Catalyst and Chemical Containers, www.hooverferguson.com/products/catalyst-chemical-containers.html
4. UOP, UOP Field Operations Services. Accessed August 22, 2017. www.uop.com/?document=uop-field-operating-services&download=1
5. Domergue, B., le Goff, P., Ross, J., Octanizing Reformer Options, PTQ Q1, 2006. Accessed August 22, 2017. www.eptq.com
6. Shell Global Solutions, About Reactor Internals, July 21, 2017. http://www.shell.com/business-customers/global-solutions/refinery-technology-licensing/reactor-internals/about-reactor-internals.html
7. Petroval, A Company Built on Innovation. Accessed August 15, 2017. http://www.petroval.com/PETROVAL-Corporate.pdf
8. Wooten, J. T., Dense and Sock Loading Compared, Catalyst Handling Service. *Oil Gas J.*, October 12, 1998.
9. Cat Tech., Cat Tech Catalyst Services, http://www.cat-tech.com/cat-tech-services.html
10. Crealyst, What's Dense Loading? www.crealyst.fr/whats-real-dense-loading/
11. Dataparc, 3 Key Benefits of Plant Information Management Systems (PIMS), http://blog.dataparcsolutions.com/3-key-benefits-of-plant-information-management-systems-pims
12. INF Systems, Plant Information Management Systems, www.infsystemsgroup.com/our-solutions/1/plant-information-management-system-pims
13. Solomon Associates, Solomon Applications, Performance Monitoring Services—Profile II. www.solomononline.com/applications/portfolio-monitoring-services
14. Olsen, T., Schodowski, E., Improve Refinery Flexibility and Responsiveness, pages 87–89. *Process Optimization, Hydrocarbon Processing*, September, 2016. Accessed August 22, 2017. http://www.emerson.com/documents/automation/improve-refinery-flexibility-responsiveness-en-38250.pdf
15. Emerson Process, Heat Exchanger Health Monitoring for Improved Reliability, Performance and Energy efficiency. Accessed August 22, 2017. www.2emersonprocess.com/siteadministrator/PM central web document
16. Gyngazova, M. S., Chekantsev, N. V., Korolenko, M. V., Iyanchina, E. D., Kravtsov, A. V., Optimizing the Catalyst Circulation Ratio in a Reformer with a Moving Bed via a Combination of Real and Computational Experiments. *Catal. Ind.*, 4(4), 284–291, 2012.
17. Emerson Process, Petroleum Refining Pump Health Monitoring, http://www2.emersonprocess.com/en-US/industries/refining/applications/Pages/EAMPumps.aspx
18. Emerson Process, Petroleum Refining Fired Heater Solutions, http://www2.emersonprocess.com/en-us/industries/refining/applications/pages/eamfiredheaters.aspx
19. US EIA, Petroleum Marketing Monthly, Petroleum and Other Liquids, July 3, 2017. www.eia.gov/petroleum/marketing/monthly/
20. Oyekan, S. O., US9371493, Low Coke Reforming, June 2016.
21. Oyekan, S. O., Robicheaux, M. G., US937149, Mixed Additives for Low Coke Reforming, June 2016.
22. Oyekan, S. O., Rhodes, K. D., Newlon, N. K., US8778823, Feed Additives for CCR Reforming, July 2014.
23. Oyekan, S. O., Robicheaux, M. G., CA2822742C, Mixed Additives for Low Coke Reforming, December 2016.

24. Doolin, P. K., Zalewski, D. J., Oyekan, S. O., Catalyst Regeneration and Continuous Reforming Issues, in *Catalytic Naphtha Reforming* by Antos, G. J., Aitani, A. M., Marcel Dekker, 1995.
25. Hammel, E. J., CCR Platforming Turnaround, https://www.scribd.com/document/161881736/CCR-Platforming-Turnarounds-CUOP-LLC-2001-pdf
26. Porocel Corporation, Length and Density Grading. Accessed July 15, 2017. http://www.porocel.com/17-length_and_density_grading/
27. Blashka, S. R., Welch, J. G., Technology, Catalyst Separation Method Reduces Platformer Turnaround Costs. *Oil Gas J.*, 93(38), September 1995.
28. MOGAS Valves, Continued Catalyst Regeneration, Benefit of MOGAS Valves, www.mogas.com/en-us/resources/media-centre/public/documents/application-notes/application-note-continued-catalyst-regeneration
29. Galvin, T., on Paper by Ganapati, M., Ding, R., Mooley, P. D., Modular Construction of Catalyst-Regen Unit Saves Time. *Costs Oil Gas J.*, 98(25), June 2000.
30. Standlee, C., Raising Gasoline Octane—The Considerations. *MSTRS Meeting*, Abengoa, May 5, 2015.

8 Special Catalytic Reforming Topics

8.0 INTRODUCTION TO SPECIAL CATALYTIC REFORMING TOPICS

In addition to catalytic reforming reactions, catalysts, and catalyst regeneration systems, there are some other important catalytic reforming topics that merit review, such as strategies and technologies for reducing reformate benzene. Oil refiners had conveniently used benzene for meeting their gasoline octane targets, especially in winter months in the United States. Starting in January 2011, the US EPA Mobile Source Air Toxics (MSAT 2) gasoline fuel program stipulated an annual average gasoline benzene content standard of 0.62 volume percent for reformulated and conventional gasoline nationwide. Successful strategies and process technologies that are used for reducing gasoline benzene are reviewed in this chapter. Several related reforming process topics are also reviewed. The list of topics include opportunity oils and recommendations for processing them profitably and effective chloride management within catalytic reforming processing units and refinery reliability. Appropriate technologies for the minimization of regenerator vent gas hydrogen chloride and volatile organic compounds to meet US EPA Refinery Maximum Achievable Control Technology regulations are reviewed. Achieving high utilization factors in the operations of catalytic reformers and oil refineries is highly dependent on the experience of the oil refiner's manpower resources for operations, engineering, and unit and equipment maintenance. Experienced process technologists and refinery reliability specialists are vital resources. These experts usually coordinate and manage oil refining strategies and activities to ensure optimal operations of crude distillation units, catalytic reformers, and other refinery process units and auxiliary systems. One vital set of activities is the monitoring and maintenance of key equipment such as heaters, pumps, recycle gas and net gas compressors, and heat exchangers. A good oil refining company is usually well staffed with excellent support and management personnel for environmental and safety, human resources and development, plant operations, planning and product control, engineering, process safety management, technical service/technology support, feedstock acquisition and purchasing, and asset development/project management. Such oil refining companies usually have high-performance central research centers and dedicated quality control laboratories at each of their refineries for timely, daily analytical support for process units and for the characterization of oil fractions, unit products, and oil refined products. Process technologists and reliability specialists should lead in the management of all key aspects of refinery operations and maintaining excellent reliability of processing assets. Process technologists and mechanical specialists should also provide annual or biannual opportunities for the training of relevant refinery personnel by process areas or sets of mechanical assets. An oil refining company of the kind described can then more efficiently use the information and data derived from the oil refiner's laboratory information management system and plant information management systems and this book for optimizing the profitability and reliability of its oil refineries. As suggested several times in this book, oil refiners who are not adequately staffed should avail themselves of the services of oil refining consultants who can aid in elevating the competitiveness and profitability of their refineries.

8.1 BRIEF REVIEW OF OPPORTUNITY OILS AND REFORMING OPERATIONS

Many oil refining companies purchase opportunity crude oils due to their discounted low prices relative to reference conventional oils with the goal of processing the oils to enhance the profitability of their refineries. For conventional and some nonconventional crude oils, detailed crude oil assays provide oil quality data such as API gravity, sulfur content, total acid number, and oil fraction yields used in assessments of potential operational and reliability challenges that may arise from processing opportunity oils.[3,4] Often only basic physical properties of opportunity oils are provided by the seller at the time of purchase. Oil refining companies are usually concerned with the quality of oils with respect to potential processing operations and reliability challenges that could occur in refineries. These challenges could start at the desalters, progress in stages, and evolve in different forms through crude distillation towers to individual process units. Superimposed on potential operational and reliability challenges are those that could also occur due to challenges with blending chemically dissimilar opportunity oils. These are because there could arise some oil incompatibility issues in blending some oils with the typical crude oil slates of refineries.[5,6,8] "Opportunity oils," for the purpose of a broader discussion, has been expanded by the author to include oil fractions that are purchased from external sources in addition to plant-produced oil fractions. Three example cases of opportunity oils from several experiences of mine over the past 25 years are reviewed and suggestions are made for effective management and processing of opportunity oils.

8.1.1 Initial Experiences with the Processing of Shale Oil

Shale oils have effectively replaced a good percentage of light sweet crude oils that were imported annually by US oil refining companies. Due to increased production of shale or tight oil, US oil production is expected to surpass that of Saudi Arabia by 2020 with continued consequent economic and environmental benefits for the United States.[42] This review of a case of my initial experience with the processing of shale oil is not intended in any way to trivialize the economic and environmental benefits of tight oils. In fact, the global oil and gas positions of the United States have been enhanced due to increased tight oil production. It must, however, be stated that despite the easy availability and cost attractiveness of shale oils, processing large fractions of shale oil in crude slates can have significant negative impact on activity and product yields in naphtha processing and similar product selectivity moderations in the other hydroprocessing units of refineries. It is important to factor in and account for the possibilities of lower product yields before processing significant percentages of shale oils in refineries. Physical oil incompatibility and reliability challenges such as precipitation of asphaltenes from the other crude oils and fouling are now well known and understood as the concentration of shale oil increases significantly in crude oil slates of oil refineries. Some oil refiners are now incorporating separate crude distillation units for shale oil distillation. Other crude oil distillation units are now used for crude oil slates with low percentages of shale oil. Representative properties of shale and other light sweet crude oils are given in Table 8.1. Bakken and Eagle Ford shale oils contain more naphthas relative to those of WTI, LLS, and Qua Iboe crude oils.[6,7] Shale oils are similar to light sweet crude oils and differ with respect to having higher fractions of naphtha and less kerosene and diesel fractions.

About 7 years ago, shale oil was processed for the first time through one of the refineries and fixed-bed catalytic reformers that I managed for technical and operational excellence. The catalytic reformer is located in a 75-MBPD oil refinery. It is a mixed topping/cracking refinery with no hydrotreating processing units. As a result, the refinery was restricted to the refining of light sweet crude oils. As shale oil was introduced at a relatively high percentage in that refinery's crude slate, naphtha from the crude unit was fed directly to a fixed-bed catalytic reformer. Reformate yield even at low-severity operation of 80 octane number declined precipitously by a significant 4 volume percent for the10-MBPD fixed-bed catalytic reformer within a few weeks of operations. Hydrogen production also declined significantly by over 20%. Reformate and hydrogen production declines suggested that there was either a naphtha feed composition change or catalyst poisoning. As the reformer operated without the benefit of a naphtha hydrotreater upstream and was producing low-octane reformate, catalyst poisoning was the first item checked. On further review, and after discounting catalyst poisoning as the causative factor, it was determined that the reformer feed had become significantly more paraffinic with the introduction of an appreciable fraction of "opportunity" Eagle Ford shale oil in the crude slate, and that

TABLE 8.1
Properties of Shale and Light Sweet Crude Oils

	Bakken	Eagle Ford	WTI	LLS	Qua Iboe
API Gravity	41	47.7	40.8	35.8	37.6
Sulfur, wt. %	0.2	0.1	0.32	0.36	0.1
TAN. Mg KOH/gm	0.1	0.1	0.1	0.4	0.26
	Fraction, wt. %				
Light Ends, (C1–C4)	3.5	1.13	3.4	2.54	2.2
Naphtha, (C5–360 F)	36.3	37.1	32.1	17.2	28.3
Kerosene, (360–450 F)	14.3	11.93	13.8	14.8	16.9
Diesel, (450–650 F)	14.1	21.08	14.1	26.4	22.3
Vacuum Gas Oil, (650–1050 F)	26.1	24.21	27.1	27.3	25.1
Residual Fuel Oil (1050+ F)	5.2	4.47	9.4	7.2	5.2

Source: Wier, M. J. et al., Optimizing Naphtha Complexes in the tight Oil boom, UOP, 2014.[9]

led to significant reformate and hydrogen yield losses for the nonregenerative fixed-bed catalytic reformer.

It is now fully known that shale oil naphtha is highly paraffinic and, as a result, high percentages of shale naphtha in the reformer naphtha feed negatively impact activities and product selectivity of reforming catalysts. Product selectivity losses should also be proportional to the fraction of shale naphtha in the reformer feed.[9]

For the profitable processing of shale oil, the fraction of shale oil in the crude slate should be optimized so as to minimize negative impacts on products yields and catalyst life in process units. Secondly, it is important that all technical and operational staff be fully informed and briefed on the expected impact of shale oil as the percentage of shale oil increases in the crude slate of refineries. That would enable refining staff to make necessary operational changes such as reducing the fraction of shale oil in the crude slate and specific process variable changes in downstream process units. Such advance planning and briefing of relevant key staff would also enable oil refiners to optimize the percentages of constituent oils in the crude slates for their refineries. The use of good PIMS and LIMS process feed and product information is vital for timely operational corrections and optimization of refinery process units when processing opportunity oils.

8.1.2 Challenges in the Processing of Opportunity Bitumen

Bitumen from oil sands is now being processed globally in high-complexity cracking and coking refineries, and the quality of the unconventional oil determines the processing assets that are required for reliable and profitable refining. The quality of bitumen-derived oils depends on the degree of upgrading of the bitumen via coking and hydrocracking, and the quality of diluents such as condensate, naphtha, and light sweet conventional oils that are added to reduce the viscosity of the bitumen and permit transportation of the oil through pipelines. An *Oil Sands* article clearly described the processes and nomenclatures such as Dilbit, Synbit, and synthetic crude oils (SCOs) used to differentiate the type of bitumen oils that oil refiners are processing as opportunity oils in their refineries. Mixed products of a diluent or condensate with bitumen are referred to as Dilbit, as shown in Figure 8.1. Heavy sour oils with API gravity in the range of 20–22 and containing greater than 1 weight percent sulfur are typically referred to as Dilbit products.[10] Such opportunity oils are most likely to be more profitably processed in complex refineries that have cracking and coking configurations. Due to the high fraction of heavy components and high sulfur, bitumen oils are not suitable for processing in hydroskimming or simple conventional refineries and have to be upgraded via decoking and hydrocracking processes to produce synthetic crude oils with API gravity in the range of 30–35.

In a survey, Weber and Yeung broadly classified opportunity crudes into three groupings, namely heavy sour, extra heavy, and crude oils that contain total acid numbers that are greater than 0.5 mg KOH/gram. They then requested the participation of oil refiners in a survey and asked refiners to list the type of opportunity crude oils that they were processing. Oil refiners' responses show that they are processing a variety of Dilbit, Synbit, SynDilBit, and synthetic crude oils in their crude slates. The survey results on opportunity crude oils are as shown in Figure 8.2. Lower crude oil costs, oil

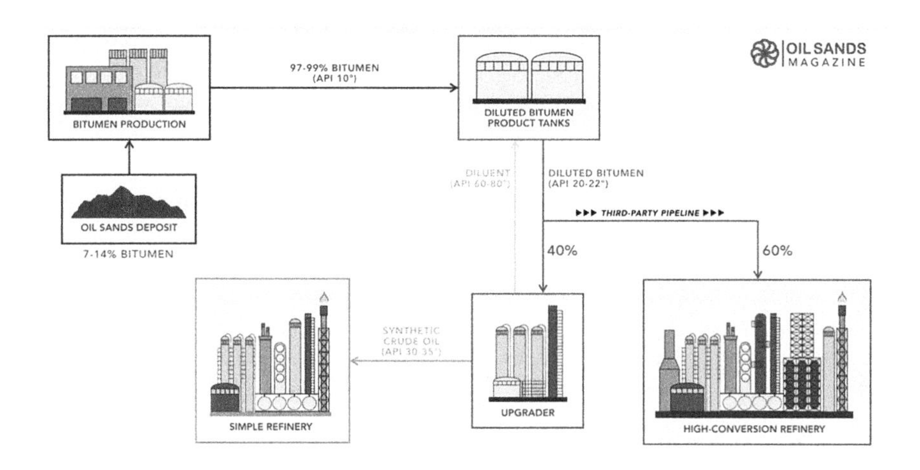

FIGURE 8.1 Bitumen, Dilbit, Synbit, and synthetic crude oil.
From Oil Sands Magazine, Products from the Oil Sands: Dilbit, Synbit & Synthetic Crude Explained, http://www.oilsandsmagazine.com/technical/product-streams.[10] (Courtesy of *Oil Sands Magazine*.)

accessibility, and supply stability are factors that are driving up the use of opportunity crude oils in refineries.[11] Additional drivers cited for processing opportunity crude oils in that survey included the need for higher diesel yields, higher propylene yields, and increased coking margins relative to cracking margins. Primary concerns and challenges with processing opportunity crude oils identified were increased rates of corrosion in refinery process units, fouling, high coke make, inferior product yields, higher energy demands, and incompatibility with other oils in crude slates. Oil refiners reported that crude oil quality variation was a concern, and that is expected from the description of how Dilbit, Synbit, and synthetic crude oils are generated from bitumen.

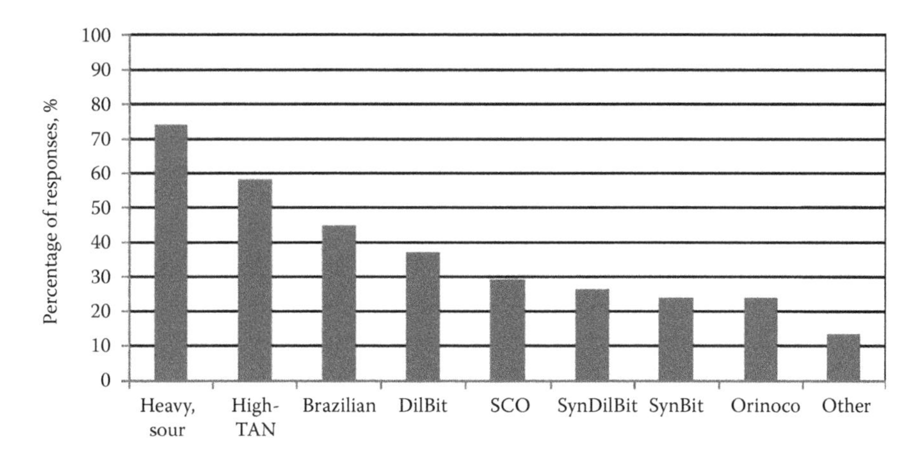

FIGURE 8.2 Opportunity crude oils processed in oil refineries.
From Weber, C., Yeung, S., Opportunity Crudes: To Process or Not to Process, *COQA Meeting*, October 27, 2011.[11]

Conventional crude oils show some quality variations, and quality variations are expected to be even more pronounced in Dilbit, Synbit, and synthetic crude oils. Western Canadian Select sold as an unconventional oil is composed of up to 20 heavy conventional oil streams, a nonupgraded bitumen from Alberta oil sands, an upgraded bitumen or synthetic oil, and a diluent or condensate. Western Canadian Select is expected to exhibit significant variations in physical and chemical properties because WCS batches essentially represent mixes of over 20 streams of vastly varying oil compositions and contaminants. Selected available properties of conventional and unconventional oils taken from a variety of sources are provided in Table 8.2.[12–19]

Hamaca Blend is a Dilbit from Venezuela. The properties of the Hamaca bitumen are 8.4 API gravity, 3.8 weight percent sulfur, and nickel and vanadium concentrations of 115 and 388 wppm, respectively.[20] Kearl is a heavy, high-sulfur diluted bitumen that is ideal for refineries with heavy sour crude oil processing capabilities. Kearl, as a synthetic crude oil, has an API gravity of 20.3, and contains 3.80 weight percent sulfur, making it an attractive coker feed.[21]

In 2012, Zuata oil, a Dilbit presumably from Orinoco in Venezuela was processed at 10 volume percent in a slate of 8 crude oils in a 270-MBPD crude distillation unit in one of the oil refineries that I supported. The resultant naphtha fraction from the crude unit

TABLE 8.2
Selected Properties of Conventional and Unconventional Oils

Crude Oil	API Gravity	S, wt. %	TAN	N, ppm	Ni, ppm	V, ppm
Maya	21.80	3.60	0.30	3200.00	53.40	298.10
WCS	21.00	3.50	0.80	2480.00	46.00	115.00
Zuata	19.40					
Zuata Sweet	30.10	0.12	<0.05		0.10	0.50
Escravos	33.40	0.17	0.06			
Bonny Light	35.40	0.24	0.27	1030.00	3.60	0.40
Hamaca Blend	26.00	1.55	0.70	2474.00	42.00	152.00
Zafiro Blend	30.00	0.25	0.77	2520.30	15.00	2.30
Kearl Blend	20.30	3.80	2.10	3261.30	47.50	128.40
Mars	29.60	1.98	0.46		15.00	41.90
WTI	40.80	0.24	<0.1	800.00	4.30	4.30
LLS	38.40	0.39	0.40		3.00	4.50
Brent	38.30	0.40	0.03		1.00	6.00
Basra Light	34.10	1.10	0.06	1040.00	8.50	30.00
Cold Lake Blend	19.20	3.95	1.10	3709.30	59.90	157.00

Source: BP Global Energy Trading, Detailed Crude Oil Assays Accessed July 10, 2017, http://www.bp.com/en/global/trading/crude-oil-and-refined-products/crudes/americas/mars.html; Scribd, Crude Oil Properties and Quality Indicators, Uploaded by Felipe Grigorio. Accessed July 15, 2017, https://www.scribd.com/document/320233512/CRUDE-OIL-PROPERTIES-AND-QUALITY-INDICATORS-pdf.[12,19]

Note: WCS, Western Canadian Select, is an unconventional oil.

was hydrotreated and combined with hydrocracker product naphtha that had been treated with a sulfur guard bed to eliminate residual mercaptans and other sulfur compounds.

Combined naphtha was processed in a 75-MBPD CCR unit. The first indication of a potential problem was an increase in the reformer feed sulfur from a 0.2 wppm level to 0.5 wppm. A review of the naphtha hydrotreater operations revealed that the hydrotreater feed sulfur concentration had increased from a reference 550 ppm to 1000 ppm and feed silicon concentration was increasing with the processing of 10 volume percent Zuata in the crude slate. Based on the need to protect the catalytic reformer and naphtha hydrotreater catalysts, operational changes were made at the source by reducing the percentage of Zuata dilbit in the crude slate to manageable levels consistent with optimizing the combined operations of the naphtha hydrotreater, CCR platformer, and other refinery assets that were negatively impacted by the high sulfur, silicon, and other metal contaminants from the Zuata oil.

Possible hypotheses for the high silicon, sulfur, and nitrogen in the naphtha derived from the Zuata were either that an unhydrotreated coker naphtha or another diluent containing silicon, sulfur, and nitrogen was used as the diluent in the Zuata dilbit. Another factor could have been the degree of upgrading achieved for the synthetic crude oil used as a key constituent of the Zuata dilbit. Preliminary oil refinery economic evaluations had indicated significant profitability benefits for processing the Zuata oil due to its attractive price differential relative to the average price of the other crude oils in that refinery's crude slate. In order to optimize refinery profitability and price benefits associated with the processing of opportunity oils, it is necessary to monitor relevant equipment and operations in the oil refinery. The active, continuous, dynamic monitoring should cover operations from the desalter through the crude unit and the other process units. Ultimately, the result of such an active monitoring could include managing the rate of use of the opportunity oil in crude slates so as not to negatively impact product yields, catalyst life in process units, and the reliability of equipment and process units. Additionally, oil refiners can also conduct comprehensive audits after processing opportunity oils at the different percentages of processing in the crude slate. Oil refiners can quantify "actual" realized benefits after deducting the costs of lower product yields and changes in process units' throughputs. Other factors that could also be compensated for are the impacts of opportunity oils on the life of catalysts in downstream processing units, cost of chemical treatment for minimizing corrosion, and fouling in the refinery associated with acquisition of opportunity oils.

8.1.3 Guidelines for Refining Opportunity Oils Can Minimize Losses

The last case example in this section is one emphasizing the need for the use of effective guidelines for trouble-free processing of opportunity oils. It is being provided to emphasize the need to ensure that guidelines are followed once they are established. A good program to ensure that guidelines are being followed is to provide or institute regular annual or biannual process technology and operational meetings. At the informational meetings, relevant technical and operational data, process concepts, and guidelines can be shared with the appropriate refinery personnel. The meetings can be used as a forum to share guidelines and best practice principles. In addition, during the recommended process technology and operations meetings, key process technologies, associated equipment, technology and equipment suppliers, and advancements in process technologies could be discussed. Information exchanges are necessary for timely incorporation of necessary innovative process technology advancements and reliability enhancements to practice. These meetings can help to foster a common resolve to achieve high utilization factors for process units and improve the profitability of oil refineries. Management goals such as the sharing and comparing of performance data for process units and benchmarking similar catalytic reformers and other processing units against competitors can be realized through meeting participants. Specific, dedicated corporate functional process and equipment groups of the kind that are alluded to are used in oil refining companies. These groups are sometimes referred to as centers of excellence and help in leveraging their technical and operational excellence in the training of relevant personnel and enhancing the profitability and reliability of oil refineries.

About 20 years ago, one of my company's oil refineries suffered significant production losses as a result of an upset due to accelerated deposition of polymerized products in a naphtha hydrotreater reactor. That naphtha hydrotreater typically produced hydrotreated naphtha feed to meet stringent reformer feed sulfur, nitrogen, and metal contaminant specifications for a 65-MBPD cyclic catalyst regenerative reformer operating at 99 reformate severity and low pressure. Platinum monometallic catalysts were used at that time in the 65-MBPD cyclic catalyst regenerative reformer. Fortunately, the refinery had a second naphtha hydrotreater and a 75-MBPD cyclic catalytic reformer operating at 98 octane severity to meet some of the demands of reformate and hydrogen of the refinery.

A cost-attractive opportunity naphtha had been purchased for processing and was segregated as per established best practice in a tank. Based on past oil processing experiences and best practice, opportunity naphthas were to be segregated from straight-run or virgin naphtha produced from refineries' crude distillation units and only processed with straight-run naphtha through naphtha hydrotreaters after reviewing necessary feed analysis for sulfur, diolefins, olefins, nitrogen, feed composition, and feed boiling ranges via the ASTM D86 method. In addition, it was stipulated in the best practice principles that opportunity oils should only be processed in low concentrations in a mix with straight-run naphtha through the naphtha hydrotreater. Naphtha processing and unit operations should be monitored closely as required until unit operations are deemed to be satisfactory and in steady state based on feed and product specifications. It was also stipulated that segregated naphtha should not be processed in "block," that is, as the only naphtha through the naphtha hydrotreater.[1,2] Despite these guidelines, an "attractive" opportunity naphtha was inadvertently processed in "block" through one of the naphtha hydrotreaters. Block processing implies that the opportunity naphtha was greater than 60% of the naphtha processed through a process unit.

With the block processing of the opportunity naphtha, catalytic performance of the naphtha hydrotreater deteriorated rapidly. It was later determined that this was due to excessive deposition of polymerized hydrocarbons and other fouling material at the top of the naphtha hydrotreater reactor from the high concentrations of olefins and diolefins in the coker naphtha feed relative to standard straight-run naphtha. The opportunity naphtha was mainly coker naphtha. In addition, the naphtha hydrotreater catalyst was deactivated due to excessive deposition of silicon. Properties of a variety of samples of naphthas are shown in Table 8.3.[22] Straight-run and unsaturated naphtha properties usually vary significantly due to the crude type and type of downstream process units that produce the naphthas. Based on the properties in Table 8.3, it was obvious why the coker naphtha caused the major upset of the naphtha hydrotreater unit. The naphtha hydrotreater exhibited an excessive high pressure drop in the reactor. That led to a decision to take the naphtha hydrotreater out of service for an unscheduled, immediate turnaround maintenance and catalyst replacement. The loss of operations of the naphtha hydrotreater led to huge negative impacts on reformate and hydrogen production in the refinery as downtime and turnaround for the naphtha hydrotreater led to a shutdown of the 65-MBPD cyclic catalyst regenerative reformer for 10 days.

Adherence to a best practice principle that stipulated review of available physical and chemical properties before purchasing and detailed in-house laboratory-generated physical and chemical characterizations before processing opportunity oils would have prevented the downtime for turnaround maintenance and catalyst replacement for the naphtha hydrotreater. In addition, downtime and loss of production of reformate and hydrogen of the 65 MBPD would not have occurred. Based on the properties of opportunity naphthas, safer, more reliable operations of naphtha hydrotreaters then would have entailed either processing opportunity naphthas in very low concentrations in a blend with straight-run naphthas or processing opportunity naphthas in dedicated two- or three-stage naphtha hydrotreaters with the first reactor operating as a diolefin and olefin saturator and the second reactor for hydrodesulfurization and hydrodenitrogenation reactions.[1,2,22–24,28,29] In another configuration, a second

TABLE 8.3
Properties of Refinery Naphthas

	Straight-Run Naphtha (ANS)	FCCU Naphtha	Delayed Coker Naphtha	Fluid Coker Naphtha
API Gravity	53	54	57	55.4
Sulfur, ppm	270	730	2500	8000
Nitrogen, ppm	2.1	38	100	150
Diolefins, vol. %	0	0.5	2	4
Olefins, vol. %	0	22.5	43	36
Silicon, ppm	0	0	10	0
Potential Gums, mg/100 ml	<1	0	300	0
Paraffins, vol. %	43	28	24	20
Naphthenes, vol. %	39	11	23	20
Aromatics, vol. %	15	40	8	20

Source: Reid, T. A., Coker Naphtha Hydrotreating Processing Solutions for Trouble-Free Operations, Akzo Nobel Catalysts.[22]

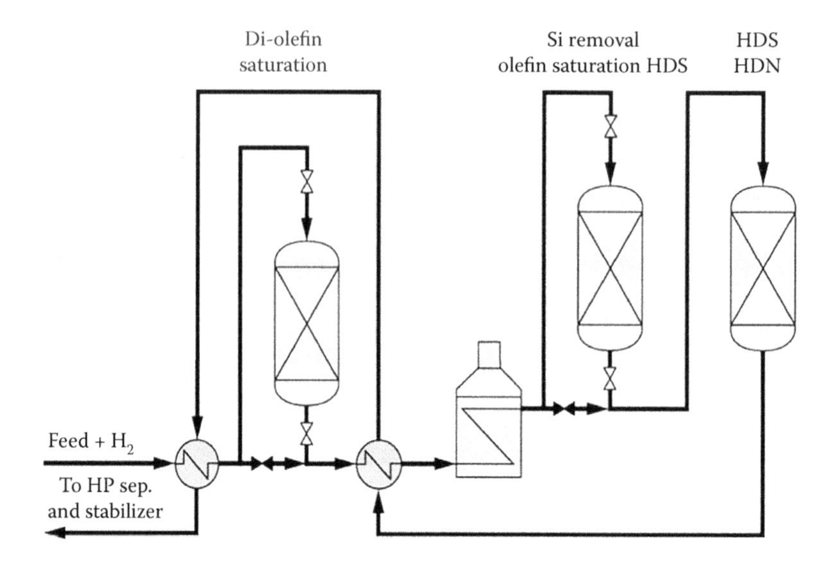

FIGURE 8.3 Coker naphtha hydrotreater with silicon guard bed.
From Haldoe Topsoe, Coker Naphtha Hydrotreating Technology, https://www.topsoe.com/products/coker-naphtha-hydrotreating-technology[24]. (Courtesy of Haldor Topsoe A/S.)

dedicated reactor placed before a third reactor is used for silicon and olefin saturation, as shown in Figure 8.3.[24] The third and last reactor is used for the hydrodesulfurization and hydrodenitrogenation of the coker naphtha. Today, processing assets available for meeting low-sulfur gasoline regulatory specifications can be used to process significant percentages of unsaturated opportunity naphthas more reliably and profitably in one of the catalytic cracker gasoline hydrotreating process units.[23,25,28,29]

8.2 STRATEGIES FOR MEETING GASOLINE BENZENE REGULATIONS

Global oil refining and energy industries are demonstrating strong resolve to meet transportation fuel product quality regulations in many countries in the combined efforts of government and energy leaders to reduce harmful emissions from the use of fossil fuels in order to protect humanity and the environment. The Clean Air Act in the United States requires that the EPA regulate fuels and fuel additives for use in motor vehicles, motor vehicle engines, or nonroad engines or vehicles if such fuel, fuel additive, or any emission product causes or contributes to air or water pollution that may endanger the public health or welfare.[37] It is with purposeful energy and drive that the Environmental Protection Agency has led the United States from the era of leaded gasoline to the use of oxygenate additives and more recently to over 90% reduction in average gasoline sulfur. In December 1999, the Tier 2 program for cleaner vehicles and gasoline was announced, which called for more protective tailpipe emissions standards for all passenger vehicles. To protect low-emissions control technologies, such as catalytic converters, and reduce harmful air pollution in vehicles, low gasoline sulfur specifications were mandated. Recent US EPA regulations in the

TABLE 8.4
History of US EPA Major Motor Vehicle Fuel Regulations

Regulation Title	Summary of Rule and Applicability to Health Effects	Regulation Publish Date (*Year/Month*)
Final Rule for Control of Air Pollution from Motor Vehicles: Tier 3 Motor Vehicle Emission and Fuel Standards	The Tier 3 standards are part of a comprehensive approach to reducing the impacts of motor vehicles on air quality and public health. The standards will reduce both tailpipe and evaporative emissions from all passenger vehicles and provide more stringent gasoline sulfur standards.	2014/04
Final Rule for Control of Hazardous Air Pollutants from Mobile Sources	In February 2007, EPA finalized this rule to reduce hazardous air pollutants from mobile sources. The rule limits the benzene content of gasoline and reduce toxic emissions from passenger vehicles and gas cans. EPA estimates that in 2030 this rule would reduce total emissions of mobile source air toxics by 330,000 tons and VOC emissions (precursors to ozone and PM2.5) by over 1 million tons.	2007/02
Final Rule for Control of Air Pollution From New Motor Vehicles: Tier 2 Motor Vehicle Emissions Standards and Gasoline Sulfur Control Requirements	The Tier 2 program is part of a comprehensive approach to reduce emissions by providing protective tailpipe emissions standards for all passenger vehicles and lowering standards for sulfur in gasoline.	2000/02

Source: US EPA, Federal Gasoline Regulations, www.epa.gov/gasoline-standards/federal-gasoline-regulations[37]

last 20 years are listed in Table 8.4.[37] By 2007, US oil refiners met the 30 wppm average sulfur and 80 wppm sulfur cap in their gasoline production. In a 1999 report, the EPA estimated the cost of implementation by US oil refiners would be about $5.3 billion and health and environmental benefits would accrue to about $25.2 billion.[26] As discussed in Section 2.4.2, European Union countries and Japan specify an average 10 wppm sulfur for their gasolines and many other countries now use 50 ppm gasoline sulfur. Gasoline sulfur specifications for selected countries are listed in Table 8.5.[27] The next US EPA gasoline quality regulation stipulated that oil refiners reduce the production of hazardous pollutants or air toxics from use of transportation fuels by effecting reductions in some of the chemical compounds in gasoline. Mobile Sources Air Toxics Phase 2 rules are intended to ensure that air toxics such as benzene and

TABLE 8.5
Gasoline Sulfur for Selected Countries

Country	2005	2010	2015	2020
Brazil	1000	1000	50	50
China	500	150	50	10
EU-27	50	10	10	10
India	500	150	150	150
Japan	50	10	10	10
Russia	500	500	50	10
Thailand	150	150	50	50
United States	30/90/300 (1)	30/80 (2)	30	10 (3)
South Africa	1000	500	500	10

Source: www.transportPolicy.net, Global Comparisons of Fuels.[27]

other hydrocarbons such as 1, 3-butadiene, formaldehyde, acetaldehyde, acrolein, and naphthalene emitted by cars and trucks are substantially reduced in the environment. MSAT 2 compounds are a subset of volatile organic compounds, which are known to contribute to ozone formation and possibly to particulate matter.[31] EPA set the date of January 1, 2011, for meeting a lower concentration of benzene, which is one of the key target compounds of the MSAT 2 rules.

Benzene is a known carcinogen and has other major health issues and environmental effects[31] and as a result, average benzene concentration in gasoline for oil refiners and importers was set at 0.62 volume percent for both reformulated and conventional gasoline starting in January 1, 2011.[30,31,40]

According to the US EPA, modest estimates suggested that it would cost "about $400 million in 2030 to implement and produce $6 billion in health benefits."[32,33] However, another cost estimate from a1978 study for the EPA had indicated a cost to the petroleum refining industry of $5.3 billion for a 94.5% reduction in gasoline benzene from 1.37 volume percent to 0.26 volume percent and that cost, including capital recovery, was about 2.2 cents per gallon of gasoline. That cost estimate did not include the cost of octane replacement due to loss of benzene, volume replacement, and the effect on the chemical industry.[34] A review of the technologies applied by oil refiners in the last decade and available for reducing gasoline benzene are discussed as well as other low-cost effective technology options. Oil refiners with coking configuration refineries, paraffin isomerization processing units, and more elaborate gasoline blending capabilities applied low-cost processing options to meet the 0.62 volume percent gasoline benzene standards.

Most of the benzene in gasoline comes from reformates with smaller contributions from light straight-run naphtha and FCC naphtha, as shown by the representative data in Table 8.6.[31] If pyrolysis gasoline is blended, it introduces a small amount of benzene, as its volume in gasoline is usually quite low. Alkylate, light hydrocracker naphtha, paraffin isomerate, and other gasoline blending components do not contribute any appreciable benzene to gasoline. There are, however, wide variations in component benzene concentrations relative to those listed in Table 8.6 due to

TABLE 8.6
Benzene Concentrations of Blending Components

Stream	Benzene, Vol. %	Stream, Vol. %	Gasoline Bz, Vol. %
Reformate	5.00	30.00	1.50
LSR Naphtha	1.50	10.00	0.15
FCC Naphtha	1.20	33.00	0.40
Other	0.00	27.00	0.00

Source: Oyekan, S. O., Personal Notes; Colwell, R. F., Benzene in Gasoline, Regulations and Remedies. Process Engineering Associates. (Source: "Final Regulatory Impact Analysis" EPA420-R-07-002), February 2007.[1,31]

Note: Where Bz. is benzene and represents the volume percent contributed by that stream in the gasoline product.

differences in refinery configurations, type of crude oil slate processed, naphtha splitting, and naphtha processing assets. Additionally, the blendstock components used, reformer naphtha type, and catalytic reforming processes that oil refiners are using impact the relative contributions of benzene in gasoline benzene.[36]

Based on the high contributions of benzene in gasoline from reformate and light straight-run naphtha of about 75%–85%, special processing of optimally fractionated naphtha fractions to produce lower benzenes in reformates and light naphtha blending components and efficient blending provide low-cost options for meeting regulated gasoline benzene concentrations. However, projects and processing options selected to meet low gasoline benzene regulation by oil refiners are dependent on the configurations of their refineries and the type and number of processing assets available for producing gasoline blend components. In addition, the percentage of ethanol and other oxygenates that can be blended also impacts gasoline benzene. A good supply of light naphtha and blend components with different octane and Reid vapor pressure characteristics could aid in minimizing capital outlay for benzene reduction technologies for oil refineries.

Two major fundamentally based methodologies are utilized for meeting low gasoline benzene concentrations. The methodologies are based primarily on chemical reactions that lead to formation of benzene and conversion of benzene to cyclohexane and paraffin compounds, as shown in Figures 8.4 through 8.6. Specific details of these reactions were discussed in Chapter 4 of this book.

Benzene is selectively produced in catalytic reformers via the dehydrogenation of cyclohexane and also via dehydroisomerization of methyl cyclopentane and other C6

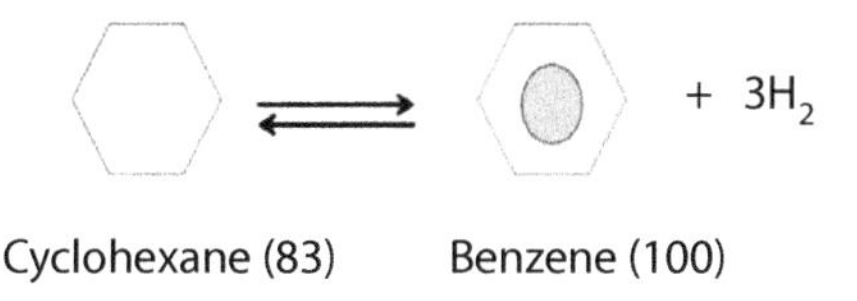

FIGURE 8.4 Cyclohexane dehydrogenation.

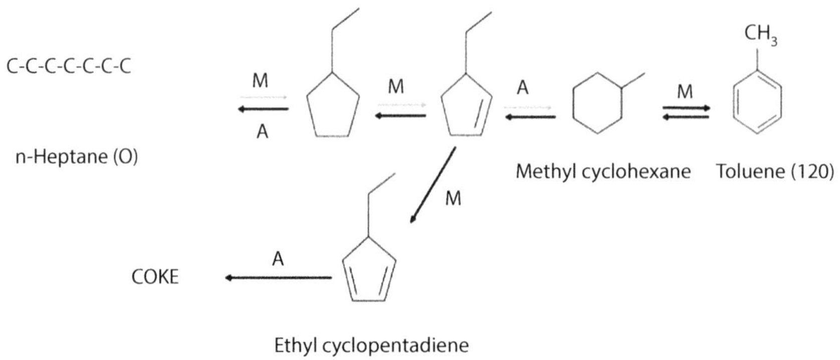

C-C-C-C-C-C-C

n-Heptane (O)

Methyl cyclohexane Toluene (120)

COKE

Ethyl cyclopentadiene

FIGURE 8.5 Paraffin aromatization reaction.

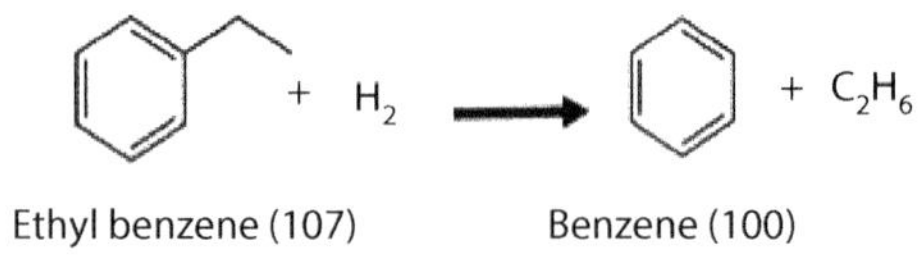

Ethyl benzene (107) Benzene (100)

FIGURE 8.6 Hydrodealkylation of alkyl aromatics.

cyclic compounds. In addition, through paraffin aromatization and hydrodealkylation reactions, benzene is produced in reformers from secondary benzene precursor compounds even after efficiently fractionating out primary benzene precursor compounds from the catalytic reformer feed.

Prefractionation and postfractionation options refer primarily to stages where separations are effected in the processing of total naphtha in oil refineries to reduce the concentrations of benzene in the ultimate reformate blending component.

In the prefractionation option, benzene reduction is initiated with fractionation of naphtha and concentrating primary benzene precursor C6 cyclic compounds in the light naphtha fraction. Depletion of primary benzene precursors from the heavy naphtha reformer feed essentially reduces benzene production in catalytic reformers. Secondly, benzene in the C6 cyclic compounds in the light naphtha can be eliminated via a number of catalytic processes. Benzene can be disposed of via hydrogenation in dedicated benzene saturation and light naphtha isomerization units to produce cyclohexane. Alkylation reactions of benzene and ethylene and propylene lead to the production of high-octane ethyl benzene and propyl benzene. This process technology option is attractive, as it does not lead to octane reductions. The prefractionation strategy for the reduction of reformate benzene has as the essential objective significant reduction or elimination of primary benzene precursor compounds such as methyl cyclopentane, hexane, and benzene from the heavy naphtha feed to the catalytic reforming unit.

The objective is achieved by the use of an appropriate naphtha splitter or fractionation unit located either at the crude distillation unit or after the naphtha hydrotreater. A simplified drawing of the prefractionation option is shown in Figure 8.7.

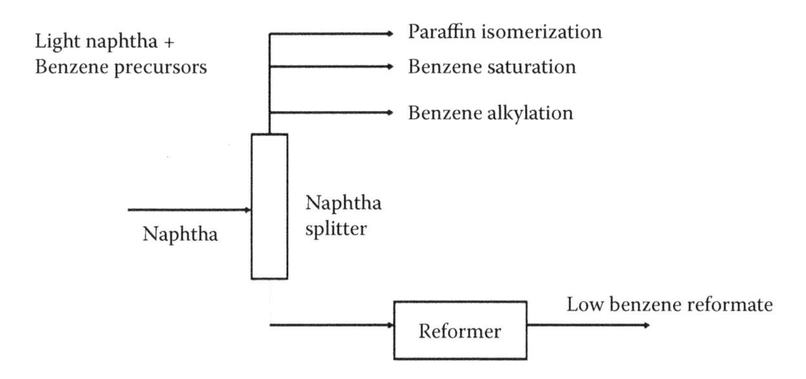

FIGURE 8.7 Concentrating benzene precursors in light naphtha.

For oil refiners that operate pentane/hexane isomerization, or Penex, units, hydrotreated light naphtha in an appropriate boiling range containing primary benzene precursor compounds is charged to the Penex unit and the heavy naphtha fraction is charged to the catalytic reformer. Benzene is saturated in the paraffin isomerization unit or in a dedicated benzene saturation unit, as shown in Figure 8.7. Benzene saturation reactions occur in the pentane/hexane isomerization process at low temperatures of 280–350 F and pressures of 400–700 psig over a platinum aluminum-chlorided catalyst.

However, it should be noted that in the prefractionation scheme and despite the removal of benzene precursor compounds from the reformer feed, there is usually about 2.0–3.5 volume percent benzene in reformates due to contributions from paraffin aromatization and alkyl aromatics hydrodealkylation reactions, as reviewed previously and as shown in Figures 8.5 and 8.6.

Postfractionation strategy involves recovering benzene from reformate produced in catalytic reformer units in an associated aromatics extraction unit. The benzene product can be disposed of by selling it to the chemical industry and also by hydrogenating benzene to produce to cyclohexane. Extracted benzene can also be alkylated with ethylene and propylene to produce high-octane ethylbenzene and propyl benzene.[(31)] Oil refiners that have catalytic reformers with aromatic compound extraction capabilities can use the postfractionation option with minimal capital expenditures. The overall goal of gasoline benzene reduction in refineries can then be achieved via the use of prefractionation or postfractionation processing strategies, and the ultimate strategies employed are highly dependent on the processing assets in refineries and gasoline blending capabilities of the oil refiner. A detailed analysis of an oil refiner's assets for meeting gasoline benzene regulations should be conducted and existing assets identified for meeting gasoline benzene specification. The oil refiner can then select from a mix of low-cost effective processing and gasoline blending options. There could be a need to study options for either adding or conducting revamps of process units and naphtha fractionation units. Oil refiners could opt to use a variety of process technology units. The process units could include pentane/hexane isomerization, benzene saturation, and aromatic extraction. Alternatively, oil refiners could also achieve low gasoline benzene via cost effective gasoline blending to reduce reformate benzene. Oil refiners can use a variety of blendstock such as raffinate, alkylate, fluid catalytic cracker naphtha, hydrocrackate, and hydrotreated light naphtha to meet gasoline octane, benzene, sulfur, and Reid vapor pressure specifications.

In one of the companies where I had responsibilities for naphtha processing in several refineries, an extensive review of refining processing assets and gasoline blending capabilities was conducted to determine specific approaches for meeting refinery and overall corporate EPA gasoline benzene concentrations and Reid vapor pressure regulations. Various low-cost effective options were identified and applied in the seven refineries. Representative performance data of reformate benzene reductions are highlighted for 2008 and 2012 before and after implementation of prefractionation strategies. Comparative data for reformate benzene and Reid vapor pressure from 2008 and 2012 operations for two continuous catalyst regenerative and a fixed-bed cyclic catalyst regenerative reformers are provided in Table 8.7. Changes made to reduce primary benzene precursor compounds in the feed to CCR 2 led to the production of reformates with 70% reduction in benzene concentration, as shown in Table 8.7. The reformate benzene reduction was sufficient for meeting gasoline benzene specifications for that refinery and, as such, negligible changes were made to the cyclic catalyst regenerative reformer naphtha feed. Changes made to the naphtha splitter operations in the refinery with CCR 1 led to about a 50% reduction in reformate benzene. The refineries referred to are of the coking configuration and have pentane/hexane isomerization units that are also used for benzene saturation. In addition, the refineries have alkylation, catalytic crackers, and hydrocracking process units. Efficient gasoline blending schemes are utilized in all of the company's refineries and these have led to meeting MSAT 2 gasoline benzene specifications with minimal capital expenditures.

Fifty to 70% reformate benzene reductions realized in the refineries were due primarily to efficient fractionation to eliminate all of the primary benzene precursors such as cyclohexane, methyl cyclopentane, and hexanes and processing the benzene precursor depleted naphtha feed in catalytic reformers. Hydrotreated light naphtha

TABLE 8.7
Pre- and Post-2011 Catalytic Reformer Data for Benzene Reduction

	Cyclic Regen		Continuous Regen 1		Continuous Regen 2	
	2008	2012	2008	2012	2008	2012
Feed Rate, MBPD	33.9	19.9	45.8	40.0	30.5	39.6
Distillation, F						
IBP	87.6	76.6	196.3	214.7	58.5	72.0
FBP	334.0	339.1	368.7	378.0	376.7	387.0
N+2A, vol. %	56.8	69.7	55.2	55.7	56.1	57.8
C5+ Octane, RON	91.9	92.6	96.6	96.1	101.1	99.7
Reformate Benzene, vol. %	2.5	2.4	2.7	1.3	4.5	1.5
Reformate RVP, psia	2.7	2.6	4.0	3.2	3.5	3.2

and paraffin isomerate contributed essentially zero benzene to gasoline blends. Selected process data for paraffin isomerization units are provided in Table 8.8.

Feed x factor and feed benzene for paraffin isomerization units increased between 2008 and 2012 for two of the paraffin isomerization units, and that reflected changes that were made in naphtha splitter operations to concentrate primary benzene precursor compounds in light naphtha feeds. The prefractionation strategy was replicated in my other refineries and 40%–70% reformate benzene reductions were realized for the continuous catalyst regenerative reforming units.

In summary, for the prefractionation options, attainable percentages of benzene reductions in refineries are dependent on the light and heavy naphtha split to eliminate benzene precursor compounds from the heavy naphtha. Reformate benzene is a function of the naphtha feed quality and processing severity of a catalytic reformer. [1,2,36,38]

Four of the seven refineries that I covered had pentane/hexane isomerization units. Processing conditions were modified in two of the units and they were used exclusively as benzene saturation units so as to minimize gasoline benzene and Reid vapor pressure. The refineries, as indicated previously, are mostly of the coking configuration type with extensive, efficient fuel blending assets and capabilities. Oil refiners that have coking configuration refineries usually have a wide variety of blendstock components to use in gasoline production, as shown in Table 8.9.[35,39] An oil refiner with a wide variety of gasoline blend components has significant advantages in gasoline production, as it can easily compensate for octane losses associated with benzene reductions in the reformate blendstock.

The postfractionation option or strategy is usually conveniently applied by oil refiners that have refinery assets such as catalytic reformers and associated aromatic extraction units. Heavy naphtha feed to catalytic reformers are typically not subjected to elimination of benzene precursor compounds. Hydrotreated heavy naphtha is processed in the catalytic reformer and a reformate splitter after the reformer separates the benzene, toluene, and xylene fraction from the reformate, as shown in Figure 8.8. Benzene is eliminated by direct removal from the reformate product.[1,36,38]

TABLE 8.8
Naphtha Splitter Impact on Isomerization Unit Feed Quality

	Isom A		Isom B		Isom C	
	2008	**2012**	**2008**	**2012**	**2008**	**2012**
Charge Rate, MBPD	20.00	25.10	14.20	12.40	12.20	16.50
Reactor Pressures, psig	480.00	476.00	450.00	400.00	661.00	675.00
Lead Reactor Inlet Temp., F	325.00	324.90	332.70	300.00	295.00	260.00
Feed × Factor	12.80	32.50	16.10	14.40	5.20	9.60
Feed Benzene, vol. %	1.59	2.60	1.86	1.10	1.44	1.62
Product Benzene, vol.%	0.00	0.00	0.00	0.00	0.00	0.00

Note: Feed × factor is the sum of benzene, cyclohexane, methyl cyclopentane and C7+ hydrocarbons in the light naphtha feed to the C5/C6 isomerization unit.

TABLE 8.9
Typical Gasoline Blend Component Octanes

No.	Component	RJP, psi	(R+M)/2	MON	RON	API
1	iC4	71.0	92.5	92.0	93.0	
2	nC4	52.0	92.5	92.0	93.0	
3	iC5	19.4	92.0	90.8	93.2	
4	iC5	14.7	72.0	72.4	71.5	
5	iC6	6.4	78.8	78.4	79.2	
6	ISR gasoline [CS-180 F]	11.1	64.0	61.6	66.4	78.6
7	ISR gasoline isomerized a nice-through	13.5	82.1	81.1	83.0	80.4
8	HSR gasoline	1.0	60.5	58.7	62.3	48.2
9	Light hydrocrackate	12.9	82.6	82.4	82.8	79.0
10	Hydrocrackate, CS-06	15.5	87.4	85.5	89.2	86.4
11	Hydrocrackate, C6-190 F	3.9	74.6	73.7	75.5	85.0
12	Hydrocrackate, 190–290 F	1.7	77.3	75.6	79.0	55.5
13	Heavy hydrocrackate	1.1	67.5	67.3	67.6	49.0
14	Cake gasoline	3.6	63.7	60.2	67.2	57.2
15	Light thermal gasoline	9.9	76.8	73.2	80.3	74.0
16	C6 light thermal gasoline	1.1	72.5	68.1	76.8	55.1
17	FCC gasoline, 200–300 F	1.4	84.6	77.1	92.1	49.5
18	Hydrag. light FCC gasoline, CS	13.9	82.1	80.9	83.2	51.5
19	Hydrag. CS-200 F FCC gasoline	14.1	86.5	81.7	91.2	58.1
20	Hydrag. light FCC gasoline, C6	5.0	80.2	74.0	86.3	49.3
21	Hydrag. CS-FCC gasoline	13.1	85.9	80.7	91.0	54.8
22	Hydrag. 300–400 F FCC gasoline	0.5	85.8	81.3	90.2	48.5
23	Reformate, 94 RON	2.8	89.2	84.4	94.0	45.8
24	Reformate, 98 RON	2.2	92.3	86.5	98.0	43.1
25	Reformate, 100 RON	3.2	94.1	88.2	100.0	41.2
26	Aromatic concentrate	1.1	100.5	94.0	107.0	
27	Alkylate, C3	5.7	89.1	87.3	90.8	
28	Alkylate, C4	4.6	96.6	95.9	97.3	70.3
29	Alkylate, C3, C4	5.0	93.8	93.0	94.5	
30	Alkylate, CS	1.0	89.3	88.8	89.7	
31	Palymer	8.7	90.5	84.0	96.9	59.5

Source: Jechura, J., Product Blending and Optimization Considerations, http://inside.mines.edu/~jjechura/Refining/11_Blending_Optimization.pdf; Gary, J., Handwerk, G., Kaiser, H., Blending Component Values for Gasoline Blending Streams. *Petroleum Refining Technology and Economics*, 5th edition, CRC Press. March 2007.[35,39]

Gasoline octane giveaway and gasoline vapor specifications are some of the gasoline challenges that can arise as oil refiners blend components to compensate for octane loss due to benzene reduction. These challenges call for significant planning of light and heavy naphtha processing, optimization of catalytic reforming process units, especially continuous catalyst regenerative reformers, and use of alternate

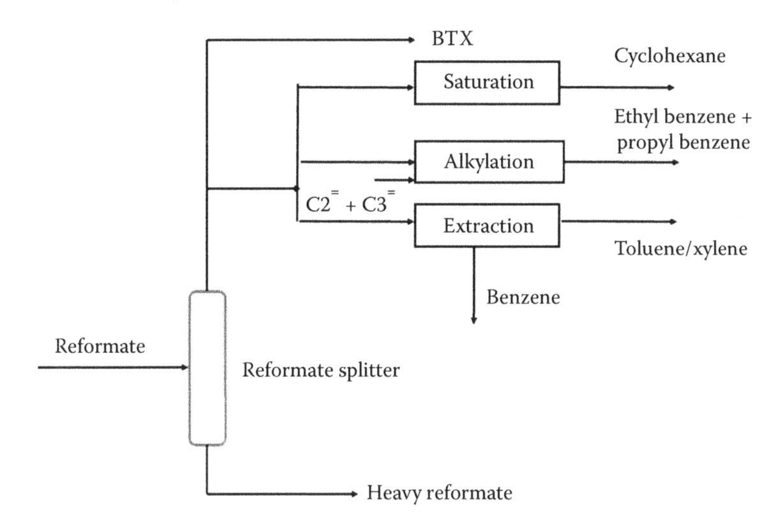

FIGURE 8.8 Postfractionation option leading to benzene reduction.
From Oyekan, S. O., Personal Notes.; Meister, M. et al., *Oil Gas J.*, 104, 38–45, September 11, 2006; Schiavone, B. J., Technology Advancements in Benzene Saturation. NPRA 2007 Q&A Technology Forum, October 11, 2007.[1,36,38]

gasoline blending components including natural gasoline. An oil refiner should audit its benzene reduction programs periodically as required. The audits should include assessing and benchmarking pentane/hexane and catalytic reformers processing and gasoline blending solutions periodically.

Oil refiners can also benchmark their processing units and reformate benzene concentrations relative to those of competitors' refineries via the use of appropriate benchmarking services such as those offered by Solomon Associates.

Examples of cost-effective prefractionation strategy programs used in process units and refineries are highlighted to share the results of timely studies, good planning, successful oil refining naphtha processing, and efficient gasoline blending for meeting fuel and environmental regulations in a proactive way.

8.2.1 Process Technologies for Meeting Gasoline Benzene Regulations

The US EPA led the January 2011 oil refining industry compliance with MSAT 2 regulations that focused primarily on reducing gasoline benzene concentrations to reduce harmful pollution and minimize the impact of a known carcinogen on people. It is expected that health benefits realized from cleaner gasoline and benzene reduction to the 0.62 volume percent concentration in the United States will spur further reductions in gasoline benzenes for the European Union and other countries relative to the currently regulated 1.0 volume percent concentration listed in Table 8.10.

Based on the outstanding work of the oil refining technologies licensing and catalysts supplier companies, the oil refining industry is equipped with the necessary portfolio of catalysts and technologies to meet gasoline benzene concentrations of 0.62 volume percent and lower. Most of the process technologies are based on

TABLE 8.10
Parameters for Selected Gasoline for Global Markets

Fuel Parameter		China V	Bharat IV (1)	Euro IV	Euro V	EPA RFG (2)	EPA Conv (3)	WFC (4)
RON, min	(5)	81/92/95	91.0	91-95	91–95	NA	NA	91/95/98
MON, min	(6)	NS	81.0	81–85	81–85	NS	NS	82.5/85/88
Sulfur,	Wppm	50		10	10	10		10
AKI	(7)	84/87/91	NS	NS	NS	87/87/91	87/87/91	NS
Aromatics (8)	vol.%	40.0	35.0	35.0	35.0	20.7/19.5	27.7/24.7	35.0
Olefin (9)	vol.%	25.0	21.0	18.0	18.0	11.9/11.2	12.0/11.6	10.0
Benzene (10)	vol.%	1.0	1.0	1.0	1.0	0.66/0.66	1.21/1.15	1.0
Lead	mg/l	5.0	5.0	5.0	5.0			NS
Density	Kg/m^3	NS	720–775	NS	720–775	NS	NS	715–770
RVP (11)	KPa	(12)	60.00	60–70	60–70	47.6/82.0	57.2/83.6	(13)

saturation of benzene after concentrating benzene precursors in the light naphtha fraction. Benzene saturation technologies include paraffin isomerization units, UOP's BenSat™, Axens BenFree™, and CD Tech CDHydro™ process. Axens conventional benzene saturation process is as shown in Figure 8.9, and the BenFree™ process is shown in Figure 8.10. UOP's simplified BenSat™ process is shown in Figure 8.11.

The CD Tech CDHydro process saturates benzene to cyclohexane in a catalytic distillation benzene-toluene column. Product benzene from the CDHydro™ process is about 0.5 volume percent for a feed stream of C5-C9 hydrocarbons. CD Tech's CDHydro™ process is shown in Figure 8.12.

Solvent liquid-liquid extraction process units and extractive distillations units are also used. UOP's Sulfolane℠, ED Sulfolane™, and Carom™ are used for the separation of benzene to less than 0.1 volume percent concentration from C6-C7 reformate in a reformate splitter overhead streams.[31] A simplified drawing of the UOP Carom™ is shown in Figure 8.13.

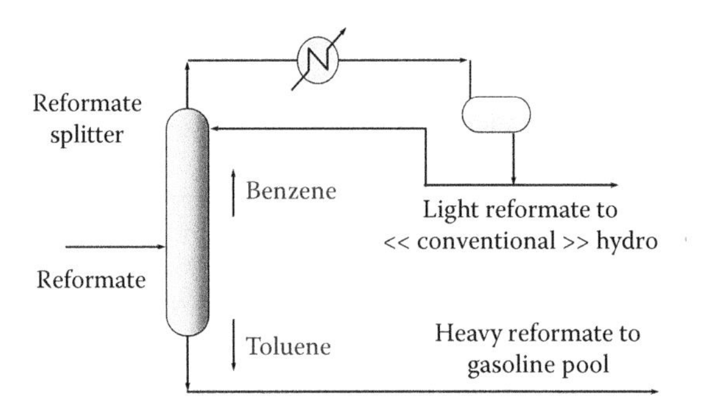

FIGURE 8.9 Postreformate splitter conventional hydrogenation of benzene.
From Oyekan, S. O., Personal Notes; Colwell, R. F., Benzene in Gasoline, Regulations and Remedies. Process Engineering Associates. (Source: "Final Regulatory Impact Analysis" EPA420-R-07-002), February 2007.[1,31] (Courtesy of Process Engineering Services.)

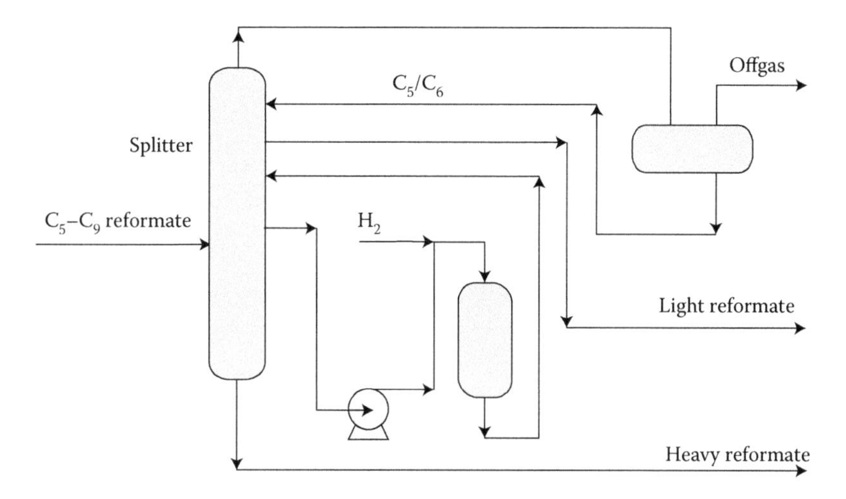

FIGURE 8.10 Axens BenFree™ process for low benzene reformate.
From Colwell, R. F., Benzene in Gasoline, Regulations and Remedies. Process Engineering Associates. (Source: "Final Regulatory Impact Analysis" EPA420-R-07-002), February 2007.[31] (Courtesy of Axens.)

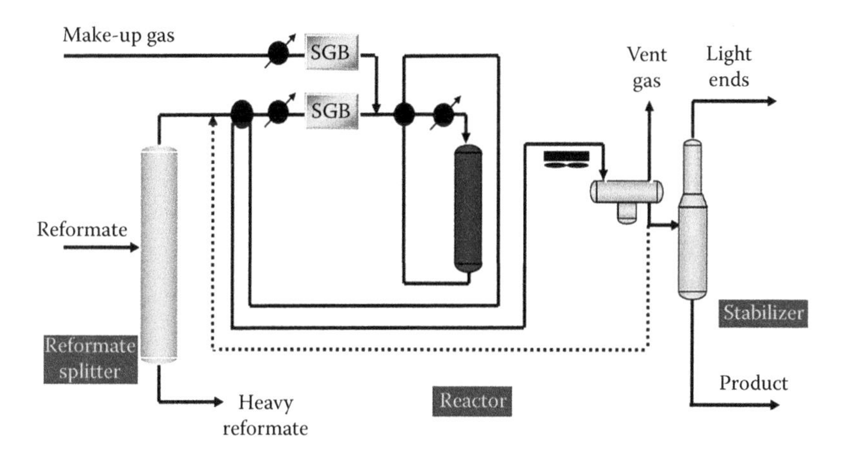

FIGURE 8.11 UOP's BenSat™ process with sulfur guard beds.
From Colwell, R. F., Benzene in Gasoline, Regulations and Remedies. Process Engineering Associates. (Source: "Final Regulatory Impact Analysis" EPA420-R-07-002), February 2007.[31] (Courtesy of Process Engineering Services.)

Other aromatic extraction technologies that are used for reformate feeds with aromatics in the concentration range of 10–95 volume percent include the GTC technology GT-BTX™, which is reported to reduce the resultant gasoline blend component benzene to less than 0.1 volume percent.

A fourth set of process technologies for reducing gasoline blend component benzene from reformate streams involves alkylating benzene with ethylene and propylene to produce high-octane ethyl benzene and propyl benzene of 120 and 124

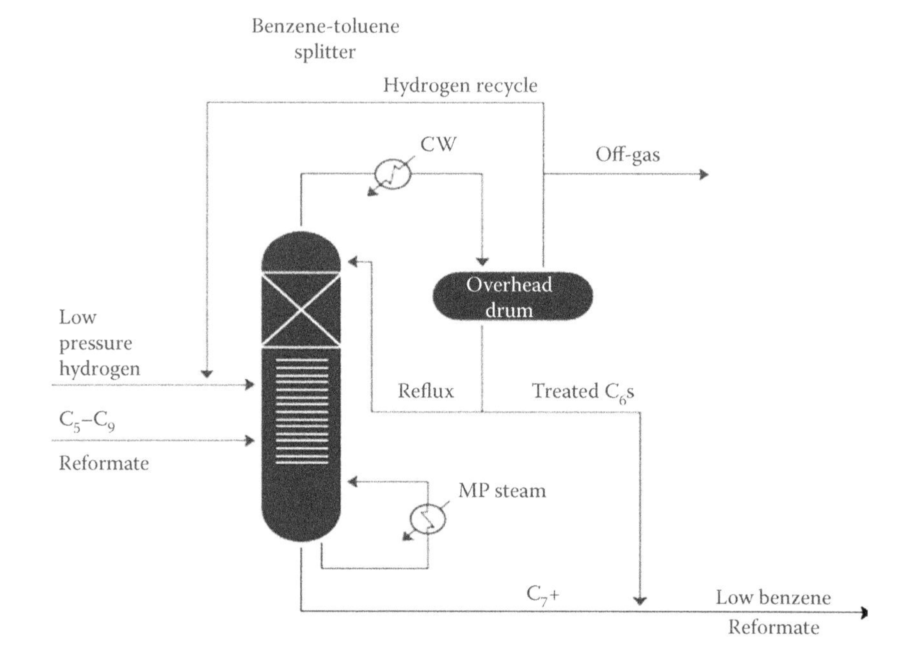

FIGURE 8.12 CD Tech catalytic distillation CDHydro™ process.
From Colwell, R. F., Benzene in Gasoline, Regulations and Remedies. Process Engineering Associates. (Source: "Final Regulatory Impact Analysis" EPA420-R-07-002), February 2007.[31] (Courtesy of Process Engineering Services.)

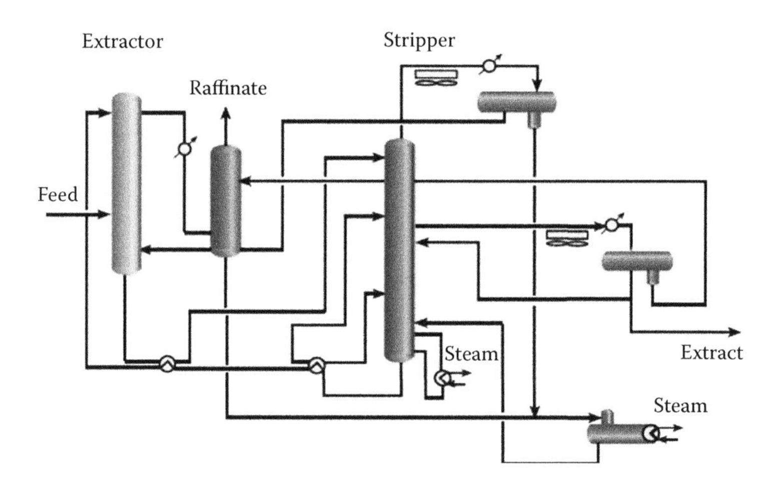

FIGURE 8.13 UOP's Carom process for benzene extraction.
From Colwell, R. F., Benzene in Gasoline, Regulations and Remedies. Process Engineering Associates. (Source: "Final Regulatory Impact Analysis" EPA420-R-07-002), February 2007.[31] (Courtesy of Process Engineering Services.)

octanes, respectively. UOP's Alkylmax™[38,43] and ExxonMobil's BenzOUT™ are some of the licensed alkylation process units offered. In the Alkylmax™ process, benzene is reacted with propylene to produce isopropyl benzene (cumene) and di-isopropyl benzene. ExxonMobil's BenzOUT™ is offered as a grassroots unit and can also be retrofitted into an existing facility such as a polyglas unit.[41] The key sections of ExxonMobil's BenzOUT process are shown in Figure 8.14.

Oil refiners, with the great assistance of process technologies licensors and catalyst supplier companies, met US EPA gasoline benzene reductions through the use of a combination of prefractionation and postfractionation strategies. The successful approaches involved the use of naphtha fractionation, benzene saturation, paraffin isomerization, and benzene extraction from reformates.

Most of the low-capital technology options that are used by oil refiners to reduce gasoline benzene are based on benzene saturation as the main reaction. Oil refiners experienced some losses of blendstock octane barrels as the saturation of benzene in oil refineries led to octane losses. Basically, via the benzene to cyclohexane saturation reaction, the octane loss per mole of benzene converted is about 30 octane numbers. In addition, substantial octane barrels losses were incurred due to conversion of benzene precursor compounds to lower octane hydrocarbons in paraffin isomerization units instead of aromatics in catalytic reformers. Environmental advocacy groups in the United States have suggested that reductions in the concentrations of toluene and xylenes in gasoline should be effected since these hydrocarbons raise similar carcinogenic health-related concerns to those that led to the reduction of benzene in gasoline. Undoubtedly, reducing toluene and xylene concentrations in gasoline would lead to major losses of octane barrel reformate blendstock at current octane operating severities of catalytic reformers. Oil refiners would be compelled to increase the average octane severity of catalytic reformers in order to compensate for octane barrels losses if toluene and xylene concentrations are reduced in gasolines.

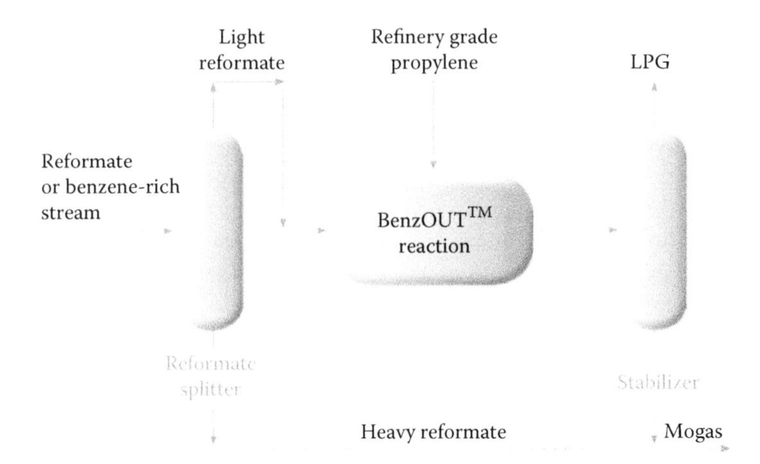

FIGURE 8.14 ExxonMobil's BenzOUT™ process for alkylating benzene.
From ExxonMobil, Cost-effective Process for Benzene Reduction in Gasoline, www.catalysts-licensing.com[41]. (Courtesy of ExxonMobil Corporation.)

Other blendstock toluene and xylene concentrations would also be reduced by such a stipulation on gasoline quality.

The US EPA is also reviewing plans that could lead to raising the Corporate Average Fuel Economy standards to 54.5 miles per gallon of gasoline for automobiles in the United States by 2025. It has been suggested that gasolines with antiknock indexes of 95+ research octane number may be required to meet the 2025 CAFÉ standard gasoline, which is being referred to as renewal super premium (RSP) gasoline.[69] Since the US EPA had also stipulated that 36 billion gallons of biofuels should be blended per year into transportation fuels by 2022, that specific renewable fuel standard (RFS) could provide some relief in terms of the incremental octane barrels of blendstock that could be required in oil refineries to produce RSP gasoline. Thus, depending on the percentage of oxygenates that is permitted for RSP gasoline, catalytic reformers and alkylation units would be expected to make up the required incremental octane barrels in oil refineries. The expectation is that catalytic reformers, especially continuous catalyst regenerative reformers, will be required for the production of the incremental octane barrels of gasoline blendstock in oil refineries.

8.3 CHLORIDE MANAGEMENT AND REFINERY RELIABILITY

There are two major operational scenarios that require critical management of inorganic and organic chlorides in oil refineries. The first is to apply good management principles to minimize or eliminate negative impacts of inorganic and organic chlorides on the performance and reliability of crude and vacuum distillation columns. Residual inorganic and organic chlorides in oil fractions from crude distillation units and continuous chloride additions that are required to boost and stabilize activity and selectivity of catalyst performances in light naphtha isomerization units and catalytic reformers must be diligently managed. Effective management of inorganic and organic chlorides in oil refineries and good routine and turnaround maintenance are necessary in order to achieve the goals of enhanced productivity and reliability of refineries in a highly competitive oil refining and marketing business.

8.3.1 Management of Total Chlorides in Crude Unit Operations

Crude oil and intermediate feedstock acquisitions are a major risk-taking business for oil refiners, as they have to contend with the qualities of a wide variety of conventional and unconventional bitumen, shale oil, and synthetic crude oils. Opportunity oils have variable qualities for some of the reasons discussed in Section 8.1. Crude oils typically have significant concentrations of inorganic chlorides and sediments that are usually removed through the operations of effective desalters before heat exchange and heating up of the oil in furnaces. Residual inorganic chlorides, nitrogen, and sulfur compounds in crude oils after the desalter can cause major equipment reliability challenges due to under deposit corrosion as a result of ammonium chloride deposition and aqueous corrosion due to hydrochloric acid in the overhead sections of crude distillation towers.[47] Crude and vacuum distillation operations were reviewed in Chapter 3. It was emphasized that good desalter operations are required to minimize the amount of residual chlorides in the treated crude oil that is carried into the furnace and crude distillation tower. The residual chlorides could cause corrosion in the first-stage overhead exchangers of the crude tower. A simplified diagram of the crude distillation tower is provided in Figure 8.15.

In a major incident that involved the processing of a crude oil containing high organic chlorides in the range of 3–3000 ppm in an oil refinery in 1997, rapid tube leaks occurred in four of the first-stage overhead shell and tube exchangers of the crude tower. Organic chlorides are usually not removed in desalter operations due to their limited solubility in water.[44,51,52] Despite the usual injection of an organic neutralizer and a filming-amine corrosion inhibitor upstream of the overhead exchangers, rapid corrosion of the exchanger tubes occurred.[44] Furthermore, naphtha from the crude oil tower that was processed in the naphtha hydrotreater caused leaks to develop in the 300 series stainless steel tubes of the reactor effluent coolers during the same period of processing the highly contaminated crude oil.

In a paper on the installation of a desalter for the atmospheric gas oil feed to an FCC unit in Sinopec, Ye reported on the inspection of a crude oil for chloride. Ye reported that despite the fact that there was a high percentage of chloride in the naphtha fraction, there was significant amount of chloride in the atmospheric gas oil to cause abnormally poor operation of the fluid catalytic cracking unit.[45] It was also reported that a chloride balance of a crude oil after the desalter and before the crude distillation column showed that organic chlorides were 96% of the residual chlorides and the organic chlorides caused operational and reliability problems in the crude distillation tower.

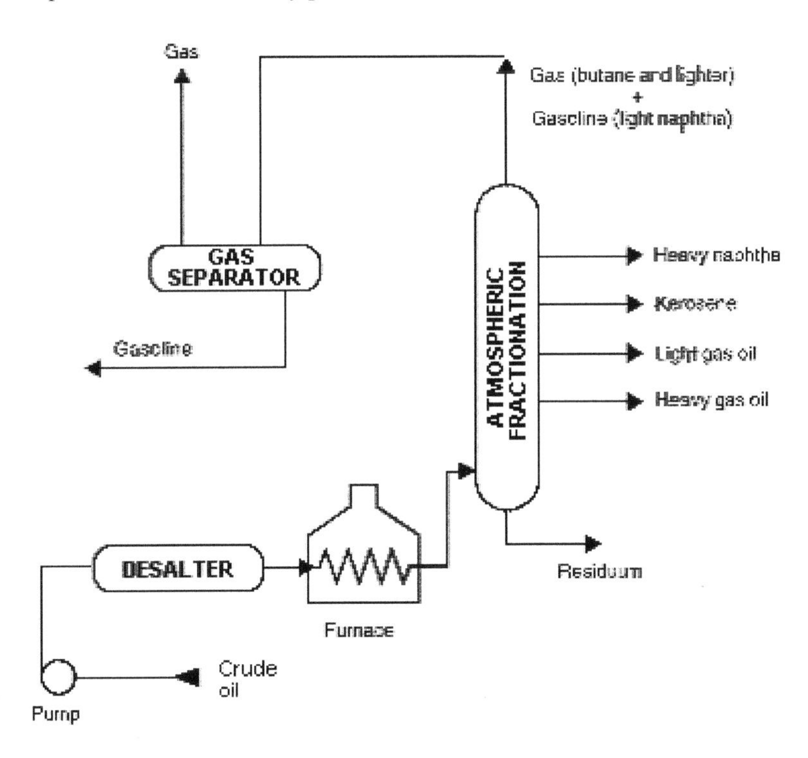

FIGURE 8.15 Crude distillation unit showing desalter and fractionation column.

A number of crude oils were identified as "problematic" oils based on the fact that significant amounts of organic chlorides were entrained in the oils. The organic chlorides were possibly due to oil contamination during production and transportation of the oils. Some of the "problematic" crude oils included Cabinda, Ratawi, Didon, Ashtart, Arab medium, Arab heavy, Illinois Basin crude, Urals Tallyn, Palanca, and Equadorian Oriente and crude oils that were transported through the Russian export pipeline.[46] It is now generally accepted that organic chlorides are not natural constituents of crude oils and are only present as a result of contamination during production and transportation.[48,50] The properties of some organic chloride compounds that could be present as contaminants in opportunity oils and used in catalytic reforming processes and light naphtha paraffin isomerization are listed in Table 8.11.[46]

Unconventional tight oils would most likely have entrained organic chlorides due to the use of hydraulic fracturing fluids containing a variety of chemical compounds. Fracturing fluids are used for physical separation and reactions with the shale structure during oil and gas production.[49] Since the projection is for greater use of opportunity oils, organic chloride compounds that are typically not removable via the crude desalter will continue to pose significant challenges with respect to adequate protection and maintenance of overhead sections of fractionation units in oil refineries.

One of the major problems with organic chloride contamination occurs when the naphtha fraction (IBP to ~400 F) is hydrotreated and hydrogen chloride is formed. Hydrogen chloride reacts with available water to form hydrochloric acid, which can cause rapid and severe corrosion even in small quantities. Piping and vessels can be rendered dangerously thin and susceptible to leaks and explosions.[50] Since nitrogen compounds are present in the naphtha, ammonium and iron chloride compounds

TABLE 8.11
Properties of Some Organic Chlorides

Compound	Molecular Weight	Boiling Point, F	Gravity	Solubility g/100 ml
Carbon Tetrachloride	154	170.6	1.59	0.10
Tetra Chloroethylene	164	249.8	1.60	0.02
Vinyl Chloride	62	8.6	0.90	0.27
Chloro Benzene	113	269.6	1.11	0.02
Chloroprene	88	138.2	0.96	0.03
Freon 113	187	118.4	1.56	0.02
Propylene Dichloride	113	204.8	1.16	0.26
Dichloro Methane	85	104.0	1.13	1.30
Trichloroethylene	131	188.6	1.50	0.10

Note: Gravity is the density of the organic compound relative to that of water. Water has a density of 1 gm/cc; Solubility is defined as solubility in water at 68 F and the unit is in grams per 100 ml of water.

$$NaCl + H_2O \longrightarrow NaOH + 2HCl$$
$$MgCl_2 + 2H_2O \longrightarrow Mg(OH)_2 + 2HCl$$
$$CaCl_2 + 2H_2O \longrightarrow Ca(OH)_2 + 2HCl$$

FIGURE 8.16 Hydrolysis of inorganic salts in atmospheric distillation unit.

are formed, which cause plugging of pumps and exchangers. It is suggested that the acceptable concentration of organic chlorides in naphtha be set at 1 wppm. Residual inorganic salts that are not removed in the desalter are hydrolyzed in the atmospheric distillation unit to form HCl, as shown in Figure 8.16. Hydrochloric acid and ammonium salts that are produced then lead to salt depositions and corrosion of refinery process overhead units and equipment, including those in the overhead sections of crude distillation columns.

Organic chlorides are converted in hydrotreating units to produce hydrogen chloride, which in turn is hydrolyzed to hydrochloric acid. As a consequence, high chlorides are usually reported for samples from the overhead sections of crude distillation towers, vacuum distillation units, hydrotreater stripper, hydrocracker unit, and catalytic reforming unit debutanizers. Similarly, high chlorides are often reported for samples from naphtha hydrotreater feed/effluent exchangers.[46]

To sustain the operations and reliability of the crude unit and minimize corrosion of crude unit overhead section, opportunity crude oils should be fully inspected for chloride types and blended with other crude oils to reduce organic chloride concentrations. Desalter operations should be improved by maintaining water wash at about 5% and at temperatures greater than 250 F to help minimize the concentrations of residual inorganic salts entering furnaces and crude distillation towers. Atmospheric distillation tower overhead sections should also be subjected

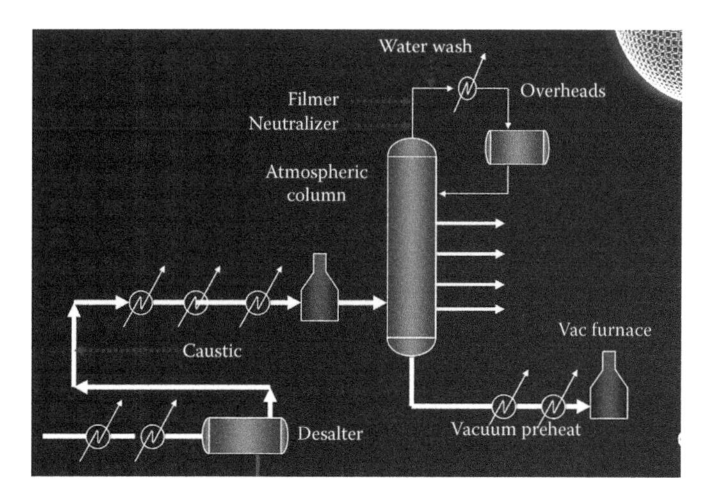

FIGURE 8.17 Corrosion control for atmospheric distillation unit overhead section. From Lordo, S. A., Primer on Organic Chlorides and Their Control, *Nalco Energy Services*.[46] (Courtesy of Nalco Champion.)

to efficient neutralizer and filmer chemical treatment programs and adequate water washes. Neutralizer, filmer, and water washes can be applied to the overhead section of the atmospheric distillation unit, as shown in Figure 8.17.[46]

8.3.2 Management of Chlorides in Oil Refining Processes

There are a number of routes for the introduction of inorganic and organic chloride compounds into a refinery, as reviewed in Section 8.3.1. Poor desalting operations can permit crude oil–borne inorganic chloride to be introduced. Purchased opportunity feeds can bring in organic chlorides. Additives used in the refinery can also add chlorides. The composition and functions of additives used in oil refineries should be reasonably understood with respect to interactions with the variety of catalysts in process units and other chemical treating agents. Cooling water used in condensers often contains some amount of chloride and other inorganic salts that can cause fouling and corrosion in refinery equipment. A small amount of organic chloride is added in light naphtha isomerization process units that are operated at low reactor temperatures of 250–360 F over platinum chlorided alumina catalysts and at pressures of 400–900 psig. The effluent gas from the stabilizer containing a small volume of hydrogen chloride is treated in a caustic scrubber before exiting the paraffin isomerization unit, as shown in Figure 8.18. Thus, hydrogen chloride in the effluent gas is fully contained within the battery limits.

An alternate paraffin isomerization technology referred to as the total isomerization process is also used. It is operated with low-activity platinum/zeolite catalysts at reactor temperatures in the range of 500–600 F and pressures of 200–300 psig and does not require chloride addition. The high-activity paraffin isomerization and catalytic reforming process units are the oil refining processes that require significant organic chloride addition to boost and stabilize catalytic performance.

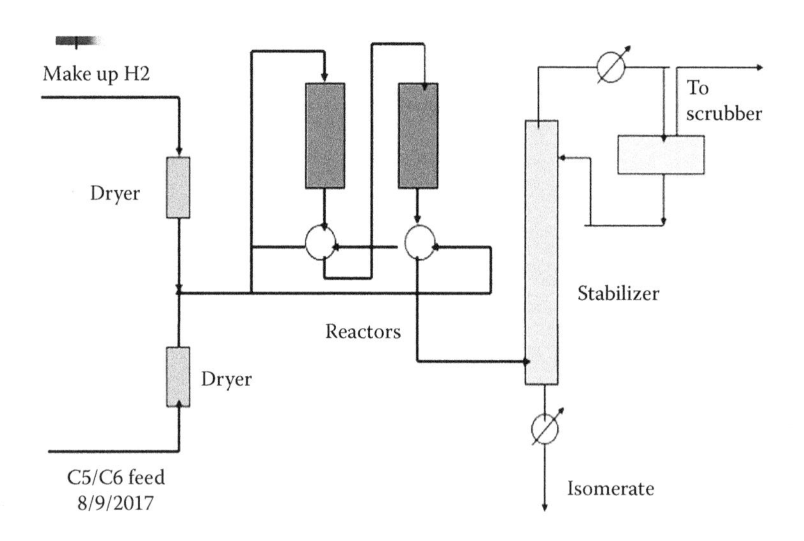

FIGURE 8.18 Paraffin isomerization unit.

By far most of the organic chlorides introduced in the refinery are through catalytic reformer operations, as organic chloride is added to promote and maintain catalyst activity, selectivity, and stability. The organic chloride is converted to hydrogen chloride within the reactors. A simplified flow diagram of the UOP CCR platformer is shown in Figure 8.19. The figure is used in this section to emphasize relevant sections of the unit, use of organic chloride, and ultimate disposition of hydrogen chloride through reformate, net hydrogen gas, light ends, and fuel gas.

Some of the hydrogen chloride reacts with ammonia to form ammonium chloride and with water to form hydrochloric acid in the cooler sections of product separation systems. Ammonium salts and hydrochloric acid cause fouling and corrosion in combined feed exchangers, recycle gas compressors, net gas compressors, recovery sections, and the stabilizer and debutanizer overhead sections.

Organic chloride addition is carefully controlled in fixed-bed semiregenerative and cyclic regenerative reformers during steady-state operations of the units and during catalyst regenerations. It is vital to diligently follow operating procedures for the three catalytic reforming processes with respect to operating at appropriate catalyst chloride concentrations so as to optimize the productivity of the reformers and minimize chloride-induced reliability challenges. Applicable water/chloride management procedures for fixed-bed semiregenerative reformers and cyclic regenerative reformers were reviewed in detail in Sections 6.1 and 6.3, respectively. The focus in this section of the review is on the continuous catalyst regenerative reforming process as the chemistries of hydrogen chloride formation, concerns with disposal of hydrogen chloride through catalytic reformer liquid and gas products, ammonium salt fouling mechanisms, and net hydrogen gas and stabilizer chloride management are essentially the same for the three catalytic reforming technologies.

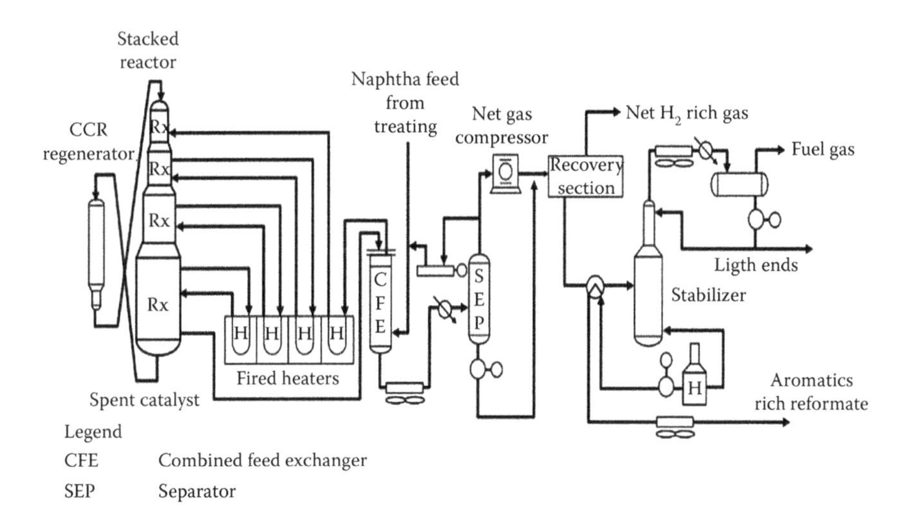

FIGURE 8.19 UOP CCR platformer with stacked reactor system.
Illustration Courtesy of Honeywell UOP.

8.3.2.1 Chloride Management for Optimizing Reforming Operations

Platinum concentrations in catalytic reforming catalysts used in CCR process units are usually in the range of 0.25–0.35 weight percent, tin concentrations of 0.2–0.30 weight percent, and about 1.0 weight percent chloride on a gamma alumina support. Catalytic reforming catalysts were reviewed in reasonable detail in Chapter 5. Reforming process units usually generate most of the inorganic chloride and precisely hydrogen chloride in oil refineries due to the continuous addition of organic chloride to the units. In one of the oil refineries and set of naphtha processing units that I managed, my conservative estimates indicated that two of the continuous catalyst regeneration reformers released about 100,000 pounds of hydrogen chloride annually into catalytic reformer equipment, chloride guard beds, regenerator vent gas caustic scrubbers, and the refinery. The estimated amount of hydrogen chloride deposited within that oil refinery did not include chloride losses associated with equilibrating fresh makeup catalyst and chloride retained on elutriated catalyst particles and fines. Effluent and entrained hydrogen chloride are deposited in a variety of equipment and streams in the refinery. It is known that hydrogen chloride is deposited on catalytic reforming equipment, in stabilizers, refinery fuel gas, light end products, regeneration vent gas, reformate tanks, net hydrogen gas, chloride beds, and caustic scrubbers. The net hydrogen gas produced by a catalytic reformer is usually treated in a chloride guard bed to remove hydrogen chloride before transporting the gas through refinery piping to various hydroprocessing equipment such as feed/effluent exchangers, low- and high-pressure separators, fin fans, and stripper overhead sections. Catalyst chloride data from a representative CCR process unit's hydrogen chloride production estimates are listed in Table 8.12.

Chloride guard beds are used specifically to remove hydrogen chloride from net hydrogen gas. For oil refiners operating with either no chloride guard bed units or poorly performing chloride guard bed adsorbents and procedures, significant amounts of hydrogen chloride are deposited in their refineries with consequent major fouling and corrosion challenges in various sections and equipment of hydroprocessing units. Typical areas of hydrogen chloride-induced fouling and corrosion in hydroprocessing are as depicted by Lordo in Figure 8.20.[46]

Since organic chlorides are introduced by opportunity oils that are a small percentage of the crude oil slates, it is reasonable to conclude that catalytic reforming units are the major contributors of hydrogen chloride in oil refineries.

Therefore, the first recommendation for controlling and managing hydrogen chloride from catalytic reformers is to use catalysts with excellent chloride retention

TABLE 8.12
Annual HCl Production from CCR Units

	Unit A	Unit B	Unit C
CCR, pounds per hour	1400	4500	1800
Regen Catalyst Cl, wt. %	1.10	1.10	1.08
Annual HCl, M pounds	25.30	81.10	32.40

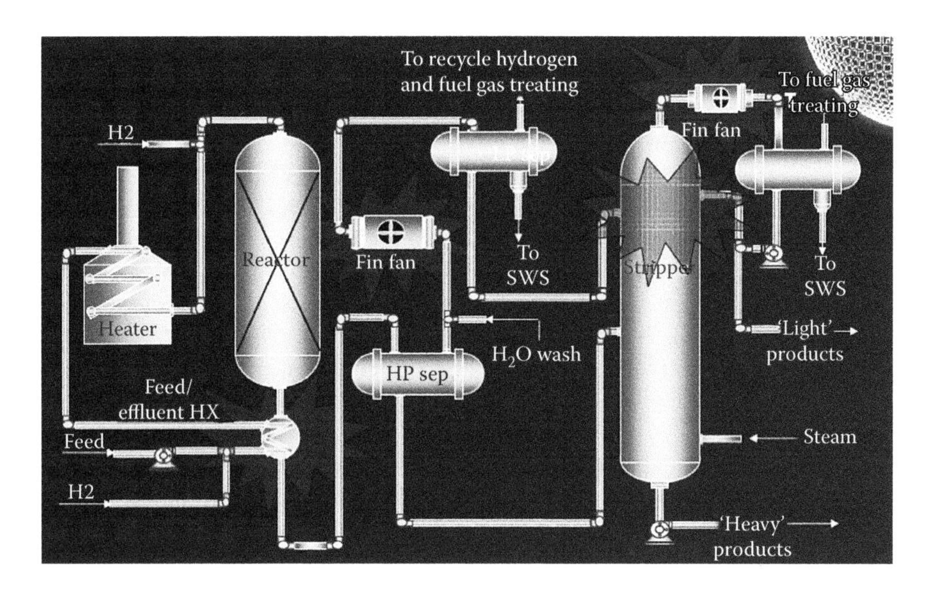

FIGURE 8.20 Areas of chloride induced fouling and corrosion in hydroprocessing units. From Lordo, S. A., Primer on Organic Chlorides and Their Control, *Nalco Energy Services*.[46] (Courtesy of Nalco Champion.)

characteristics, as that would reduce organic chloride addition rates and hydrogen chloride in catalytic reformers and in the net hydrogen gas. Secondly, it is important to monitor recycle gas water/chloride to reduce excessive chloride stripping from the catalysts within the reactors.

8.3.2.2 Mitigating HCl-Induced Fouling in Catalytic Reformers

The first significant negative impact of hydrogen chloride after the reactors is fouling and corrosion of combined feed effluent exchanges, condensers, piping, recycle gas compressors, and stabilizers. Fouling is due to ammonium chloride, ammonium sulfide, and bisulfide salt deposition in separators, condensers, debutanizer trays, debutanizer overhead, and condensers. Ammonia is formed from the residual organic nitrogen compounds in the feed to the catalytic reformer. Hydrogen chloride and hydrogen sulfide react with ammonia to form salts in the cooler areas of the product separation section. Ammonium salt formation reactions are shown in Figure 8.21.

Salt deposition and corrosion occur and often result in unscheduled downtime and processing interruptions of some days for stabilizer washes and as long as a few

$$NH_3(g) + HCl(g) \longrightarrow NH_4Cl(\text{solid})$$

$$NH_3(g) + H_2S(g) \longrightarrow NH_4HS(\text{solid})$$

$$2NH_3(g) + H_2S(g) \longrightarrow (NH_4)_2S(\text{solid})$$

FIGURE 8.21 Ammonium salts formation reactions.

weeks for tray replacements. Additionally, ammonium salt deposits accumulate and cause ineffective fractionations, losses in the yields of reformates and hydrogen from catalytic reformers, and curtailment of rates of downstream hydroprocessing units due to reduced hydrogen supply.

Ammonium salt deposition and corrosion may occur over a period of a few months and is dependent on chloride addition rates, turnaround cycles, and catalytic reformer operating conditions. Fouling may be significant in heat exchangers and distillation tower trays and cause poor fractionation, loss of revenues, and higher severity catalytic reformer operations. Operating costs usually increase due to greater heater duty and higher severity reformer operations.[53]

Critical equipment and stabilizer areas that are negatively impacted by ammonium chloride fouling and hydrogen chloride induced corrosion are shown in Figure 8.22.[54,65]

The recommended approach for managing refinery reliability issues due to fouling and corrosion is to proactively determine the contributing causative compounds and chemistries and minimize the impact of the contributing compounds. It is highly recommended that reliability problems be mitigated as much as possible at the source. Ammonium salt-induced fouling and corrosion problems in the catalytic reformer can be controlled or minimized via changing the composition of the crude slate, adequate naphtha hydrotreater and stripper operations, proper use of water wash procedures, effective use of additives, and efficient management of reformer operations.

Where feasible, the naphtha hydrotreater should be operated to achieve <0.2 ppm nitrogen and <0.2 ppm sulfur in the reformer feed. Use of high-performing naphtha hydrotreater catalysts should enhance hydrodenitrogenation and hydrodesulfurization activities. Discussions from past National Petrochemicals and Refiners Association (NPRA, and now AFPM) Q&A conferences have emphasized that reductions in platformer feed nitrogen and sulfur and good water/chloride management in the

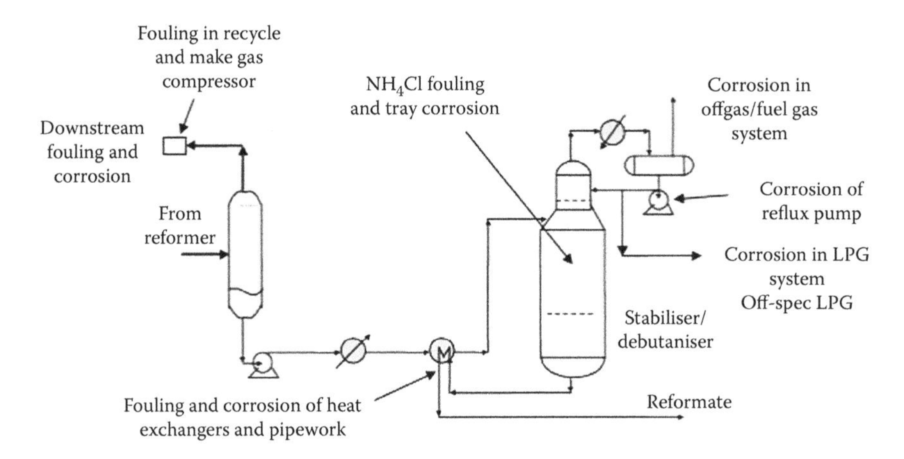

FIGURE 8.22 Hydrogen chloride related fouling and corrosion areas in reformers.
From Broadhurst, P. V., Hawkins, G. B., Young, N. C., Issues in the Removal of Chloride Compounds from Catalytic Reformer Product Streams. PTQ Chloride Guard Bed Paper, 2003.[65] (Courtesy of Johnson Matthey PLC.)

reformer would lead to significant improvements in the performance and reliability of recycle gas compressors and stabilizers.[56–60] Effective recycle gas water/chloride management, alluded to in the preceding section and in Chapter 6, would aid in reducing the amount of water deposited in the cooler sections of catalytic reformers.

One effective method used for removing ammonium salt deposits is good water washes of the stabilizers during turnarounds. Use of the services of special equipment cleaning companies such as Hydrochem is recommended.

Another approach used in managing and minimizing the negative effects of hydrogen chloride in the product separation section is to use adsorbents to remove hydrogen chlorides in the product separation section. Chloride adsorbent technology suppliers suggest several chloride guard bed locations that could help mitigate ammonium salt deposition on critical equipment and eliminate HCl in reformate, LPG, fuel gas, and net hydrogen gas or make gas.[61,64,65] A diagram of a simplified catalytic reformer unit and possible locations of chloride guard beds is provided as Figure 8.23.[65]

Chloride guard beds are used by some oil refiners to treat the feed to the stabilizer and the refinery fuel gas and possibly the LPG. Most oil refiners use dedicated chloride guard beds to treat the net hydrogen gas.

The Kurita Company has introduced an ammonium chloride-free (ACF) technology that uses additives of chemical liquid formulations that are based on a strong organic base. The strong organic base reacts with strong acids such as hydrochloric acid to form a liquid ACF chloride salt. ACF salts have a neutral pH, are very hygroscopic, and can be easily removed with water. According to the Kurita Company, the organic chemical neutralizes hydrogen chloride, thereby eliminating

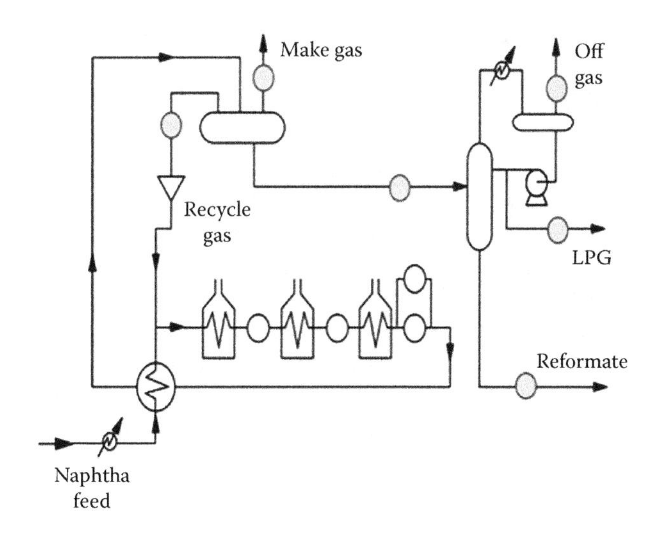

FIGURE 8.23 Possible chloride guard bed locations for HCl.
From Broadhurst, P. V., Hawkins, G. B., Young, N. C., Issues in the Removal of Chloride Compounds from Catalytic Reformer Product Streams. PTQ Chloride Guard Bed Paper, 2003.[65]. (Courtesy of Johnson Matthey PLC.)

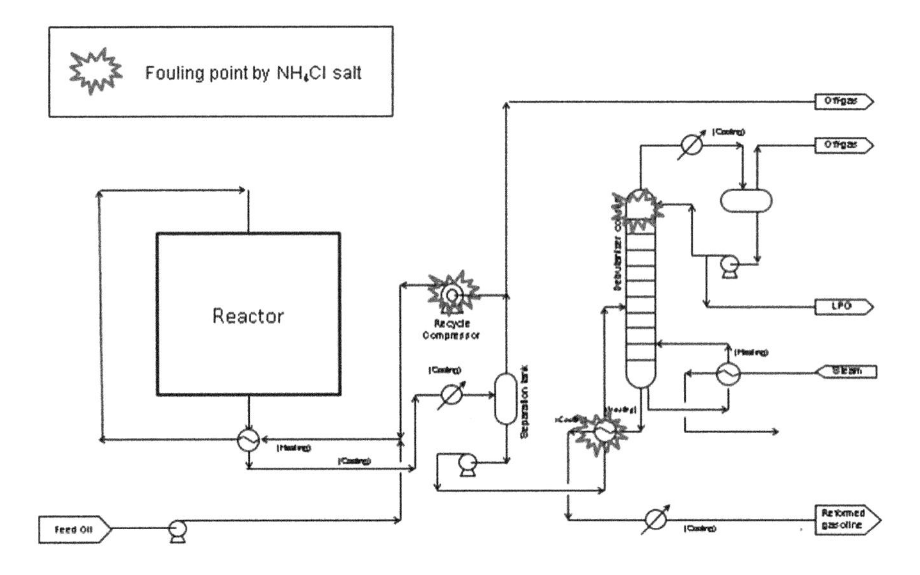

FIGURE 8.24 Kurita's ACF technology application in a reformer debutanizer. Courtesy of Kurita Company.

the reaction of ammonia and hydrogen chloride. The ACF technology also reacts in aqueous medium to remove some of deposited ammonium chloride salts. Kurita's ACF technology is excellent for online removal of and prevention of ammonium salts formation in catalytic reformer systems.[55] An application scheme for ACF is shown for the reformer system in Figure 8.24.

8.3.2.3 Eliminating HCl in Net Hydrogen Gas from Catalytic Reformers

Net hydrogen gas is used in naphtha, diesel, cat feed hydrotreaters, hydrocracking, selective hydrogenation, and paraffin isomerization units. Hydrogen is one of the key products of catalytic reforming units. As reviewed in Section 8.3.2.1, organic chloride is added during the oxy-chlorination step in the catalyst regeneration process for the redispersion of oxides of platinum and promoter metals during catalyst regeneration. Sufficient chloride is added to make up for chloride that is stripped from the catalyst during coke burn. In addition, as discussed in Chapter 6, chloride is also added to the reactor section in fixed-bed regenerative reformers to maintain about 1.0 weight percent chloride on the catalyst. For CCR process units, chloride is added in the catalyst regenerator and specifically in the chlorination zone. Excess hydrogen chloride that is generated in the reactors is distributed between reformate, net hydrogen gas, and regeneration vent gas. During steady-state catalytic reformer operations, hydrogen chloride in the recycle gas is about 1 ppm for semiregen and cyclic regenerative reformers, whereas that of CCR platformers is usually in the 2–8 ppm range.

Net hydrogen from catalytic reformers should, therefore, contain about 1–8 ppm of hydrogen chloride.[1] Some high configuration refineries have additional hydrogen supplied via pipelines from gas companies and dedicated refinery-located hydrogen plants based on steam methane reforming technology. Pipeline hydrogen and steam methane reformer-supplied hydrogen do not contain hydrogen chloride and, therefore, most of the hydrogen chloride in the refinery is from untreated net hydrogen gas.

Chloride guard bed adsorbents are recommended for eliminating hydrogen chloride in the net hydrogen gas as well as for other products from the catalytic reformer, as shown in Figure 8.19, for refinery fuel gas, reformate, and light ends. Chloride guard beds on light ends protect the vessels and equipment in the saturated gas plant, and those on the refinery fuel gas protect burner tips in fired heaters. Developments in alumina-based adsorbents for hydrogen chloride adsorption have led to the current promoted alumina and mixed oxide adsorbent systems. The high acidity of earlier alumina adsorbents was modified and promoted alumina with basic metal oxides such as sodium oxide are currently used. Basic metal oxides react with hydrogen chloride to form salts, and salt formation enhances the adsorptive capacity of the promoted alumina relative to the nonpromoted alumina. In addition, basic metal oxides decrease the acidity of the alumina substantially.[61] The modification of the alumina adsorbent delays acidification of the promoted alumina and formation of organic chlorides and polymerized, high molecular weight organic compounds that are referred to as "green" or "red" oils in oil refineries. Eventually, promoted alumina sites are acidified, leading to formation of undesired products, as shown in Figure 8.25.[62,63]

With acidification of the promoted alumina adsorbent and mixed adsorbents, organic chloride and green oil compounds that are formed go essentially undetected. Organic chloride reacts with hydrogen and decomposes at high temperatures downstream of the chloride guard beds in piping, equipment, and hydroprocessing units. Green or red oils that are generated on spent adsorbents cause reliability challenges due to contamination and fouling of units.

Because a chloride guard bed adsorbent performance can be drastically and negatively degraded by acidification, a two-stage, lead/lag chloride guard bed system with sampling points after each bed is recommended. The lag chloride guard bed should be replaced after HCl breakthrough is detected. After replacement, the lag becomes the lead bed and the freshly loaded adsorbent is placed in the lag position. This arrangement and operation should provide greater protection and improved reliability for the oil refinery from the fouling and corrosive consequences of net hydrogen gas-borne HCl, organic chloride, and green oil. In addition, a reasonable frequency of monitoring of the lead and lag chloride guard beds for hydrogen chloride and timely adsorbent replacements is required. A simple diagram of the two-stage lead/lag chloride guard bed is shown in Figure 8.26.

A single chloride guard bed as shown in Figure 8.27 can also be used as long as spent chloride adsorbents are replaced expeditiously as required. In addition,

$$\mathrm{R{=}R' + HCl \longrightarrow R'Cl}\ \text{(organic chloride)}$$

$$\mathrm{XR_2C{=}CR_2 + H^+ \longrightarrow (R_2C{=}CR_2)_{X^-}}\ \text{(green oil)}$$

FIGURE 8.25 Undesirable reactions on acidified HCl adsorbents.

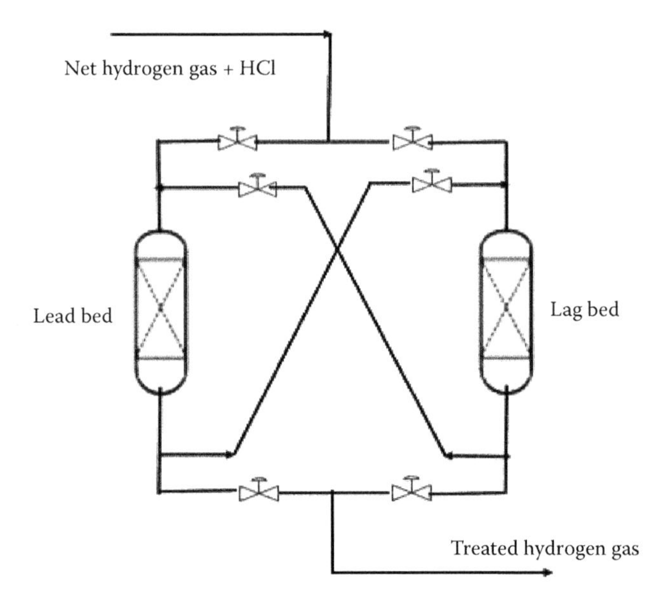

FIGURE 8.26 Lead/lag chloride guard bed arrangement.

special considerations should be made to sample the treated gas for organic chloride breakthrough. Draeger, Sensidyne, and Matheson-Kitigawa can all supply special tubes for the detection of paraffinic chlorides such as methylene chloride, tri-chloroethane, and olefinic chlorides such as vinyl chloride and tri-chlorotheylene. If oil refiners are unable to effectively monitor the chloride compounds for their single chloride guard bed, they could elect to replace their adsorbents at about 70%–80% of estimated hydrogen chloride coverage based on past spent adsorbent chloride concentrations.

Promoted alumina, zeolitic materials, and mixed oxide adsorbents are available suitable adsorbents depending on the performance objectives of the oil refiner. Oil refiners can choose the adsorbent that is suitable for their net hydrogen chloride removal service. Currently promoted alumina adsorbents are favored by many oil refiners.

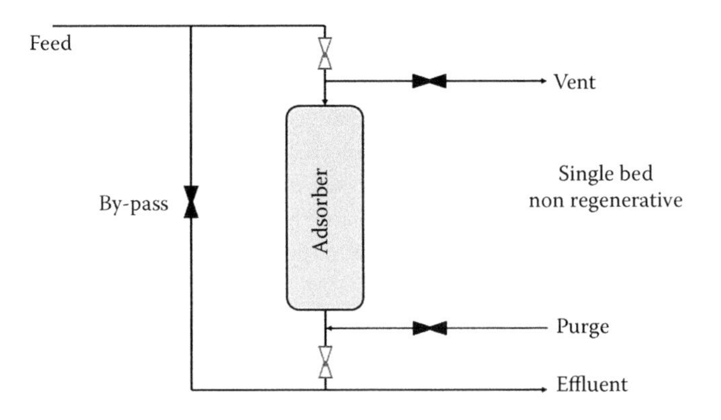

FIGURE 8.27 Single chloride guard bed.

The use of good efficient chloride guard bed management systems is vital, and equally vital are good monitoring programs for oil refining process unit operations, especially catalytic reformer operations. Additionally, there should be regular auditing of process unit operations and of the operations of key refinery equipment, burners in fired heater, fouling and corrosion in saturated gas plants, and data of monitored items compared to expected targets and benchmarked against productivity and reliability of other oil refineries, if available. Modifications should be made as required and best practices established for monitoring, inspections, and regular routine and turnaround maintenance of various refinery equipment as required. Operating a refinery in the top echelon for refinery reliability and safety requires dedication, monetary and personnel resources, and a highly conscious corporate health, safety, and environmental culture.

8.4 CATALYTIC REFORMER OPERATIONS AND RELIABILITY CHALLENGES WITH REFINERY MAXIMUM ACHIEVABLE CONTROL TECHNOLOGY II

The US EPA published an RMACT II regulation that was contained in US EPA 40 CFR Part 63 National Emission Standards for Hazardous Air Pollutants (NESHAP) for petroleum refineries on April 11, 2002. Fluid catalytic cracking (FCCU), sulfur recovery units (SRUs), and catalytic reforming units (CRUs) were the oil refining units that were impacted by the proposed regulation for a variety of hazardous air pollutants.[68] For catalytic reforming units, RMACT II regulation mandated that oil refiners meet regeneration effluent or vent gas hydrochloride chloride limits of either less than 30 ppm or 97% reduction relative to reference untreated emission levels and less than 10 ppm toxic organic compounds by April 11, 2005. Regeneration vent gas HCl from fixed-bed regenerative reforming units is limited to less than 20 ppm, and that for TOC is less than 10 ppm.

Oil refiners in the United States are meeting the US EPA RMACT II regulations limiting regeneration effluent gas hydrogen chloride and toxic organic compounds, and the technologies in use are covered here for the benefit of national oil refining companies operating in countries that could use similar technologies for improving the quality of their atmospheric environment. It is expected that some form of the USA RMACT II regulations will be applied globally in future programs to reduce hazardous air pollutants. US oil refiners can use some of the technical and operational information reviewed for improved operations and reliability of their regeneration vent gas hydrogen chloride and toxic organic compound removal and enhance the overall profitability and reliability of their catalytic reformers and refineries.

Fixed-bed semiregenerative reformer regeneration procedures include continuous circulation of caustic/water wash flow treatment to neutralize any acidic gases, especially hydrogen chloride, and removal of toxic organic compounds during the coke burn step. The caustic neutralizing wash is contacted with the effluent gas after the last reactor and before the effluent coolers, as was reviewed in Section 6.1.4 and as shown in Figure 8.28.[66] Spent caustic/water solution effluent from the separator is sent to a wastewater treatment unit.

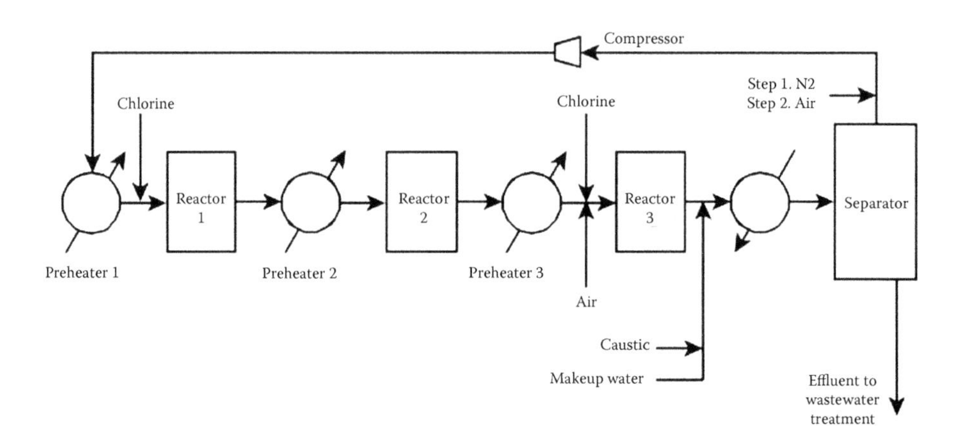

FIGURE 8.28 Caustic neutralization of regeneration effluent gas.
From Leavitt, M. et al., Technical Support Document for the 2004 effluent Guidelines program Plan. EPA-821-R-04-014.[66]

Fixed-bed semiregenerative catalytic reformers with appropriate caustic neutralization of the regeneration effluent gas for HCl and TOC removal typically meet the RMACT II regulations limits. As discussed previously, oil refiners can also use treatment services of companies such Hydrochem and others for their turnaround services support and catalyst regenerations.

Due to the frequent reactor regenerations of cyclic regenerative reformers, a chloride guard bed was determined to be sufficient to meet RMACT II-mandated HCl and TOC limits. A chloride guard bed with appropriate monitoring and timely replacement of spent adsorbents is a cost-effective technology for meeting RMACT II HCl and TOC limits for cyclic regenerative catalytic reforming process units.

Continuous catalyst regenerative reformers are also required to meet US EPA RMACT II regeneration vent gas (RVG) HCl and TOC limits continuously while operating the catalyst regeneration section. United States oil refiners met RMACT II regulations with moderate capital expenditures and opted for one or more of the three available chloride elimination technologies for their regeneration effluent gases. Chloride guard beds, the Honeywell UOP Chlorsorb system, and the traditional caustic scrubber vent gas wash tower technologies are available to non-US oil refiners that may be called upon to reduce regeneration effluent gas hydrogen chloride and toxic organic compounds. Selection criteria for the CCR reformers that were in my sphere of responsibilities included use of cost-effective technologies, moderate capital expenditures, moderate operational and maintenance costs, and good reliability, with negligible impacts on CCR catalytic performance and productivity.

Typical regeneration effluent gases from catalytic reformers contain nitrogen, nitrogen oxides, sulfur oxides, oxygen, carbon dioxide, water, and hydrogen chlorides in the range of 500–2500 ppm for CCR units. Due to the mix of hydrogen chloride, water, and other gaseous compounds in the regeneration vent gas, a corrosive environment is expected in areas of low temperature where condensation of regeneration vent gas water and hydrogen chloride occurs. It is necessary to emphasize at this point that available technologies for meeting RMACT II regulations pose either reliability or catalytic reforming operational challenges.

According to UOP, the Chlorsorb system is an effluent treating technology that recovers chloride from regenerator vent gas streams and recycles the chloride back to the reactor section of the CCR process unit. Catalytic reforming catalyst is essentially used as the adsorbent. The process with the spent catalyst operating as a chloride adsorbent is shown in Figure 8.29.[67] The other available technology involves scrubbing the regeneration vent gas with caustic solution and completing disengagement of gas and caustic solution in a vent gas wash tower (CS/VGWT). Caustic scrubbing of acidic gases, especially hydrogen chloride, has been used for over 50 years to protect effluent coolers and separators during catalyst regenerations in fixed-bed semiregenerative reformers. A variation of caustic scrubbing has been applied successfully in neutralizing hydrogen chloride in platinum chlorided-alumina catalyzed paraffin isomerization unit off gas for several decades.

The major concern with respect to use of caustic scrubber technology in the treatment of process effluent gases is corrosion of the scrubber metallurgy due to poor management of circulating caustic solution quality and the cost of spent caustic disposal. The Venturi scrubber or vent gas wash tower may have to be replaced as a result of excessive corrosion. Concerns with potential equipment corrosion have led to recommendation of Hastelloy 2000, a better corrosion resistant metallurgy, for use in manufacturing Venturi scrubbers and carbon steel for vent gas wash towers. Some US oil refiners have elected to use Incolloy for the wash tower to improve the overall reliability of their caustic scrubber and vent gas wash tower systems. A simple illustration of a Venturi scrubber and vent gas wash tower system is provided in Figure 8.30.

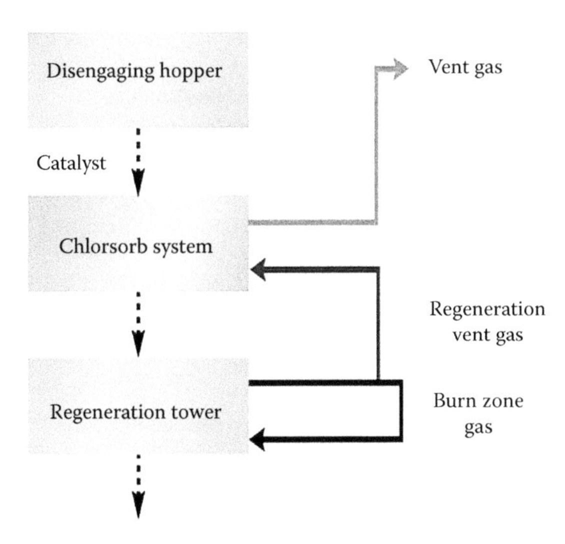

FIGURE 8.29 UOP's chlorsorb process for vent gas HCl removal. Illustration Courtesy of Honeywell UOP.

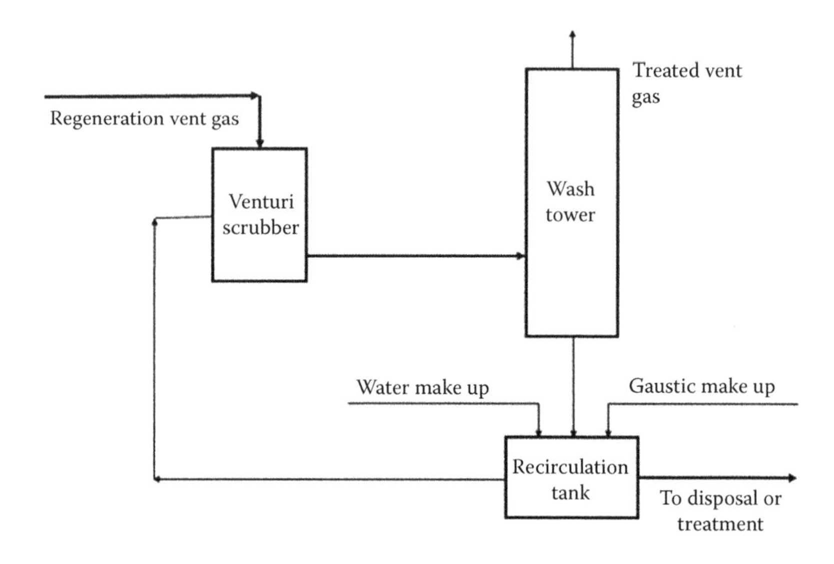

FIGURE 8.30 Caustic scrubber vent gas wash tower unit.

Honeywell UOP listed a number of benefits for use of the Chlorsorb technology relative to the caustic scrubber vent gas wash tower system in a 2013 paper, and the technologies are compared in Table 8.13.[67] Despite the cited benefits of Chlorsorb relative to CS/VGWT with respect to costs associated with caustic use and disposal and utilities, there are some operational concerns with respect to the reforming catalyst essentially functioning as the adsorbent for chloride elimination

TABLE 8.13
Comparison of CS/VGWT and Chlorsorb

Effluent Treating Technology	Venturi Scrubber/Vent Gas Wash Tower	Chlorsorb Systems
RMACT II Compliance	Yes	Yes
Make-up Chloride Injection, MTA	33.6	6.5
Make-up Caustic Injection, MTA	312.3	0
Spent Caustic for Disposal, MTA	13,164	0
	Utilities	
Electricity, 1000 KW-hr/year	213	10
Cooling Water, MTA	249,000	0
LP Steam, MTA	0	1901
Condensate Consumed, MTA	13,120	0

Source: Shakur, M., Lok, K., Gaicki, R., Go Green with UOP's Chlorsorb™ System by Eliminating Caustic Usage from CCR Vent Gas, Honeywell UOP. www.uop.com/go-green-with-uops-chlorsorb-system-by-eliminating-caustic-usage-from-ccr-vent-gas-treating/.[67] Illustration Courtesy of Honeywell UOP.

in the Chlorsorb technology. Based on reviews of catalytic reforming catalysis and reforming processes in Chapters 4, 5, and 6, care should be exercised in catalytic reforming operations so as to maintain catalyst chloride and the necessary metal/acid functionality balance of catalysts. As reviewed previously, catalyst chloride and metal/acid functionality balance are critical for achieving optimal performance with respect to reformate and hydrogen production and catalytic performance stability. Operating with current catalysts that have excellent chloride retention characteristics poses special challenges with use of the Chlorsorb as catalyst chloride, and total system hydrogen chloride could substantially increase.

The possible ramifications of the spent catalyst performing as a chloride getter were reviewed and the US EPA set spent catalyst and Chlorsorb-treated catalyst chloride limits of 1.35 and 1.8 wt. %, respectively, due to concerns with catalyst chloride accumulations. For CCR reforming operations, regenerated catalyst chlorides are usually capped at between 1.0 to 1.2 weight percent for optimal catalytic performance. High catalyst chlorides obviate the need for adding any more chloride and conducting oxy-chlorination of catalysts. Not conducting adequate oxy chlorination and platinum redispersion for catalysts could lead to subpar-performing catalysts, significant degradation of catalytic performance, and resultant reformate and hydrogen yield losses. It is necessary to emphasize that significant CCR productivity losses could be incurred due to poor catalyst regeneration operations, as shown in Figure 8.31.

Reformate yield differences for good continuous catalyst regeneration and poor regeneration to no regeneration for catalyst in a 20-MBPD CCR process unit are provided Figure 8.31. At higher octane severities, the reformate yield differences increased significantly, and it is about 4 volume percent at a reformate octane of 100.

Additionally, high chlorides on catalyst and high hydrogen chlorides in CCR systems exacerbate challenges due to increased fouling and corrosion in product separation sections; higher chlorides in reformates, light ends, and fuel gas products; and higher frequency of chloride guard bed adsorbent replacements.

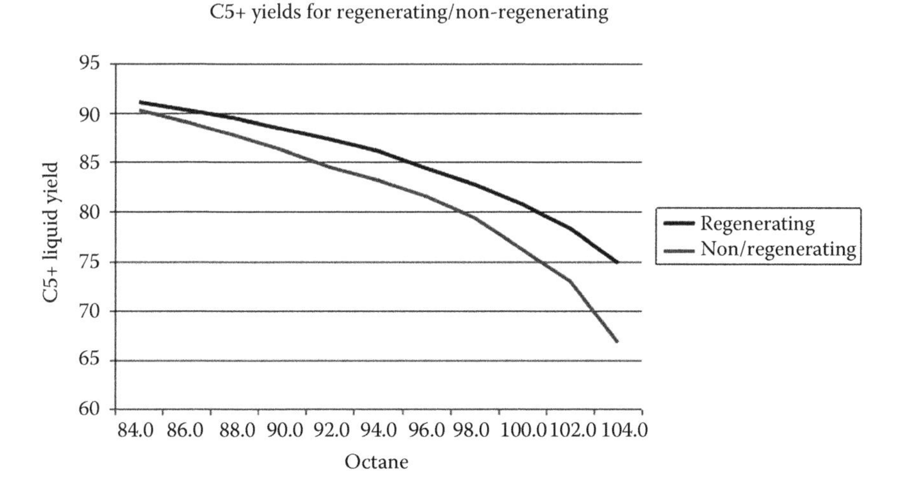

FIGURE 8.31 Differences in reformate yields with octanes.

Caustic scrubber/vent gas wash tower technology can also create major challenges that were reviewed earlier and include high rates of corrosion of metallurgy in the caustic scrubber and vent gas wash tower units. They also include costs of fresh caustic and disposal of spent caustic solution. Monitoring and inspection frequencies have to be increased and matched with diligence in maintaining adequate caustic solution quality to meet RMACT II regeneration vent gas HCl and TOC limits. Failure to maintain effective management of caustic scrubber vent gas wash tower operations could lead to increased corrosion of metallurgy and replacements of piping, scrubber, and vent gas wash tower systems. Frequent inspections of the caustic vent gas water systems should be conducted, and replacements of key equipment should be conducted as required during planned CCR process unit turnarounds.

Three technologies are available for substantial reductions of hydrogen chloride and toxic organic compounds in regeneration vent gas. Chloride guard bed technology has been found to be convenient and cost effective for meeting regeneration vent gas HCl and TOC reductions for fixed-bed cyclic regenerative reformers with its intermittent reactor regenerations. In the case of fixed-bed semiregenerative reformers, the use of continuous circulation of caustic/water for neutralization of HCl during catalyst regenerations has been found to be cost effective. Chlorsorb and caustic scrubber wash tower technologies are suitable for continuous catalyst regenerative reformers. The short review provided in this section could assist oil refiners in making a decision on an appropriate technology for their specific CCR naphtha processing requirements. It is also necessary to emphasize that a continuous monitoring, frequent inspections, and proactive routine and turnaround maintenance are required for technologies that the oil refiner selects. The replacement costs for corroded sections of regeneration vent gas equipment and sections could be high, and downtime for equipment maintenance should be minimized.

8.5 SUMMARY

Several special catalytic reforming and related topics are covered in detail. Key learning or lessons of the oil industry show that good planning and use of good process technologies and catalytic reforming processes can aid in meeting a variety of oil refiners' challenges. Processing unconventional and opportunity oils can boost refinery profitability, and it can also generate processing and reliability challenges. The roles of catalytic reformers for the production of gasoline blendstock and hydrogen, and as "gate keepers" with respect to the monitoring of oil contaminants, are emphasized. Prefractionation and postfractionation strategies for meeting gasoline benzene requirements and the vital impact of catalytic reforming on benzene production are reviewed. Some digression was considered necessary to review process technologies that are used for meeting the MSAT II gasoline benzene concentration of 0.62 volume percent in the United States. Process technologies for gasoline benzene reductions are expected to be of great use globally as other countries consider it necessary to reduce the concentrations of benzene in gasolines. The management of inorganic and organic chlorides in oil refining is extremely important, as ammonium salts and corrosive acidic solutions formed can lead to major reliability challenges in the crude distillation units, catalytic reformers, and refineries. Technologies for managing hydrogen chloride generated via catalytic reforming operations can be applied successfully by following operating and maintenance guidelines. It is necessary to follow guidelines provided for managing single and lead/lag chloride guard bed systems for treating net hydrogen gas. Potential operational and reliability challenges are reviewed for the Chlorsorb and caustic scrubber vent gas wash tower systems to enhance oil refiners' decision making in their selection of technology for meeting RMACT II regulations for regeneration vent gas hydrogen chloride and toxic organic compounds. Guidelines are provided for moderating the expected catalytic reforming operational challenges with the use of the Chlorsorb technology. Similarly, corrosion-induced reliability challenges with use of the caustic scrubber vent gas wash tower are so severe that oil refiners are selecting more corrosion-resistant metallurgy for their caustic scrubber and vent gas wash tower sections.

REFERENCES

1. Oyekan, S. O., Personal Notes.
2. Oyekan, S. O., Catalytic Naphtha Reforming Lecture Notes in Catalytic Processes, in *Petroleum Refining* by Chuang, K. C., Kokayeff, P., Oyekan, S. O., Stuart, S. H., AIChE Today Series, 1992–1998.
3. Weber C. Yeung, S., Opportunity Crudes. *COQA Meeting*, October 27, 2011.
4. Joint UNDP/World Bank Energy Study Management Assistance Programme, Crude Oil Price Differentials and Differences in Oil Qualities; A Statistical Analysis, ESMAP Technical Paper 81, October 2005 Accessed July 10, 2017, www.esmap.org.
5. Garrett, T., Rathanakhambay, A., Robbins, N., Wunder, M., Yeung, T., *The Challenges of Crude Blending – Part A*. Hydrocarbon Publishing, February 2016. Accessed July 6, 2017, http://www.digitalrefining.com/article/1001216,The_challenges_of_crude_blending___Part_1.html.
6. ExxonMobil, assays Available for Download, Worldwide Operations. Crude Oil Trading, http://corporate.exxonmobil.com/en/company/worldwide-operations/crude-oils/assays.
7. Capline Pipeline, Most Current Complete Assay for Crude Type of Choice, http://www.caplinepipeline.com/Reports2.aspx
8. Benoit, B., Zurlo, J., Overcoming the Challenges of tight Shale Oil Refining: Processing Shale feedstocks, www.eptq.com.
9. Wier, M. J., Sioui, D., Metro, S., Sabitov, A., Lapinski, M., Optimizing Naphtha Complexes in the tight Oil boom, UOP, 2014.
10. Oil Sands Magazine, Products from the Oil Sands: Dilbit, Synbit & Synthetic Crude Explained, http://www.oilsandsmagazine.com/technical/product-streams
11. Weber, C., Yeung, S., Opportunity Crudes: To Process or Not To Process, *COQA Meeting*, October 27, 2011.
12. BP Global Energy Trading, Detailed Crude Oil Assays Accessed July 10, 2017, http://www.bp.com/en/global/trading/crude-oil-and-refined-products/crudes/americas/mars.html
13. What, When, How, In Depth Tutorials and Information, Assay of Crude Oils. Accessed July 5, 2017, what-when-how.com/petroleum-refining/assay-of-crude-oils-petroleum-refining/
14. Rhodes, A. K., Benchmark West Texas Intermediate Crude Oil Assayed Oil & Gas Journal, 92(33), August 15, 1994.
15. Capline Pipeline, Specifications for LLS Crude oil, http://www.caplinepipeline.com/reports1.aspx.
16. Crudemonitor.ca, Western Canadian select, www.crudemonitor.ca/crudes/index.php?acr=WCS.

17. Stratnor, Western Canadian Select whole Crude Properties, stratnor.com/wp-content/uploads/2015/07/Western-Canadian-Select1.xlsx.
18. Rhodes, A. K., Brent Blend, UK., North Sea Marker Crude Assayed. *Oil & Gas Journal*, 93(6), February 6, 1995.
19. Scribd, Crude Oil Properties and Quality Indicators, Uploaded by Felipe Grigorio. Accessed July 15, 2017, https://www.scribd.com/document/320233512/CRUDE-OIL-PROPERTIES-AND-QUALITY-INDICATORS-pdf.
20. Oyekan, S. O., Torrisi, S., Opportunities and Challenges in Transportation Fuels Production. *AIChE Regional Process Technology Conference*, Houston. October 2, 2009.
21. ExxonMobil, Kearl Crude Oil, Worldwide Operation, Crude Trading Accessed July 10, 2017, http://corporate.exxonmobil.com/en/company/worldwide-operations/crude-oils/kearl.
22. Reid, T. A., Coker Naphtha Hydrotreating Processing Solutions for Trouble-free Operations, Akzo Nobel Catalysts.
23. Debuisschert, Q., Nocca, J. L., *Prime-G+ TM: Commercial Performance of FCC Naphtha Desulfurization Technology*, Axens IFP Group Technologies.
24. Haldoe Topsoe, Coker Naphtha Hydrotreating Technology, https://www.topsoe.com/products/coker-naphtha-hydrotreating-technology.
25. Sanghavi, K., Schmidt, J., Achieve Success in Gasoline Hydrotreater, Refining Developments, 2011.
26. US EPA, Regulatory Announcement: EPA's Program for Cleaner Vehicles and Cleaner Gasoline. December 1999, nepis.epa.gov/Exe/ZyPDF.cgi/P1001Z9W.PDF?Dockey=P1001Z9W.PDF.
27. www.transportPolicy.net, Global Comparisons of Fuels.
28. Kalyanaraman, M., Smyth, S., Greeley, J., Pena, M., SCANfining Technology: A Proven Option for Producing Ultra-Low Sulfur Clean Gasoline. *LARTC 3rd Annual Meeting*, Cancun, Mexico, April 9–10, 2014.
29. Korpelshoek, M., Rock, K., How clean is your Fuel? Hydrocarbon Engineering, July 2009.
30. US EPA, Benzene. 71-43-2, www.epa.gov/sites/production/files/2016-09/documents/benzene.pdf.
31. Colwell, R. F., Benzene in Gasoline, Regulations and Remedies. Process Engineering Associates. (Source: "Final Regulatory Impact Analysis" EPA420-R-07-002), February 2007.
32. Potash. S., Benzene research in China Informs EPA Regulation, www.fic.nih.gov/News/GlobalHealthMatters/september-october-2015/Pages/china-benzene-studies-inform-epa.aspx.
33. Eilperin, J., New EPA Rules for Gasoline Limit Benzene, a Carcinogen Washington Post. February 10, 2007
34. Turner, F. C., Felton, J. R., Kittrell, J. R., Cost of Benzene Reduction in Gasoline to the Petroleum Industry, EPA-450/2-78-021. April 1978.
35. Jechura, J., Product Blending and Optimization Considerations, http://inside.mines.edu/~jjechura/Refining/11_Blending_Optimization.pdf.
36. Meister, M., Crowe, T., Keesom, W., Stine, M., Meister, M., Crowe, T., Keesom, W., Stine, M., Study Outlines US Refiners Options to reduce Gasoline Benzene Levels. *Oil Gas J.*, 104, 38–45, September 11, 2006.
37. US EPA, Federal Gasoline Regulations, www.epa.gov/gasoline-standards/federal-gasoline-regulations.
38. Schiavone, B. J., Technology Advancements in Benzene Saturation. NPRA 2007 Q&A Technology Forum, October 11., 2007.

39. Gary, J., Handwerk, G., Kaiser, H., Blending Component Values for Gasoline Blending Streams. *Petroleum Refining Technology and Economics*, 5th Edition. CRC Press. March 2007.
40. Nispel, D., Varraveto, D., *How Refineries Meet the New EPA Benzene Regulations*, Burns & McDonnell Company Presentation. 2008 Accessed July 6, 2017, http://docplayer.net/41149531-How-refineries-meet-the-new-epa-benzene-regulations-presented-by-dave-nispel-and-dominic-varraveto.html.
41. ExxonMobil, Cost-effective Process for Benzene Reduction in Gasoline, www.catalysts-licensing.com.
42. Merrill, T. W., Schizer, D. M., The Shale Oil and Gas Revolution, Hydraulic Fracturing and Water contamination: A Regulatory Strategy. *Minn. L. Re.*
43. Oil and Gas Journal, New and Modified Processes and Catalysts Needed for New Fuels. *Oil and Gas J.*, June 18, 1990.
44. Gutzeit, J., Effect of Organic Chloride Contamination of Crude Oil on Refinery corrosion, NACE-00694, Corrosion 2000, Orlando, Florida, March 26-31, 2000.
45. Ye, X., Gas Oil Desalting Reduces Chlorides in Crude. *Oil & Gas J.* October 16, 2000.
46. Lordo, S. A., Primer on Organic Chlorides and Their Control, *Nalco Energy Services.*
47. Chambers, B., Yap, K. M., Srinivasan, S., Yunovich, M., Corrosion in Crude Distillation Unit Overhead Operations: A Comprehensive Review paper 11360. *NACE Corrosion Conference & Expo*, 2011.
48. Cox, J., Chevron Finds Buyer for Tainted Oil, but Questions Remain. June 7, 2012.
49. McElreath, D., Comparison of Fractionation Fluids Composition with Produced Formation Water Following Fracturing – Implication of fate and Transport. *Chesapeake Energy.*
50. Crude Oil Quality Group, Crude Oil, Contaminants and Adverse Chemical Compounds and Their Effects of Refinery Operations, May 27, 2004.
51. Bagdasarian, A., Feather, J., Hull, B., Stephenson, R., Strong, R., Crude unit Corrosion and Corrosion Control. Corrosion/96. Paper 96615 NACE International, Houston, TX 1996.
52. Gutzeit, J., Controlling Crude Unit Overhead Corrosion – Rules of thumb for Better Crude Desalting, Paper No. 07567, NACE International Houston, TX 2007.
53. Alverson, F., Mullenix, A. J., Technology, New Process Removes Inorganic Chlorides from Reformates. *Petrolite Corp.* May 1996.
54. Hawkins, G. B., Vulcan Chloride Guard Technology, GBH Enterprises.
55. Kurita, Kurita's Technology Prevents Fouling and Corrosion, www.kurita.eu/en/ACF_technology.
56. NPRA Q&A Session on Reforming, 133, 1989.
57. NPRA Q&A Session on Reforming, 150, 1992.
58. NPRA Q&A Session on Hydrotreating, 98, 1997.
59. NPRA Q&A Session on Hydrogen Processing, 133, 2000.
60. NPRA Q&A Session on Light Oil Processing, 185, 2001.
61. Hawkins, G. B., Third Generation High Capacity Chloride Removal from Catalytic Reformer Product Streams, ICI Katalco.
62. Matsuoka, J., Aizawa, S., Fujiwara, K., Takase, T., Shioya, Y., Brown, R. S., New Chloride removal Catalyst in CCR Unit. AM-00-50. July 2003.
63. Matsuse, H., New Chloride removal Catalyst in CCR Unit July 2003.
64. Honeywell UOP, Adsorbent Solutions in Refining, www.uop.com/?document=adsorbents-solutions-in-refining&download=1.
65. Broadhurst, P. V., Hawkins, G. B., Young, N. C., Issues in the removal of Chloride Compounds from Catalytic Reformer Product Streams. PTQ Chloride Guard Bed Paper, 2003.

66. Leavitt, M., Grumble, B., Grubbs, G. H., Smith, N., Wall, T., Gooden, J., Johnson, C., Matuszko, J., Technical Support Document for the 2004 effluent Guidelines program Plan. EPA-821-R-04-014.
67. Shakur, M., Lok, K., Gaicki, R., Go Green with UOP's Chlorsorb™ System by Eliminating Caustic Usage from CCR Vent Gas, Honeywell UOP, www.uop.com/go-green-with-uops-chlorsorb-system-by-eliminating-caustic-usage-from-ccr-vent-gas-treating/.
68. EPA, National Emission Standards for Hazardous Air Pollutants For Petroleum Refineries: Catalytic Cracking Units, Catalytic Reforming Units, and Sulfur Recovery Units. 40CFR Part 63, FRL-7163-71, RIN 2060-AF28.
69. Standlee, C., Raising Gasoline Octane – The Considerations. *MSTRS Meeting*, Abengoa, May 5, 2015.
70. Schnepf, R., Yacobucci, B. D., Renewable Fuel Standard (RFS), Overview and Issues, CRS Report for congress, CRS 7-5700. www.crs.gov.

9 Special Topics—Turnaround and Platinum Management

9.0 INTRODUCTION TO TURNAROUND AND PLATINUM MANAGEMENT

Oil refiners should be staffed appropriately so that they can have the necessary manpower resources to conduct timely routine maintenance on equipment when required and execute well-planned, capital-effective turnaround maintenance for catalytic reforming process units satisfactorily. Satisfactory turnaround maintenance for a catalytic reformer is completed on schedule and within the capital budget. In addition, turnaround maintenance should be completed safely with no injuries to refinery personnel and contract workers. Another desired objective is that turnaround and startup of the catalytic reformer or other process units be conducted with no incidents of excessive emissions into the environment. Turnaround planning and maintenance and cost-effective precious metal management principles are covered in this chapter.[1,2] As discussed in Chapter 8, experienced process technologists and refinery reliability specialists are vital resources with respect to turnaround planning and execution. The experts usually interface with contracted companies with respect to catalyst handling and coordinate critical repairs on heaters, compressors, and fractionation units.

The management of platinum is important, and oil refiners should strive to operate cost-efficient platinum catalyzed processes such as light naphtha and butane isomerization processes, benzene saturation processes, and catalytic reformers. The cost of platinum required to manufacture the required amount of catalysts for a catalytic reformer could be in the range of 15–50 million dollars. An overview of the manufacturing process for platinum-containing catalysts is provided as well as strategies and procedures for continuous monitoring of the catalysts through the stages of handling, use, and platinum recovery from spent catalysts.

9.1 TURNAROUND PLANNING AND MAINTENANCE

Good planning and execution of incident-free, cost-effective, successful process unit and refinery turnaround maintenance projects are of utmost importance. Turnarounds conducted at the required optimal frequency are vital for maintaining the reliability, productivity, and profitability of oil refiners' assets in a highly competitive, global oil refining and marketing business. Refining of high percentages of opportunity crude oils with widely varying properties and those with high sulfur, high TAN, and organic chloride contaminants exacerbates challenges with respect to maintaining operations and reliability excellence. The challenges are even greater in refineries with high Nelson complexity indices. Processing of substantial fractions of tight oils in crude slates has also led to incompatibility issues with significant asphaltene precipitation and increased maintenance of crude and vacuum distillation units. With the need to stay competitive and profitable, aggressive turnaround schedules and production pressures have also led to major accidents, fires, and explosions in oil refineries either during the turnaround maintenance period or in the transition to startups for process units. Some notable recent refinery accidents include the BP Texas City refinery explosion in March 2005, the ENI Sannazzaro refinery fire of December 2016, and the Reliance Jamnager refinery fire in November 2016. Most oil refinery accidents and fires have occurred during turnaround maintenance of process units.[8–11]

In addition to increased routine and turnaround maintenance costs, there are high costs of lost production during process unit downtimes for turnarounds that should be accounted for. Tie-ins of revamps of process units to incorporate process enhancements and process technologies to meet fuel and environmental quality regulations such as MSAT 2 and RMACT II are usually accomplished during turnaround maintenance periods. As a result of the myriad of challenges, oil refiners should avail themselves of automation technologies for refinery asset health monitoring, wireless sensing, and statistical monitoring.[4,6,7] Performance data of heaters, rotating equipment, fractionation units, and piping of the catalytic reforming units should be monitored and trended to determine appropriate times for taking shutdowns of the catalytic reforming unit for turnaround maintenance. Turnaround planning and organizational structures, planning cycles, management steering committees, and multifunctional groups of corporate and refinery personnel are ideal representatives that can incorporate some of the good practices in turnaround maintenance project activities.[3,5–7] A multifunctional turnaround group should include representatives from relevant functional departments of an oil refinery. Turnaround groups should consist of representatives from the refinery's technical service, engineering and maintenance, product control and planning, operations, health, safety and environmental, purchasing, warehousing, corporate rotating equipment and heater reliability technologists, and corporate process technologists.

Group meetings should be held to create preliminary work scope, purchase catalyst and inert materials and chemicals, acquire equipment, and secure turnaround companies' services at least 6 months to 1 year before process unit outage and start of turnaround activities.

Planning for routine and turnaround maintenance should ideally commence either immediately after startup of a grassroots process unit or after the turnaround of a process unit. Using information and data gathered from plant information management systems and laboratory information management systems during past operations of the process unit, turnaround groups can then plan systematically for scheduled outages and mechanical work. The cost of the turnaround should have been determined and a budget allocated at least 1 or 2 years before the start date of the turnaround. Turnaround groups should use established best practice principles to determine the work scope. Best practice guidelines should enable the turnaround group to establish a schedule of key milestones for completion of different phases of the planning and execution of the turnaround maintenance. Critical milestones should include getting turnaround service companies on board at least 1 or 2 months before process unit outages, and lease agreements should be completed for cranes and other equipment. Catalysts, chemicals, and other necessary replacement materials should have been purchased. The group should secure the services of process and turnaround maintenance companies several months before process unit outage and begin to hold regular turnaround meetings for the sharing of project status updates and training of relevant personnel.

9.1.1 Turnaround Planning for Catalytic Reforming Units

Planning for the next turnaround maintenance of catalytic reforming process units should start after the completion of the last turnaround and startup. Catalytic and equipment performance data should be tracked and drops in performances analyzed. Based on results of the analyses of performance declines, a comprehensive list of items to be worked on during turnaround can be determined and prioritized. A list of equipment and various systems in the catalytic reforming process unit that are to be subjected to detailed inspections and maintenance should also be compiled. The list could include the following key items.

- Fired heaters
- Reactors based on catalytic performance and catalyst poison episodes (all catalytic reformers) and reactor coke burn profiles during catalyst regenerations (fixed-bed catalyst regenerative)
- Product separation section, which should cover combined feed effluent exchangers, recycle gas compressors, net gas compressors, stabilizer operations
- Catalyst circulation and catalyst fines management systems (CCR)
- Catalyst reduction zone section (CCR)
- Catalyst regeneration system
- Regeneration effluent gas technologies for meeting RMACT II HCl and TOC systems
- Chloride guard bed for net hydrogen gas

Critical reviews of the list of items above would enable turnaround groups to select special items for maintenance and companies that can provide necessary services and products for the turnaround and startup operations of the catalytic reformers. At the top of that list of service providers should be inspection experts who can provide detailed, comprehensive inspections of catalytic reforming reactors, catalyst circulation, and catalyst regeneration systems and excellent catalyst handling companies. Oil refiners typically contact Axens and Honeywell UOP for recommendations of appropriate CCR process unit inspectors for licensed reforming units. Other key support companies that should be used include those for reactor internal material fabrication and services, chemical treatment, heaters, turnaround maintenance, and technical support from catalyst and adsorbent supplier companies. A short review of possible findings and suggested actions is provided below. Due to work scope and the need to be focused on safety, schedule, and communication, brief turnaround group meetings should be scheduled daily with leading members of the diverse support groups to review status of the work in each section, safety concerns, logistics, and progress as soon as deemed necessary based on best practice guidelines for turnaround projects.

a. Heaters
   - Inspections should be conducted for the heaters and maintenance conducted as required.
   - Depending on the age of the heater tubes and processing service, structural integrity of heater tubes should be checked for damage such as metal carburization. Tube repairs and replacements can then conducted as required.[12–14]
   - Feed sulfur addition systems should be installed for metal surface passivation of the tubes and protection of the heater tubes.[15,16] Metal surface passivation aids in minimizing the rate of metal carburization and tube damage.
   - Fired heater operations and energy efficiency are monitored via infrared inspection and tube wall temperature determinations. Oxidation of heater tubes leads to scale deposits on the external surfaces of the heater tubes. Heat transfer to the liquid feed decreases as a result of the scale deposits. Based on reviews of performance data, a decision could be made to apply ceramic coating to the heater tubes. Application of ceramic coating to heater tubes has been shown to result in substantial revenue gains due to increased catalytic reformer productivity and energy cost savings.[17] IGS Cetek has demonstrated that higher energy efficiencies can also be achieved for heaters with their high emissivity refractory coating.

b. Reactors
   - When reactors are opened, the catalyst bed and reactor internals should not be disturbed until a preliminary unobtrusive visual inspection is conducted. This first critical visual inspection is conducted to assess any indication of blockage of catalyst and hydrocarbon flows and accumulation of debris on the top sections of scallops and centerpipes. For CCR process units, it is recommended that the corporate process technologist, refinery inspector, expert inspectors from a catalytic reforming process inspector service company, and personnel from the process licensor field services group inspect the undisturbed reactor bed and reactor internals.
   - Usually, the catalyst in the reactors and systems is dumped and screened if the turnaround plan included reloading the "equilibrium" catalyst instead of fresh catalyst. This is an ideal time to eliminate catalyst fines from the equilibrium catalyst despite the continuous elutriation of catalyst fines conducted during CCR reformer operations. Additionally, catalyst density grading separation can be applied to remove highly coked "heel" catalyst so that the highly coked catalyst particles do not negatively impact CCR reforming process performance after startup. A representative data set from the separation of heel-contaminated catalyst particles is shown in Figure 9.1.[18]

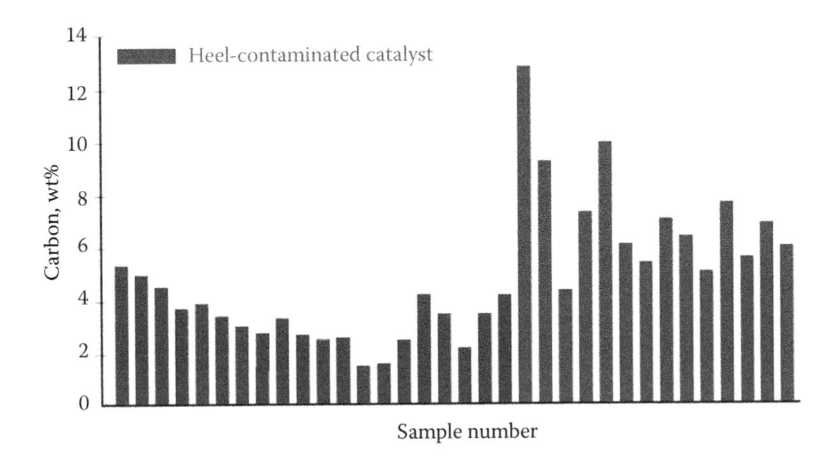

FIGURE 9.1 Data from catalyst density grading of CCR reformer catalyst.
From Porocel, Density Grading, http://www.porocel.com/25-density_grading/.[18] (Courtesy of Porocel International, LLC.)

- After the dumping of catalyst and inert catalyst spheres from reactors, reactor internals are inspected in detail. The information gathered is reviewed and reactor section maintenance work scope is established. Work scope should include checking the integrity of scallops and centerpipe and measuring the sizes of inner screen slots and monitor during turnarounds. Larger sizes of the inner screen slots negatively impact hydrocarbon flow and catalyst and platinum containment.
- Complete mechanical work scope, inspect reassembled internals, and reload inert catalyst spheres and screened catalyst (CCR) or load fresh catalyst as per catalyst loading procedure and turnaround plan. Use catalyst container bins, as reviewed previously in Chapter 6. Inert material and screened or fresh catalyst are loaded as per work scope and catalyst loading diagram. For all catalytic reformers, the reactors should all be inspected before closing the reactors after catalyst loading.
- For technical assistance on developing work scope for UOP CCR process units, review UOP's recommended list for inspection items and expected turnaround work.[19] Johnson Screens, now a subsidiary of Aqseptence, manufactures inner screens and scallops and is available to conduct screen repairs and scallops replacements.[20]

c. Product Separation Section
- Inspect and conduct maintenance as per work scope. Wash stabilizers and clean CFE as required. Recycle gas compressor is usually maintained and cleaned during each fixed-bed semiregenerative reformer. Chemical treatment companies are recommended if the oil refiner does not have necessary in-house capabilities, as turnaround maintenance periods present excellent opportunities for effective cleaning and repairs to stabilizers and rotating equipment.

d. Catalyst Circulation and Catalyst Fines Systems (UOP CCR)
    - Use expert inspectors, field service support team from process licensing company, and in-house company experts.
    - Inspect catalyst collector and the transfer lines between the last reactor and the catalyst collector. Also check transfer pipes between catalyst collector and lock hopper.
    - Inspect piping for thinness possibly due to erosion of the piping by catalyst; repair or replace sections as required.
    - Inspect lift engagers as per process licensor's guidelines.
    - Check interreactor transfer lines, other transfer lines, and lift pots (Axens).
    - Conduct maintenance and complete scheduled work for the section.

e. Catalyst Reduction Section (CCR)
    - Use expert inspectors, field service support from process licensing company, and in-house company experts.
    - Inspect the catalyst reduction system and the hydrogen gas coalescer. The hydrogen or booster gas coalescer is used for removing hydrocarbon liquid from the hydrogen gas used in catalyst platinum and tin reduction.
    - Conduct maintenance and complete scheduled work for the section.

f. Catalyst Regeneration System
    - Use expert inspectors, field service support from process licensing company, and in-house company experts.
    - Inspect outer and inner screens and measure screen slots of inner and outer screens.
    - Information from the screen slots can enable the oil refiner to determine remaining life of a screen. Measured average slot dimensions could give an indication of changes that may be occurring in the screens. Smaller average slot size may be due to screen blockage with catalyst chips and larger average slot size could be due to age and thermally induced expansions of screen slots. Specifications for inner and outer regenerator screen slots are as listed in Table 9.1.[19]

TABLE 9.1
Standard Specifications of Screen Slots

Specification	Inner Screen	Outer Screen
Average Slot Width	0.483 ± 0.051 mm	0.635 ± 0.051 mm
	0.019 ± 0.002 inch	0.025 ± 0.002 inch
Maximum Slot Width	0.635 mm	0.890 mm
	0.025 inch	0.035
Minimum Slot Width	0.330 mm	0.380 mm
	0.013 inch	0.015 inch

Source: Van den Bosch, A., Improve Operational Availability of CCR Reformers. *ERTC Conference*, Lisbon, November 2016.[19]

- Check oxygen analyzers and calibrate.
- Check transfer lines between disengaging hopper and regeneration tower.
- Inspect chloride addition and air dryer systems.

g. Regenerator Effluent Gas Unit for Meeting RMACT
- Caustic scrubber vent gas wash tower system used for meeting RMACT II HCl and TOC limits should be inspected and required maintenance conducted.
- If the oil refiner is using the Chlorsorb technology for meeting RMACT II regulation, the system, including heat exchangers, should be inspected.
- Conduct required maintenance and scheduled work for the section.

h. Chloride Guard Bed System
- Inspect chloride guard bed system.
- If required, replace one of the chloride guard bed adsorbents in a lead/lag chloride guard bed system and as required for the single chloride guard bed unit. Catalytic reformer revamp or addition of technology for meeting RMACT II HCl and TOC limits project.

i. Other Turnaround Related Capital Projects
- A significant capital project would be better worked on and managed by another dedicated team so as not to unduly extend the scope of the turnaround project team.
- A few key technical personnel, including the corporate process technologist, should be members of the special capital project team. The special capital team would then focus on their primary responsibilities with respect to catalytic reformer revamp or add-on capital project progress.
- The turnaround and capital project teams should interface and share updates of turnaround plans and the capital project.
- The capital project equipment should be assembled and ready several weeks before tie-in to the catalytic reforming process unit. The two groups should be aligned with requirement and schedule for tie-in of the capital project unit to the catalytic reformer.

At the completion of the turnaround maintenance work, the process unit is turned over to the refinery operations group. Key process operations group members and the corporate process technologist should also be members of the startup process operations team for a smooth transition between the turnaround group's work and that of the process unit startup team. A major essential factor for safe and successful turnaround maintenance and startup is that the maintenance work be conducted based on effective communication through group meetings, frequent reviews, safety and progress updates, and reinforcement of best practice principles for turnaround maintenance and process unit startup for all the members of the turnaround and startup teams. Startup of the process units can then occur with maximum supervision of all phases by relevant personnel and with special concerns for the safety of personnel and protection of refinery assets and the environment.

9.2 OVERVIEW OF PLATINUM SUPPLY AND MANAGEMENT

Platinum, palladium, ruthenium, and rhodium are elements of the platinum group metals (PGMs). The most important and useful element in the group is platinum. About 45% of the annual demand for platinum is from the auto catalyst business. Platinum is combined with palladium and other metals as catalysts in catalytic converters. Hazardous compounds such as carbon monoxide, nitric oxide, and hydrocarbons in exhaust gases from the combustion of gasoline and diesel in automobiles are converted to harmless carbon dioxide, water, and nitrogen in catalytic converters. Platinum, palladium, and other metals are the usual components of automotive catalytic converter catalysts. Supply and demand for platinum is as provided in Table 9.2 courtesy of the Johnson Matthey Company.(21) Platinum is used in jewelry, investments, and manufacturing industries, as shown in Figure 9.2.(22)

A number of oil refining processes use platinum-containing catalysts. Platinum catalysts are used in catalytic reformers, light naphtha isomerization, benzene hydrogenation or saturation, and in other selective hydroprocessing units. Catalytic reformers use most of the platinum in an oil refinery. It is a valuable asset that should be managed efficiently based on a good understanding of the catalytic reforming process and the life of the catalyst from manufacturing of fresh catalysts to platinum recovery from spent catalysts.(23–25)

There are several factors that contribute to the total cost of precious metal catalysts, especially platinum-containing catalysts, in oil refining. Relevant concepts reviewed

TABLE 9.2
Supply and Demand for Platinum

	2013	2014	2015
	Supply, Thousands of Pt oz.		
South Africa	4205	3546	4237
Russia	725	715	696
Others	871	864	874
Total Supply	5801	5125	5807
	Gross Demand		
Auto Catalyst	3147	3354	3695
Jewelry	3028	2900	2862
Industrial	1652	1784	1836
Investment	871	272	−88
Total Gross Demand	8698	8310	8305
Recycling	−2029	−2073	−2213
Total Net Demand	6669	6237	6092

Source: Johnson Matthey, PGM Market Report, May 2015 Summary of Platinum, Supply and Demand in 2014. Accessed August 15, 2017. http://www.platinum.matthey.com/documents/new-item/pgm%20market%20reports/pgm%20market%20report%20may%202015.pdf.(21)

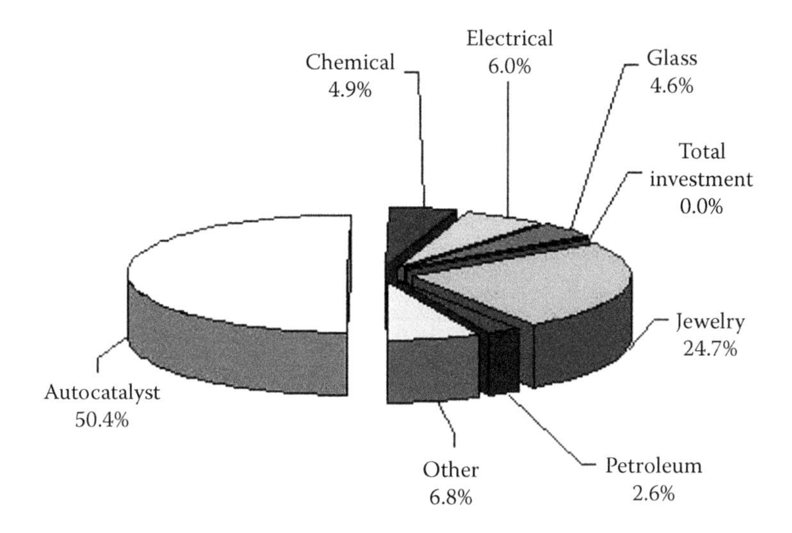

FIGURE 9.2 Platinum is used in a variety of industries.
From Total Materia, Platinum and Its Uses. February 2010. Accessed July 10, 2017, http://www.totalmateria.com/page.aspx?ID=CheckArticle&site=ktn&NM=237.[22]

for platinum and platinum-containing catalyst management can be extended for use in the management of precious metals such as palladium and silver and catalysts containing those metals. For effective management of the cost of platinum and platinum-containing catalysts, a good understanding of the key factors that contribute to the costs is required. Cost factors should include the prices of catalysts, catalyst acquisition and logistics costs, catalyst handling costs, platinum prices, platinum lease prices, laboratory services costs, platinum losses, and platinum recovery costs. Platinum-containing catalysts costs have to be continuously managed from the fresh catalyst from the catalyst supplier through catalyst life in oil refineries and platinum recovery via use of the services of a platinum reclamation company.

9.2.1 Platinum Acquisition Cost

The price of platinum has varied widely over the past 25 years. From a traditional low price of \$365.00 per troy ounce in December 1992, platinum price skyrocketed to \$2150.00 by March 2008 and stayed above \$1400.00 between 2009 and 2014, as shown in Figure 9.3.[26] The platinum price was at about \$990.00 per troy ounce in August 2017. Depending on the time of platinum acquisition for fresh catalyst use by an oil refiner, the cost of platinum acquisition could be three to six times the cost in 1992.

The alternative option for acquiring platinum by an oil refiner is leasing platinum, and platinum lease rate volatility is a concern, as shown in Figure 9.4.[27] As shown in Figure 9.4, the 3-month platinum lease rate was as high as 30% in April 2001, and the 12-month rate also peaked at 17.0%. Similarly, even with lower platinum prices in the range of \$425–440 per troy ounce in October 2001, the 3-month lease rate peaked at

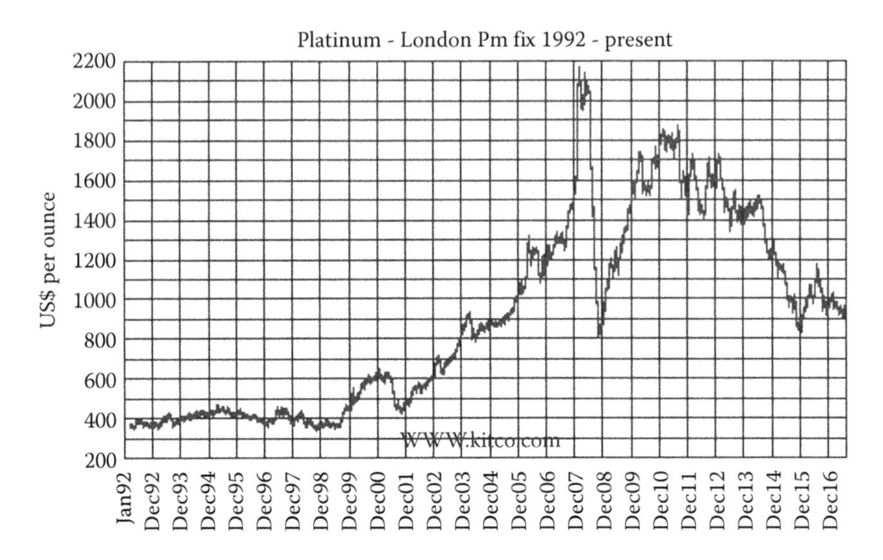

FIGURE 9.3 London platinum prices 1992–2017.
From Kitco, Platinum Prices 1992–2017. Accessed August 23, 2017, http://www.kitco.com/scripts/hist_charts/yearly_graphs.plx.[26] (Courtesy Kitco, http://www.kitco.com.)

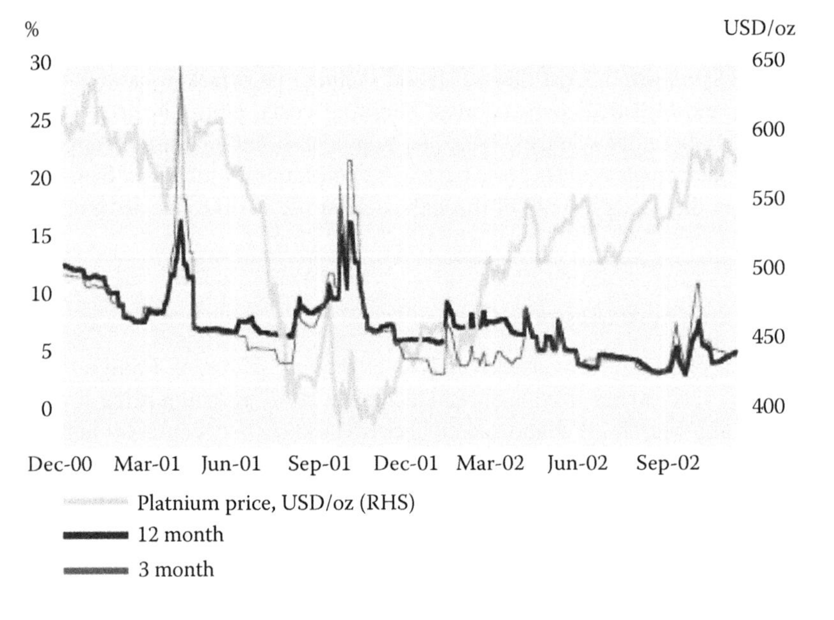

FIGURE 9.4 Platinum price and lease rates.
From Marr-Johnson, M., The Effect of Lease Rates on Precious Metals Market. The London Bullion Market Association. Accessed September 15, 2017, http://www.lbma.org.uk/assets/blog/alchemist_articles/Alch29Marr-Johnson.pdf.[27]

23% and the 12-month lease rate was 17.5%. Timing is a critical factor with respect to the opportune time to buy or lease platinum. Unfortunately, oil refiners sometimes do not have the flexibility to postpone platinum acquisitions and have to acquire platinum at exorbitant prices and lease rates when they need it.

9.2.2 Fresh Catalyst Purchase

Purchasing fresh catalyst for a grassroots catalytic reformer or replacement catalyst requires a good understanding of the elements of catalyst evaluation and selection discussed in Chapter 5. Based on past experiences with catalysts used in catalytic reformers and catalyst evaluations, an oil refiner can make a reasonably good selection of a high-performance catalyst. Key fresh catalyst properties for projected good catalytic performances are good surface area and chloride retention characteristics and good platinum, promoter metal, and chloride distributions. In reforming operations, desired performance characteristics are good reformate and hydrogen production, high activity, and low rates of catalytic coking. Good platinum distributions and attrition resistances are critically important with respect to the minimization of catalyst usage and platinum losses in catalytic reformers, especially in CCR reformers.

With respect to fresh catalyst and platinum acquisition cost, a reference case example of a 75-MBPD CCR platformer operating at 4500 pounds per hours is used for estimation of initial reformer catalyst fill cost. Catalyst requirement was 480,000 pounds for a grassroots 75-MBPD CCR reformer on the US Gulf Coast in September 2008. The catalyst inventory is 400,000 pounds for the reformer, and 80,000 pounds was the amount of catalyst estimated for makeup and initial potential unit upsets. To get an idea of the huge financial investment required, the September 2008 platinum price of $1800 per troy ounce from Figure 9.3 is used with an assumed platinum lease rate of 6%. The cost of acquiring the fresh catalyst and logistics without the required platinum is estimated at $25 per pound of catalyst. Included in the estimated price are the cost of logistics, inert materials, transportation, catalyst transfer from drums to larger catalyst bins, and use of a warehouse.

For the example reviewed, catalyst acquisition cost ex platinum is $12.0 million, and that for platinum only, based on the 0.28 weight percent platinum in the reforming catalyst, is $35.3 million. The total initial expenditure for catalyst and platinum is $47.3 million. If the oil refiner elected to lease platinum at favorable annual rate of 6%, it would be saddled with paying about $2.2 million per year for 6–8 years and would still be obligated to return all of the troy ounces of platinum. Platinum losses typically occur over the life of a catalyst, and the oil refiner would either be obligated to replace with makeup platinum or via monetary compensation. Based on this example, the oil refiner could opt to purchase the necessary platinum for the fresh catalyst.

The purchased catalyst is expected to perform as advertised by the catalyst supplier. It is expected that the catalyst meet the good projected catalytic performance and excellent physical properties as offered by the catalyst supplier. Uniform platinum and chloride distributions of the catalyst should provide good catalytic performance and minimize platinum losses via catalyst fines. Platinum management for an oil refiner starts for each of its catalytic reformers with the application of prudent platinum acquisition strategies and diligent tracking of platinum from catalyst plants through catalyst handling phases. The process continues with expert management of the operations of catalytic reformers and optimization of every aspect of the operations including catalyst regenerations. All of the listed management goals are achieved while maintaining the reliability of the processing unit. A necessary element of catalyst management is close supervision and knowledge of the process of recovering platinum from spent catalytic reforming catalysts, including tracking of the platinum at the platinum reclamation site and the ultimate return of recovered platinum either to oil refiners' pool accounts or the pool accounts of platinum leasing companies.[25] The total value of platinum in catalytic reforming and light naphtha isomerization units and in the pool accounts that I managed for one of the US oil refiners was about $150 million in 1997. Successful cost-effective management of such platinum accounts requires that the individual phases of the platinum management program be conducted with due diligence, as reviewed in the next section. Audits of the platinum management process for each of the catalytic reformers and light naphtha isomerization units should be conducted to determine percentage platinum losses for the life of the catalysts in each of the platinum catalyzed process units. In addition, comprehensive quarterly and annual audits of the corporate platinum management program should be conducted to determine the amount and percentages of platinum losses for the company. The efficacy of the platinum management system of the oil refiner can then be benchmarked against those of other companies and against financial goals. At American Fuel and Petrochemical Manufacturers (AFPM) Question and Answer conferences over the years, meeting attendees from oil refining companies have stated that their audits have shown platinum losses in the range of 6–12-plus percent. Recommendations offered in this book should enable oil refiners to effect substantial reductions of their platinum losses and improve overall management of their catalytic reformers, especially their CCR process units.[23,24]

9.2.3 Catalyst Platinum Agreements

Two different platinum settlement agreements are typically used. The first platinum agreement is between the oil refiner and the catalyst supplier. This is established so that companies can settle on the actual amount of platinum used in the manufacture of the fresh catalyst. Actual fresh catalyst platinum concentrations typically differ from those posted as nominal platinum concentrations for the catalysts. The ultimate platinum concentrations of fresh catalysts usually reflect the efficacy of the platinum and promoter metal impregnation process used during the manufacturing of catalysts. The agreements could reflect potential catalyst platinum manufacturing losses. Depending on the manufacturing process, the difference in platinum concentrations between nominal and actual for the fresh catalyst could be in the range of 1%–2% based on assays conducted on specially designated catalyst settlement lot samples.

The second set of platinum settlement agreements are those established for determining the amount of platinum that oil refiners could expect from the platinum recovery by the recovery company for their loads of spent catalysts. Agreements could stipulate a platinum recovery loss of 1%–20% for the oil refiner depending on the quality of the spent catalyst. Expected platinum loss is related to the platinum reclamation company's estimated yield of platinum from the catalyst sent by an oil refiner in the platinum recovery process. Penalties are assessed for the quality of the spent catalyst based on the concentrations of residual coke, metal contaminants, and chlorides and on the state of the alumina of the spent catalysts supplied by oil refiners. Thermally transformed alumina support to the alpha, delta, and theta phases are likely to yield less platinum via the wet chemistry methodology used in the standard platinum recovery process. A pyrometallurgical precious metals recovery process is typically used for recovering platinum from thermally damaged spent catalytic reforming catalysts and light naphtha isomerization process catalysts containing high chlorides. Platinum recovery agreements for the pyrometallurgical process usually provide for lower percentages of recovered platinum for oil refiners. It is important that oil refiners be fully aware of the broad implications of poor operations of catalytic reformers and that higher costs of platinum-containing catalysts and platinum management can lead to some refinery profitability losses.

9.2.4 Catalyst Selection and Platinum Management

The first element in platinum management for an oil refiner is to select an appropriate process technologist with good understanding of the catalytic processes in refineries. A good process technologist should lead and assist the accounting group in managing platinum and other precious metals via use of sound process and catalyst management principles. The platinum management team under the technical leadership of the process technologist, including the accounts department and laboratory personnel, would then drive a good understanding in the refinery of what is required to properly use platinum-containing catalysts in refinery processes and also ensure efficient use and protection of the platinum asset.

As reviewed several times, for catalytic reforming process units, especially for CCR process units, good fresh catalysts with excellent surface area, attrition resistance characteristics, and good platinum distribution provide a good starting point for the minimization of platinum losses during handling and naphtha reforming operations. Attrition resistance for catalysts has been improving with the introduction of new high-performance CCR reforming catalysts as percentages of catalyst fines collected for seven CCR process units operating on three different CCR catalysts ranged from 0.01% to 0.1% of catalyst circulation rates with an average of about 0.05% operating with an average catalyst circulation rate of 1460 pounds per hour.[(1)]

Catalyst handling starts with using cost-effective, catalyst loss–free procedures for loading the reactors by using catalyst container bins that can hold up to 10 drums of catalyst and are equipped with a catalyst dumping chute at the bottom. The services of catalyst container companies such as Hoover Ferguson were recommended in Chapter 7.[(28)] The use of larger catalyst bins for loading aids in minimizing catalyst and platinum losses. Catalyst particles that are not recoverable are usually found strewn on the ground beneath platformer reactors and catalyst regeneration sections of CCR process units after catalyst unloading and loading activities. These are usually unmistakable clear evidence of catalyst and platinum losses.

Similarly, dumping of spent catalysts and catalyst handling should be conducted as efficiently as feasible so as not to lose catalyst and platinum. Areas behind reactor center screens, scallops, and behind regenerator tower screens should be inspected for catalyst particles and fines. If significant amounts of catalyst particles and fines are found, the screens should be repaired and possibly replaced at the earliest opportunity after adequate planning. It is also a good time to determine and track the dimensions of the screen slots for the reactor center screen and regeneration tower inner and outer screens.

Continuous monitoring of catalytic reformers, especially of CCR platformers, has been emphasized for optimizing catalytic performances. The use of plant information management systems, laboratory information management systems, and routine inspections has been emphasized for monitoring so as to preserve the integrity of piping, transfer pipes, reactor internals, and catalyst regeneration operations. Diligent monitoring of catalytic reforming operations would aid in achieving high performance in units, especially in CCR units. It will also aid in effective management of catalyst and platinum, as the operations are typically monitored for pressure drops in reactors, obstructions and holes in piping in the circulation system, and poor catalyst regeneration operations. It is well known that larger slots on the center screens and holes in the scallops can cause significant catalyst and platinum losses, as significant amounts of catalyst particles and platinum have been recovered from product separation sections of catalytic reformers and reformate tanks. It also known that high temperature upsets in catalyst regenerations could lead to significant catalyst damage and losses. Phase transition of the catalyst alumina in high-temperature operations can lead to formation of chemically insoluble alpha alumina. Chemically insoluble alpha alumina can occlude platinum and lead to losses.

Catalyst balances should be conducted periodically to determine if there are high amounts of unaccounted-for catalyst, and if that is the case, troubleshooting should be conducted to determine the factors that could be causing catalyst and platinum losses.

Some aspects of the platinum management program proposed for catalytic reforming unit catalysts can also be replicated for use with platinum in light naphtha and butane isomerization units. For oil refiners operating units that are using palladium catalysts for the selective hydrogenation of alkylation feed and silver catalyzed petrochemical processes, the elements of platinum management that are suggested in this book can be applied for successful, efficient management of palladium and silver.

9.2.5 Overview of Platinum Management and Recovery

A number of decisions and procedures could define the overall performance of platinum management teams of oil refiners. One major decision is to select one or two good platinum recovery companies whose plants and practices meet environmental regulations. This is a necessary criterion so as to protect the oil refiner from lawsuits and liabilities stemming from the processing of the oil refiner's catalysts. A set of the platinum recovery companies would be those who use wet chemistry recovery processes. Platinum reclamation companies who use pyrometallurgical processes to recover platinum and metals from thermally damaged catalysts and catalysts that are not amenable to wet chemistry reclamation processes should also be considered. Thirdly, the services of a good, reliable analytical laboratory company should be secured, as the oil refiner needs to use accurate, reliable assays for platinum and solid contents of catalyst samples in platinum settlement discussions with platinum recovery companies.

In the process of managing a load of catalyst that will eventually arrive at the chosen platinum recovery company, some critical procedures need to be followed to ensure maximizing recovered platinum and minimizing the overall cost of platinum recovery. Before sending the load of catalyst, inert materials, and catalyst fines to a platinum reclamation company, the total material in catalyst container bins should be weighed and labeled and results compiled in a database. If the catalyst was regenerated, laboratory characterization tests indicate that it is free of toxic organic compounds such as benzene, and it meets the US Department of Transportation (DOT) requirements, the catalyst load can then be labeled as nonhazardous and transported at much lower shipping costs. However, if the catalyst is labeled as hazardous, the cost of transportation increases significantly. Thus, if feasible, it is recommended that the oil refiner regenerate catalysts before dumping. This is useful for safety reasons and for maximizing the troy ounces of platinum recovered for the oil refiner.

Density grading could be used to separate highly "coked" heel catalyst and catalyst dumped from catalyst regeneration towers.[18] The catalyst that is dumped from regeneration towers could contain unregenerated catalyst particles. Regeneration of coked catalyst from catalytic reformers can be conducted offsite by a catalyst regeneration company before submitting catalyst load to the platinum recovery company, which could add extra cost to the overall platinum recovery cost. As indicated, the minimization of coke, iron, silicon, and other contaminants during catalytic reforming operations leads to reduced cost of platinum recovery.

Another key consideration is the selection of a good witnessing agency company that can supply reliable, experienced witnessing agents to represent the oil refiner during the initial weighing and separation of catalyst from catalyst fines and noncatalyst inert materials. The witnessing agent is also required to observe catalyst sampling at the platinum recovery company's plant.

The screening of catalyst, catalyst fines, and inert materials can be conducted either at the refinery by a catalyst handling company or offsite by a catalyst handling and regeneration company. Platinum recovery companies can also conduct catalyst screening at their plants. If it is deemed necessary to conduct catalyst screening via density grading and catalyst regeneration, a suitable catalyst services company could be used. Oil refiners sometimes use the services of platinum recovery companies for conducting all facets of the recovery process, which could include catalyst regeneration and catalyst and materials screening.

If more than two companies provide the catalyst services, witnessing agents then have to work at more than one location. If catalyst weighing, screening, and separation are conducted at the platinum recovery company's plant, the work of the witnessing agents can then become more focused. AHK, Inspectorate, SGS, Ledoux, and other analytical laboratories offer excellent, accurate precious metal assays services. The typical screening of spent catalyst and inert materials at the platinum recovery site usually results in three different fractions. The fractions are whole pill catalysts, catalyst fines, and larger inert materials. Three separate sets of materials from each lot are weighed and the data input into files for the lots. At the completion of the catalyst sampling, a material balance is conducted and compared with bulk material of catalyst, catalyst fines, and large particles and extraneous materials that are bigger than catalyst particles.

9.2.6 Platinum Recovery Processes

This section on platinum recovery processes is provided to ensure that oil refiners' platinum management team members fully understand the two major precious metal recovery processes. An understanding of the factors that determine which of the processes is used for precious metal recovery leads to a better understanding of the terms of agreement usually entered into and the lower percentages of precious metals recovered for the pyrometallurgical process relative to those from the wet chemicals process.

Wet chemistry processes involve separating the alumina from platinum by reacting the catalyst with sulfuric acid and sodium hydroxide to produce aluminum sulfate and sodium aluminate, as shown by the first two reactions in Figure 9.5.[30] After the chemical elimination of the alumina support as soluble aluminum salts, the insoluble platinum is then oxidized in a chloride media to form hexa-chloroplatinic acid, as shown by the third reaction in Figure 9.5. The platinum metal is produced after purification from the hexa-chloroplatinic acid.[30]

Spent catalytic reforming catalysts that have undergone significant alumina phase transitions due to high thermal excursions during catalyst regenerations are usually insoluble in acids and caustics used in the wet chemistry platinum recovery process. An alternative platinum recovery process is used, as the guaranteed platinum recovery via wet chemistry could be much lower than 85% due to higher insolubility of the substrate or catalyst support. In this case, platinum-containing oil refining catalysts consisting of insoluble alumina support such as delta, theta, and especially alpha alumina are not subjected exclusively to the wet chemistry process. Pyrometallurgical recovery processes are used and involve high-temperature treatment to melt the catalyst containing platinum and extract the platinum after the smelting process as shown in Figure 9.6.

Pyrometallurgical processes can also be used to recover platinum from standard spent catalyst. As indicated, the percentage of precious metal or platinum recovered will be low if there are insoluble constituents. To recover as much of the precious metal or platinum as feasible, the residual insoluble catalyst portion after the wet chemistry process is then subjected to the pyrometallurgical process to recover residual platinum or precious metals. The wet chemistry process is favored, however, due to the higher percentages of platinum recoveries that can be achieved relative to the pyrometallurgical process with soluble catalyst supports or carriers.

$$Al_2O_3 + 3H_2SO_4 \longrightarrow Al_2(SO_4)_3 + 3H_2O$$

$$Al_2O_3 + 2NaOH \longrightarrow 2NaAlO_2 + H_2O$$

$$Pt + 6HCl + O_2 \longrightarrow H_2PtCl_6 + 2H_2O$$

FIGURE 9.5 Chemical reactions in wet chemistry platinum recovery process.

FIGURE 9.6 Electric arc furnaces for platinum recovery.
From Jacobsen, R.T., Catalyst Recovery–Part 1: Recovering Precious Metals From Spent Catalysts–The Basics. *Chem. Eng. Prog.*, 101(2), 20–23, February 2005.[(31)] (Courtesy Sabin Metal Corporation.)

9.2.7 Catalyst Sampling and Settlement

The activities reviewed in this section are most critical for the success of catalyst sampling and successful, equitable platinum settlement between the platinum recovery company and the client. The client could be a producer of chemical products using platinum-containing catalysts, and in this book, the focus is on oil refiners and catalytic reforming catalysts. The catalyst sampling process conducted at the platinum recovery company leads to the generation of four representative samples, as described in detail by others.[(29,30)] Jacobsen, in his elegant paper, discussed the benefits of burning catalyst before sampling and how that treatment improves the accuracy of the sampling process by impacting the percentage loss of volatile materials from the catalyst in the loss on ignition (LOI) determination.[(31)] The loss on ignition data is used in calculating the percentage of solids in catalyst samples by subtracting the determined LOI percentage from 100%.

Usually, four sets of samples are derived from representative lots of catalysts. Catalyst lots could be set at between 15,000 and 20,000 pounds. The generated samples are: whole catalyst pills, which is the main catalyst sample; ground samples of "whole" catalyst pills; and samples of catalyst fines and ground catalyst "fines." All the lot samples are then shared. One set is retained by the platinum reclamation company, a set is given to the oil refiner, and a set is kept for umpiring purposes. The set of catalyst samples kept for possible umpiring purpose is used if differences in estimated platinum for the lots exceed an agreed-upon percentage per the platinum settlement agreement between the two parties. A catalyst separation and sampling area is shown in Figure 9.7.[(31)]

FIGURE 9.7 An enclosed catalyst sampling area.
From Jacobsen, R.T., Catalyst Recovery–Part 1: Recovering Precious Metals From Spent Catalysts–The Basics. *Chem. Eng. Prog.*, 101(2), 20–23, February 2005.[31] (Courtesy of Sabin Metal Corporation.)

Necessary analytical data are generated by the platinum recovery company and the oil refiner for the catalyst lot samples. The solid content of a catalyst lot sample is calculated from the loss on ignition data. Loss on ignition data for a catalyst lot sample represent the percentage of volatile compounds in the catalyst sample such as water, residual hydrocarbons, residual coke, and gases that are released during calcination. The loss on ignition data for catalyst samples are highly dependent on the calcination temperature, amount of samples, crucible size used, and time, and as a result, there are significant variations between analytical laboratories for the same samples. Typical calcination temperatures for catalyst samples are 900 and 1000°C. The variation in data for laboratories for the same samples has been determined to be in the range of 4.5–10 weight percent.

Platinum concentrations are determined for representative ground samples of the whole catalyst particles for the lots and for representative catalyst fines samples. The platinum assay data and solid contents derived from the LOI data are used in the determination of the amount of platinum in each lot. The calculations are also repeated for the catalyst fines for each catalyst fines lot. At an appropriate previously agreed-upon time, the representatives of the oil refiner and platinum recovery company simultaneously exchange their platinum data and proceed to complete the platinum settlement process.

If there is a significant difference in platinum between the data of the oil refiner and platinum recovery company, the samples that were reserved for use by an umpire are then sent to the umpire company, which conducts required analysis and reports its data. Analytical data generated by the umpire company are then used to complete the platinum settlement as per the platinum settlement agreement. After successful completion of the platinum settlement process, the agreed amount of troy ounces of platinum is returned to the oil refiner. Currently, a number of companies recover platinum via the wet chemistry and pyrometallurgical processes, including Sabin Metal Corp, Gemini, Heraeus, Hindustan Metal, and Umicore.

In summary, effective platinum management should have as its operational underpinnings sound financial criteria and good informed capabilities of oil refiners' personnel to select catalysts with excellent physical and chemical characteristics to ensure as much platinum retention on catalyst as feasible. The oil refiner should also select good companies for reliable, expert witnessing agents; for analytical laboratory services; and for platinum and precious metal recovery. Cost-effective platinum management and excellent catalytic reformer productivity are also based on the need to maintain excellent, continuous equipment and reformer monitoring, benchmarking, and the use of key performance indicator data to identify equipment and reformer performance declines and debits. The identified reliability, productivity, and profitability drainages should be addressed as soon as feasible and definitely during routine and turnaround maintenance. As reviewed earlier, damage to reactor internals and regeneration tower screens could lead to poor reformer performance and platinum losses. It is should be apparent that there are several key mechanical and operational activities that are required to operate a catalytic reformer reliably and profitably.

Catalytic reformers will continue as one of the premier catalytic conversion units in oil refineries. As oil refiners plan for the projected mandated 2025 CAFÉ 54.5 mile per gallon consumption rate for vehicles, catalytic reformers, especially continuous catalyst regenerative reformers, are going to be relied on to help make up octane barrel shortfalls in the production of RSP gasoline of 95+ research octane number.

REFERENCES

1. Hammel, E. J., Turnaround Update, UOP, 1998. Accessed June 15, 2016. https://www.scribd.com/document/342885681/CCR-Platforming-Turnarounds
2. Blashka, S. R., Nite, K., Furfaro, A. P., Catalyst Separation Reduces Platformer Turnaround Costs. *Oil Gas J.*, 93(38), September 18, 1995.
3. Oliver, R., Complete Planning for Maintenance Turnarounds will Ensure Success. *Oil Gas J.*, 100(17), April 29, 2002.
4. Funkhouser, J., Manley, F., *Improving Refinery Reliability Performance and Utilization.* Special Report: Process Control and Informational Systems. Hydrocarbon Processing, October 2007.
5. Mohan, S. R., Five Best Practices for Refining: Maximizing Profit Margins Through Process Engineering AspenTech. White paper 2016. Accessed July 2, 2018. https://www.aspentech.com/en/resources/white-papers/five-best-practices-for-refineries-maximizing-profit-margins-through-process-engineering
6. Olsen, T., Schodowski, E., Improve Refinery Feasibility and Responsiveness, Hydrocarbon Processing, September 2015.
7. Bishop, N., Olsen, T., Get more from your Next Turnaround, Hydrocarbon Processing, November 2016.
8. Price, T., Aulds, T. J., What Went Wrong: Oil Refinery Disaster. Popular Mechanics. September 13, 2005. Accessed July 2, 2017. http://www.popularmechanics.com/technology/gadgets/a295/1758242/
9. Pinchuk, Y., The Dangers of Refinery Turnarounds. Accessed July 1, 2017. www.safersystem.com/2017/02/14/the-dangers-of-refinery-turnarounds/
10. Morris, J., Oil Refinery Workers Face Dangerous Conditions, Deadly Explosions. Accessed July 2, 2017. http://www.huffingtonpost.com/entry/oil-refineries-pollution_us_584f29f0e4b0bd9c3dfe53c2

11. Roy, D. D., 2 Dead, 8 Injured in Fire at Reliance Jamnager Refinery. November 2016. Accessed August 7, 2017. http://www.ndtv.com/india-news/fire-at-reliance-jamnagar-refinery-put-out-eight-injured-1629566
12. Grabke, H. J., Carburization, Carbide Formation, Metal Dusting, Coking. Materiali in Technolgije, 36, 297–303, 2002. Accessed July 2017. http://mit.imt.si/Revija/izvodi/mit026/grabke.pdf
13. Danis, J., API RP 571, Damage Mechanisms in the Refining Industry. Accessed July 10, 2017. https://www.equityeng.com/library/api-rp-571-overview-damage-mechanisms-in-the-refining-industry/
14. Ravestein, R., Metal Dusting in Catalytic Reforming Furnace Heaters. NACE Report 97496. Accessed August 10, 2017. https://store.nace.org/97496-metal-dusting-in-catalytic-reforming-furnace
15. Poparad, A., Ellis, B., Glover, B., Metro, S., Reforming Solutions for Improved Profits in an Up-Down World, Paper AM-59. *National Petrochemicals & Refiners Association (NPRA), Annual Meeting*, 2011.
16. Domergue, B., le Goff, P., Ross, J., Octanizing Reformer Options, Petroleum Technology Quarterly. Quarter 1, 2006.
17. Schaefer, F., Cherico, S. D., Cetek "Matrix Coating System" and It's Use at MiRO-Karlsruhe, May 2007.
18. Porocel, Density Grading. http://www.porocel.com/25-density_grading/
19. Van den Bosch, A., Improve Operational Availability of CCR Reformers. *ERTC Conference*, Lisbon, November 2016.
20. Aqseptence Group, Johnson Screens. Accessed July 1, 2017. https://www.aqseptence.com/app/en/keybrands/johnson-screens/
21. Johnson Matthey, PGM Market Report, May 2015 Summary of Platinum, Supply and Demand in 2014. Accessed August 15, 2017. http://www.platinum.matthey.com/documents/new-item/pgm%20market%20reports/pgm%20market%20report%20may%202015.pdf
22. Total Materia, Platinum and Its Uses. February, 2010. Accessed July 10, 2017. http://www.totalmateria.com/page.aspx?ID=CheckArticle&site=ktn&NM=237
23. Chaudhuri, P. M., Industry Needs Standardization of Precious Metals Management. *Oil Gas J.*, 98(44), October 30, 2000.
24. Oyekan, S. O., Management of Platinum in Catalytic Reforming Systems. *AFPM, AFPM Q&A-1: Safety, Gasoline Processing Questions Addressed at 2014 Annual Conference*, August 2015.
25. Oyekan, S. O., Management of Platinum in Catalytic Reforming Systems. www.prafises.com
26. Kitco, Platinum Prices 1992 to 2017. Accessed August 23, 2017. http://www.kitco.com/scripts/hist_charts/yearly_graphs.plx
27. Marr-Johnson, M., The Effect of Lease Rates on Precious Metals Market. The London Bullion Market Association. Accessed September 15, 2017. http://www.lbma.org.uk/assets/blog/alchemist_articles/Alch29Marr-Johnson.pdf
28. Hoover Ferguson, Catalyst and Chemical Containers. www.hooverferguson.com/products/catalyst-chemical-containers.html
29. Meyer, H., Grehl, M., Precious Metal Recovery from Spent Reforming Catalysts, in *Catalytic Naphtha Reforming* by George, J. A., Abdullah, M. A., 2nd edition, Marcel Dekker, New York, 2004, 459–475.
30. Rosso, J. P., El Guindy, M. I., Recovery of Pt and Re from Spent Reforming Catalysts, in *Catalytic Naphtha Reforming* by George, J. A., Abdullah, M. A., Parera J. M., Marcel Dekker. New York, 1995, 395–407.
31. Jacobsen, R. T., Catalyst Recovery–Part 1: Recovering Precious Metals From Spent Catalysts–The Basics. *Chem. Eng. Prog.*, 101(2), 20–23, February 2005.

Glossary

Absolute Pressure: A measure of the total pressure defined as the sum of gauge pressure and 14.7 lb/sq. in at sea level.

Absorption: The disappearance of one substance into another so that the absorbed substance loses its identifying characteristics, while the absorbing substance retains most of its original physical aspects. Used in refining to selectively remove specific components from process streams.

Acid Treatment: A process in which unfinished petroleum products such as gasoline, kerosene, and lubricating oil stocks are treated with sulfuric acid to improve color, odor, and other properties.

Acidity: The presence of acid-type constituents whose concentration is usually defined in terms of neutralization number. The constituents vary in nature and may or may not markedly influence the behavior of the oil.

Additive: Chemicals that are added to petroleum products in small amounts to improve quality or add special characteristics.

Adsorption: The process of adhering of the molecules of gases or liquids to the surface of solid materials.

AFPM: American Fuel and Petrochemical Manufacturers. The name, AFPM, was used after the change from National Petrochemical and Refiners Association (NPRA) in 2012.

AIChE: American Institute of Chemical Engineers.

Air Fin Coolers: A special radiator-like device used to cool or condense hot hydrocarbons; they are also referred to as fin fans.

Alicyclic Hydrocarbons: Cyclic (ringed) hydrocarbons in which the rings are made up only of carbon atoms and the carbon bonds are saturated.

Aliphatic Hydrocarbons: Hydrocarbons characterized by open-chain structures: ethane, butane, butenes, acetylene, and so on.

Alkylation: A process using sulfuric or hydrofluoric acid as a catalyst to combine olefins (butylene, pentene) and isobutane to produce a high-octane product known as alkylate. It is also a process wherein olefins are reacted with aromatics such as benzene.

Antiknock Index (AKI): A measure of the antiknock quality of gasoline, blend component, and hydrocarbons. It is the average of the research and motor octane numbers of gasoline. It is defined as equal to (RON + MON)/2. It is the same as the road octane number of gasoline, where RON is the research octane number and MON the motor octane number.

API: American Petroleum Institute.

API Gravity: An arbitrary scale expressing the density of petroleum products. The measuring scale is calibrated in terms of degrees API; it may be calculated in terms of the following formula:

$$\text{Degrees API} = \frac{141-5}{\text{Sp. gravity @ 60 F}} - 131.5$$

The higher the API gravity, the lighter the compound. Light crudes generally have greater than 38 degrees API and heavy crudes are commonly labeled as all crudes with an API gravity of 22 degrees or below. Intermediate crudes fall in the range of 22 to 38 degrees API gravity.

Aromatic: Organic compounds with one or more benzene rings.

Asphalt: A dark brown to black oily material. It could be solid, semisolid, or liquid in consistency, with bitumen as the predominating constituent. Petroleum asphalt is refined from crude petroleum into commercial grades of widely varying consistency. Asphalt is a natural constituent of asphaltic-base crude oils, which can be refined to yield high-value fuel products.

Asphaltenes: The asphalt compounds soluble in carbon disulfide but insoluble in paraffin naphthas.

Assay: A special analysis to get an accurate concentration of a precious metal in a material is referred to as an assay. The data generated are necessary for precious metal settlements. Assay also refers to crude oil analysis.

ASTM: American Society for Testing and Materials.

Atmospheric Column: A distillation unit operated at atmospheric pressure, also referred to as atmospheric tower.

Atmospheric Tower: A distillation unit operated at atmospheric pressure, also referred to as an atmospheric column.

Aviation Gasoline: A complex mixture of relatively volatile hydrocarbons with or without small quantities of additives, blended to form a fuel suitable for use in aviation reciprocating engines. Fuel specifications are provided in ASTM Specification D 910.

Barrel: A volumetric unit of measure for crude oil and petroleum products and it is equivalent to 42 US gallons.

Benzene: An unsaturated, six-carbon ring, aromatic compound.

Biodiesel: A blend of diesel and a biofuel. Biodiesel 20 refers to a biodiesel fuel containing 20% non fossil fuel components.

Biofuel: Non fossil fuel components such as ethanol, methyl tertiary butyl ether (MTBE), and others that can be blended into fuels.

Bitumen: Brown or black viscous residue from the vacuum distillation of crude oil or from propane extraction of shortened atmospheric residue. Bitumen also occurs in nature as asphalt lake and tar sands.

Bleeder Valve: A small-flow valve connected to a fluid process vessel or line for the purpose of bleeding off small quantities of contained fluid. It is installed with a block valve to determine if the block valve is closed tightly.

Blending: The process of mixing two or more petroleum products with different properties to produce a finished product with desired characteristics.

Block Valve: A valve used to isolate equipment.

Blowdown: The removal of hydrocarbons from a process unit, vessel, or line on a scheduled or emergency basis by the use of pressure through special piping and drums provided for this purpose.

Blower: Equipment for moving large volumes of gas against low-pressure heads.

BOB: Blendstock for oxygenate blending.

Boiling Range: The spread of temperatures, usually expressed in degrees Fahrenheit, over which an oil starts to boil or distill vapors and proceeds to complete evaporation. Boiling range is determined by ASTM test procedures for specific petroleum products.

Bottled Gas: Trade term for liquefied petroleum gas (LPG) or LP-gas.

Bottoms: Tower bottoms are residue remaining in a distillation unit after the highest-boiling-point material to be distilled has been removed. Tank bottoms are the heavy materials that accumulate in the bottom of storage tanks, usually composed of oil, water, and foreign matter.

BSD: Operating capacity of a refinery expressed in barrels per day the process unit is operating or "on stream."

BS&W: Bottom sediment and water.

BTU: British Terminal Unit.

BTX: Benzene, toluene, and xylenes.

Bubble Tower: A fractionating (distillation) tower in which the rising vapors pass through layers of condensate, bubbling under caps on a series of plates.

Bunker C Fuel Oil: A heavy residual fuel oil used by ships and industry and for large-scale heating installations, similar in requirements to No. 6 grade fuel oil.

CAFÉ: Corporate average fuel economy standards are established in the United States to specify the required miles per gallon of gasoline consumption for automobiles. The CAFÉ standard for 2025 was projected to be 54.5 mpg for some automobiles.

Calorie: The amount of heat required to raise the temperature of 1 gram of water by 1 degree centigrade, at or near maximum density.

Catalyst: A material that aids or promotes a chemical reaction between other substances. Catalysts increase reaction rates and can provide control by increasing desirable primary reactions and decreasing undesirable secondary reactions.

Catalytic Cracking: The process of breaking up heavier hydrocarbon molecules into lighter hydrocarbon fractions by the use of heat and catalysts.

Catalytic Reforming: A catalytic process to improve the antiknock quality of low-grade naphthas and virgin gasolines by the conversion of naphthenes and paraffins to produce high-octane reformate; benzene, toluene, and xylenes; and hydrogen.

Caustic Wash: A process in which distillate is treated with sodium hydroxide to remove acidic contaminants that contribute to poor odor and stability. It is also used in treating regeneration vent gas in catalytic reforming.

CBOB: Conventional blendstock for oxygenate blending.

CCR: Conradson carbon residue.

CCR: Also used for continuous catalyst regenerative reformers.

C/H: Carbon to hydrogen ratio.

Celsius: A measure of temperature also referred to as centigrade. The freezing point of water is 0 degrees C and the boiling point of water at one atmosphere is 100 degrees C.

Cetane Index: An empirical measure of ignition quality, defined as the percentage by volume of cetane in a mixture of cetane and methyl naphthalene, which has the same ignition quality when used in an engine as a fuel under test.

CFR: The Cooperative Fuel Research engine is a standard test engine used in determining the octane number of motor fuels.

CHD Unit: Catalytic Hydrodesulfurization unit.

Chemisorption: The specific chemical interaction of an adsorbate with a surface.

Cloud Test: The ASTM D97 method used in determining the temperature, known as cloud point, at which paraffin wax or other solid substances begin to crystallize out or separate from solution when an oil is chilled under specified conditions.

Coalescer: Equipment used in the separation of entrained liquids from gases, also used in the separation of emulsions into their components via various processes.

Coke: A high carbon-content residue remaining from the destructive distillation of petroleum residue and defined by the molecular formula C_xH_y.

Coking: A process for thermally converting and upgrading heavy residual into lighter products and byproduct petroleum coke. Coking also is the removal of lighter distillable hydrocarbons that leaves a residue of carbon in the bottom of units or as buildup or deposits on equipment and catalysts.

Coking Refinery: A full conversion refinery that is capable of processing a variety of oils. The refinery typically would have catalytic reforming, cracking, coking, hydrocracking, and alkylation process units.

Commercial Sector: An energy-consuming sector that consists of service-providing facilities and equipment of businesses; federal, state, and local governments; and other private and public organizations, such as religious, social, or fraternal groups. The commercial sector includes living quarters. Common uses of energy associated with this sector include space heating, water heating, air conditioning, lighting, refrigeration, cooking, and running a wide variety of other equipment.

Complexity Factor: A measure of the complexity of operations of a process unit, determined for the process unit as a function of oil conversion and cost relative to the crude oil distillation unit.

Complexity Index: An overall relative sum of the complexity factors of a process unit as developed by W. L. Nelson, which provides a measure of the number of secondary conversion units that are contained in the refinery. A measure of the processing capabilities of a refinery.

Condensate: A highly gaseous liquid coming from gas condensate wells, from which the gas is separated, the liquid remaining being shipped with crude oil in pipelines to refineries. It is also used to describe the liquid material coming from the condensers in a refinery.

Condenser: A heat-transfer device that cools and condenses vapor by removing heat via a cooler medium such as water or lower-temperature hydrocarbon streams.

Condenser Reflux: Condensate that is returned to the original unit to assist in giving increased conversion or recovery.

Cooler: A heat exchanger in which hot liquid hydrocarbon is passed through pipes immersed in cool water to lower its temperature.

Corrosion: Detrimental change in the size or characteristics of material under conditions of exposure or use. It usually results from chemical action either regularly and slowly, as in rusting (oxidation), or rapidly, as in metal pickling.

Cracking: The breaking up of heavy-molecular-weight hydrocarbons into lighter hydrocarbon molecules by the application of heat and pressure, with or without the use of catalysts.

Cracking Refinery: A medium conversion refinery with catalytic reforming, cracking, hydrocracking, and alkylation processing units.

Crack Spread: A gross indicator of the potential profitability of oil refining and marketing companies.

Crude Assay: A procedure for determining the general distillation and quality characteristics of crude oil.

Crude Oil: A naturally occurring mixture of hydrocarbons that usually includes small quantities of sulfur, nitrogen, and oxygen derivatives of hydrocarbons as well as trace metals. A mixture of hydrocarbons that exists in liquid phase in natural underground reservoirs and remains liquid at atmospheric pressure after passing through surface separating facilities. Depending upon the characteristics of the crude stream, it may also include:

Cycle Gas Oil: Cracked gas oil returned to a cracking unit.

CycleMax™: A type of a continuous catalyst regenerative process technology offered by Honeywell UOP.

Cycle Stock: Unfinished product taken from a stage of a refinery process and recharged to the process at an earlier period in the operation.

Cycling: A series of operations in petroleum refining or natural-gas processing conducted so that the steps are periodically repeated in the same sequence.

Deasphalting: The process of removing asphaltenes from reduced crude using liquid propane to dissolve nonasphaltenes.

Debutanizer: A fractionating column used to remove butane and lighter components from liquid streams.

Deethanizer: The fractionating column in a natural gasoline plant in which ethane and lighter components are removed overhead. The resulting gas stream is then deethanized.

Dehydrogenation: A reaction in which hydrogen atoms are eliminated from a molecule. Dehydrogenation is used to convert ethane, propane, and butane into olefins (ethylene, propylene, and butenes). It is also the reaction used to convert naphthenes into aromatics.

Depentanizer: A fractionating column used to remove pentane and lighter fractions from hydrocarbon streams.

Depropanizer: A fractionating column for removing propane and lighter components from liquid streams.

Desalting: Removal of mineral salts (most chlorides, e.g., magnesium chloride and sodium chloride) from crude oil.

Desulfurization: A chemical treatment to remove sulfur or sulfur compounds from hydrocarbons.

Dewaxing: The removal of wax from petroleum products (usually lubricating oils and distillate fuels) by solvent absorption, chilling, and filtering.

Diethanolamine: A chemical compound used to remove hydrogen sulfide from gas streams.

Dilbit: Diluted bitumen, and the term is used for bitumen that has been diluted with a liquid to facilitate transportation of the very viscous bitumen.

Distillate: The products of distillation formed by condensing vapors.

Downflow: Process in which the hydrocarbon stream flows from top to bottom.

Dry Gas: Natural gas with little natural gas liquids that is nearly all methane with some ethane.

EDC: Equivalent distillation capacity.

EIA: US Energy Information Administration.

End Point: In the distillation tests for oil fractions and products, the highest thermometer reading during the distillation, which is indicative of when the sample has been entirely vaporized. It is also referred to as the final boiling point.

Engine Oil: Generic term applied to oils used for the bearing lubrication of all types of engines, machines, and shafting and for cylinder lubrication other than steam engines. In internal combustion engines, it is synonymous with motor oils and crankcase oils.

EOR: End of run or end of cycle, which are used for catalytic processes to indicate end of cycles.

EPC: Engineering, procurement, and construction.

Euro V: The designation of high-quality gasoline as specified by the European Community of countries.

Extractive Distillation: The separation of different components of mixtures that have similar vapor pressures by flowing a solvent that is selective for some of the components in the feed down the distillation column as the operation proceeds. The less soluble component exits overhead while the soluble component is scrubbed from the vapors. The solvent with the dissolved component is deposited at the bottom of the column and withdrawn for separation.

Fahrenheit: A measure of temperature where 32 F is the freezing point of water and 212 F is the boiling point of water at atmospheric pressure.

FEED: Front end engineering design.

Feed X Factor: A measure of the ease of upgrading light naphtha to high-octane isomerate or product. × factor is defined as the sum of the concentrations of benzene, methyl cyclopentane, cyclohexane, and hydrocarbons with seven or greater carbon atoms. The higher the x factor, the lower the achievable octane upgrade.

Feedstock: Stock from which material is taken to be fed (charged) into a processing unit.

Flashing: The process in which a heated oil under pressure is suddenly vaporized in a tower, reducing its pressure.

Flash Point: Lowest temperature at which a petroleum product will give off sufficient vapor so that the vapor-air mixture above the surface of the liquid will propagate a flame away from the source of ignition.

Flux: Lighter petroleum used to fluidize heavier residual so that it can be pumped.

Fouling: Accumulation of deposits in condensers, exchangers, vessels, towers, piping, and so on.

Fractions: Oil refiners' term for the portions of oil containing a number of hydrocarbon compounds within certain boiling ranges, separated from other portions in fractional distillation. The fractions are distinguished from pure compounds, which have specific boiling temperatures.

Fractional Distillation: Separation of the components of a liquid mixture by vaporizing it and collecting the fractions, which condense in different temperature ranges.

Fractionating Column: Process unit that separates various fractions of petroleum by simple distillation, with the column tapped at various levels to separate and remove fractions according to their boiling ranges.

Freezing Point: The temperature at which a substance freezes and determined by the ASTM D1015 test procedure. It is also used for determining the degree of purity of high-purity hydrocarbon compounds.

Fuel Gas: Refinery gas used for heating.

Gas Oils: A fraction derived in refining petroleum with a boiling range between kerosene and lubricating oil. Derives its name from having originally been used in the manufacture of illuminating gas. The oils are used in the production of distillate-type fuel oils, diesel fuel, and gasoline.

Gasoline: A blend of reformate and other refinery products with sufficiently high octane and other desirable characteristics to be suitable for use as fuel in internal combustion engines. ASTM D439 test procedure specifies three grades for various types of motor vehicle operations.

GHG: Greenhouse gases. Nitrogen oxides, carbon dioxide, methane, hydrofluorocarbons, and perfluorocarbons are gaseous compounds that are referred to as greenhouse gases. They contribute to the climate change of the Earth.

Header: A manifold that distributes fluid from a series of smaller pipes or conduits.

Heart Cut: In refining, a narrow boiling range fraction, usually taken near the middle portion of the stock being processed.

Heat Exchanger: Equipment to transfer heat between two flowing streams of different temperatures. Heat is transferred between liquids or liquids and gases through a tubular wall.

Heating Oils: Trade term for distillate fuel oils used in heating homes and buildings and distinguished from residual fuel oils used in heating and power installations. Both are burner fuel oils.

Heavy Ends: The highest boiling portion of a gasoline or other petroleum oil. The end point as determined by the distillation test reflects the amount and character of the heavy ends present in a gasoline.

High Pressure: The term is used to describe the pressure of a processing unit and equipment. High-pressure (100 psi) gas from cracking unit distillate drums that is compressed and combined with low-line gas as gas absorption feedstock.

Hydrocarbons: Any chemical compound that is made up exclusively of carbon and hydrogen atoms. Hydrocarbons form the principal constituents of petroleum.

Hydrocracking: A process used to convert heavier feedstocks into lower-boiling, higher-value gasoline and diesel products. The process employs high pressure, high temperature, catalysts, and hydrogen.

Hydrodesulfurization: A catalytic reaction that converts organo-sulfur compounds to hydrogen sulfide and a hydrocarbon. In a broader definition, it is the process used to remove sulfur, primarily from petroleum fractions in the presence of hydrogen. Typically, in oil refining, this process also removes nitrogen and contaminant metals and saturates certain compounds.

Hydrofining: A process for treating petroleum fractions and unfinished oils in the presence of catalysts and substantial quantities of hydrogen to upgrade their quality.

Hydrofinishing: A catalytic treating process conducted in the presence of hydrogen to improve the properties of low viscosity-index and medium viscosity-index naphthenic oils. It is also applied to paraffin and microcrystalline waxes for the removal of undesirable components. This process consumes hydrogen and is used in lieu of acid treating.

Hydroforming: The catalytic reforming of naphtha at elevated temperatures and moderate pressures in the presence of hydrogen to produce high-octane BTX aromatics for motor fuel or chemical manufacture. This process results in a net production of hydrogen and rendered thermal reforming obsolete. It represents the total effect of numerous simultaneous reactions such as cracking, dehydrogenation, isomerization, and aromatization.

Hydrogenation: A refinery process in which hydrogen is added to the molecules of unsaturated hydrocarbon fractions.

Hydrogen Sulfide: An objectionable impurity present in some natural gas and crude oils and formed during the refining of sulfur-containing oils. It is removed from products by various treating methods at the refinery.

Hydroskimming: A simple refinery with crude distillation unit and few processing units such as a catalytic reformer.

Induction Period: A period under given conditions in which a petroleum product does not absorb oxygen at a substantial rate to form gum.

Inhibitor: An additive substance that, when present in a petroleum product, prevents or retards undesirable changes taking place in the product, particularly oxidation and corrosion.

Isomerization: A reaction that catalytically converts straight-chain hydrocarbon molecules into branched-chain molecules of substantially higher octane number. The reaction rearranges the carbon skeleton of a molecule without adding or removing anything from the original material.

Iso-octane: A hydrocarbon molecule (2, 2, 4-trimethylpentane) with excellent antiknock characteristics on which the octane number of 100 is based.

Jet Fuel: Kerosene-type fuels or blends of gasoline, distillate, and residual oils that are used as fuels for gas turbine–powered aircraft.

Knock: The sound associated with autoignition in the combustion chamber of an automobile engine of a portion of the fuel-air mixture ahead of the advancing flame front.

Knockout Drum: A vessel wherein suspended liquid is separated from gas or vapor.

KPI: Key performance indicator. Once established, KPIs can be used to track process units, projects, safety, and other programs and financial and economic activities.

Lean Oil: Absorbent oil fed to absorption towers in which gas is to be stripped. After absorbing the heavy ends from the gas, it becomes fat oil. When the heavy ends are subsequently stripped, the solvent again becomes lean oil.

LHSV: Liquid hourly space velocity is defined as the volumetric flow rate of the feed divided by the volume of catalyst in reactors and process units.

Light Ends: The lower boiling components of a mixture of hydrocarbons, usually C1 through C4 hydrocarbons.

Lime Treatment: The process of introducing lime into the still during the distillation of petroleum to reduce the acidity of the distillate. Other forms of distillate treatment are often preferred.

LIMS: Laboratory information management system.

Liquefied Petroleum Gas, LP-Gas, LPG, and Bottled Gas: Industry terms for any material composed predominantly of the following hydrocarbons or mixture of them: propane, propylene, butanes, and butylenes. Recovered from natural and refinery gases and kept under pressure in a liquid state, they are marketed in liquid form for industrial and domestic use as gas.

LOI: The loss on ignition represents volatile compounds that are removed at high temperatures of 900°C and higher in an analytical determination. LOI is subtracted from 100% to get a measure of the solid concentration of a catalyst sample.

Low Pressure Gas: Low-pressure (5-psi) gas from atmospheric and vacuum distillation recovery systems that is collected in the gas plant for compression to higher pressures.

Lube Stocks: Refinery term for fractions of crude petroleum of suitable boiling range and viscosity to yield lubricating oils when further processed and treated.

Mercaptans: Compounds of sulfur having a strong, repulsive garlic-like odor and a molecular formula of RSH. A contaminant of sour crude oil and products.

Metal Settlement: The process that culminates in the agreed-upon amount of troy ounces of precious metals to be transferred from one company to another. This process involves discussions of comparative assay data for lots of precious metals.

Motor Octane Number: The octane number or the antiknock quality is the octane number determined at higher motor temperatures and speeds for a hydrocarbon, blendstock, and gasoline.

Motor Oils: Lubricating oils designed for used in the oil circulating systems of automotive, aircraft, and diesel engines.

MSAT: Mobile source air toxics are compounds emitted from highway vehicles and nonroad equipment that are known or suspected to cause cancer or other serious health and environmental effects.

MSAT 2 Regulations: Mobile source air toxics phase 2 limits annual average benzene to 0.62 volume percent and annual maximum average benzene to 1.3 volume percent.

MTBF: Mean time between failures. A measure used to rate the operation and reliability of plant equipment.

Naphtha: A general term used for low-boiling hydrocarbon fractions that are a major component of gasoline. Paraffinic naphtha refers to those naphthas containing naphthenes and aromatic compounds such that their N+2A sums are less than 60, and naphthenic naphthas have much higher N+2As, where N and A are the volume percent concentrations of naphthenes and aromatics in the naphtha.

Naphthenes: Hydrocarbons (cycloalkanes) with the general formula C_nH_{2n}, in which the carbon atoms are arranged to form a ring.

Naphthenic Crudes: A type of crude petroleum containing a relatively large proportion of naphthenic-type hydrocarbon.

Nelson Complexity Index: An index developed by Nelson and used for determining the processing capabilities of oil refineries.

NPRA: It is the acronym for the National Petrochemical and Refiners Association. This name was used to replace the National Petroleum Refiners Association (NPRA) in 1998. As indicated previously, AFPM replaced NPRA in 2012.

NYMEX: New York Mercantile Exchange.

Octane Number: The antiknock quality of motor and aviation gasoline below 100 octane is expressed by a numerical scale that is based on the knocking tendencies of two pure hydrocarbons. Normal heptane, has an assigned value of zero in the knock rating scale. The second, iso-octane or 2, 2, 4-trimethyl pentane, has an assigned octane number of 100.

Octane Severity: In catalytic reforming, octane severity refers to the target reformate octane selected for the process unit.

Olefins: A family of unsaturated hydrocarbons with one carbon-carbon double bond and the general formula C_nH_{2n}.

Overhead: In a distilling operation, that portion of the charge that leaves the top of the distillation column as a vapor.

Paraffin Aromatization: A major series of reactions in catalytic reforming that culminate in the conversion of paraffins to aromatic compounds. Starting with paraffins, the reactions include dehydroisomerization of paraffins to C5 naphthenes, dehydrogenation of the C5 naphthenes, and dehydroisomerization to C6 naphthenes, followed by dehydrogenation to aromatic compounds.

Paraffin Base Crude: A type of crude oil containing predominantly paraffin hydrocarbons as distinguished from asphaltic and naphthenic-base crudes. It is a source of high-quality lubricating oils.

Paraffins: A family of saturated aliphatic hydrocarbons (alkanes) with the general formula C_nH_{2n+2}.

PIMS: Plant information management system. An automated system used for continuous monitoring and management of data for process units.

PIN: Paraffin isomerization number is a measure of the performance of a light naphtha isomerization process unit. PIN is defined as the sum of the product iso-pentane (iC5), 2,2 dimethyl butane (2,2 DMB), and 2,3 dimethyl butane ratio (23 DMB) ratios.

Platinum Dispersion: The degree of how platinum sites are located on a catalyst for better utilization and facilitation of chemical reactions by the sites. Platinum dispersion can be maintained via reactivation procedures during catalyst regenerations.

Platinum Distribution: The extent and uniformity of platinum penetration that are achieved within catalyst particles through the platinum impregnation process and during catalyst production.

Polyforming: The thermal conversion of naphtha and gas oils into high-quality gasoline at high temperatures and pressure in the presence of recirculated hydrocarbon gases.

Polymerization: The process of combining two or more unsaturated organic molecules to form a single (heavier) molecule with the same elements in the same proportions as in the original molecule.

Pour Point: The lowest temperature at which an oil will pour or flow when chilled, without disturbance, under test conditions in ASTM D97.

Pour Stability: The ability of a pour-depressant-treated oil to maintain its original pour point when in storage at low temperatures approximating winter conditions.

Precipitate: A substance separated in solid form from a liquid as the result of some physical or chemical change, differing from a substance held only mechanically in suspension, which is known as sediment.

Preheater: Exchanger used to heat hydrocarbons before they are fed to a unit.

Pressure Regulating Valve: A valve that releases or holds process-system pressure (that is, opens or closes) either by preset spring tension or by actuation by a valve controller to assume any desired position between fully open and fully closed.

Pyrolysis Gasoline: A byproduct from the manufacture of ethylene by steam cracking of hydrocarbon fractions such as naphtha or gas oil.

Pyrometallurgical Process: An alternative process used for precious metal recovery from materials that are not completely amenable to the wet chemistry process due to carrier or support insolubility in reagents. This procedure involves smelting at high temperatures for recovery of the precious metals.

Pyrophoric Iron Sulfide: A substance formed inside tanks and processing units by the corrosive interaction of sulfur compounds in the hydrocarbons and the iron and steel in the equipment. On exposure to air (oxygen), it ignites spontaneously.

Quench Oil: Oil injected into a product leaving a cracking or reforming heater to lower the temperature and stop the cracking process.

Raffinate: The product from a solvent extraction process, consisting mainly of those components that are least soluble in the solvents. The product recovered from an extraction process is relatively free of aromatics, naphthenes, and other constituents that adversely affect physical parameters.

Reactor: The vessel in which chemical reactions take place during a chemical conversion type of process.

Reboiler: An auxiliary unit of a fractionating tower designed to supply additional heat to the lower portion of the tower.

Recycle Gas: High hydrogen-content gas returned to a unit for reprocessing.

Reduced Crude: The residual product remaining after the removal by distillation of an appreciable quantity of the more volatile components of crude oil.

Refining: A series of processes to convert crude oil and its fractions into finished petroleum products, including thermal cracking, catalytic cracking, alkylation, hydrocracking, hydrogenation, hydrodesulphurization, and isomerization.

Reflux: The portion of the distillate returned to the fractionating column to assist in attaining better separation into desired fractions.

Reformate: An upgraded high octane naphtha resulting from catalytic or thermal reforming.

Reforming: The thermal or catalytic conversion of petroleum naphtha into more volatile products of higher octane number. It represents the total effect of numerous simultaneous reactions such as cracking, polymerization, dehydrogenation, and isomerization.

Reforming Severity: A combination of process conditions and reformate octane target for a catalytic reforming unit.

Regeneration: In a catalytic processes, deactivated coked catalysts are subjected to a process of burning off the coke deposits under carefully controlled conditions of temperature and oxygen content of the regeneration gas stream and reactivating the catalysts for reuse in more cycles of processing operations.

Reid Vapor Pressure, RVP: A test for determining the vapor pressure of volatile hydrocarbon products (gasolines) under controlled conditions.

Research Octane Number: The octane number or the antiknock quality is the octane number determined at relatively moderated engine test conditions for a hydrocarbon, blendstock, and gasoline.

RFS: Renewal fuel standards that refer to the use of cleaner gasoline and diesel fuels.

RMACT II: Refinery maximum achievable control technologies identified for reducing hydrogen chloride and toxic organic compounds (TOCs) to meet RMACT II regulations. Caustic scrubbers and adsorbents were used for capturing hydrogen chloride and TOCs.

Road Octane Number: The octane number posted at gasoline filling stations is the average of the research and motor octane numbers, that is, RON is to (R+M)/2, where R and M are research octane number and motor octane number, respectively.

Routine Maintenance: Ongoing preventative maintenance of equipment, plant, building, and structures.

RSP: Renewal super premium gasoline is the proposed gasoline that automobiles could use to meet a 54.5 mile per gallon CAFÉ standard by 2025.

SAGD: Steam-assisted gravity drainage oil production process used in the recovery of bitumen from tar sands.

Scallops: Reactor internals that are used in radial reactors instead of outer baskets. Scallops permit hydrocarbon and gas flow in the radial direction though the annular zone of the catalyst bed and through the centerpipe.

Scrubbing: Purification of a gas or liquid by washing it in a tower.

Smelting: A process used in the recovery of metals from hard, insoluble materials or ores. The process uses heat and a reducing agent to recover metals such as platinum.

Solvent Extraction: The separation of materials of different chemical types and solubilities by selective solvent action.

SOR: Start of runs or start of cycles for catalytic processes.

Sour Gas: Natural gas that contains corrosive, sulfur-bearing compounds such as hydrogen sulfide and mercaptans.

Specific Gravity: The ratio of the weight of a given volume of material to the weight of an equal volume of water. The standard reference material is distilled water, and the temperature of both the hydrocarbon product and the water is 60 F.

Stabilization: A process for separating the gaseous and more volatile liquid hydrocarbons from crude petroleum or gasoline and leaving a stable (less-volatile) liquid so that it can be handled or stored with less change in composition.

Straight-Run Gasoline: Gasoline produced by the primary distillation of crude oil. It contains no cracked, polymerized, alkylated, reformed, or visbroken stock. It is also referred to as straight-run naphtha (SRN).

Stripping: The removal by steam-induced vaporization or flash evaporation of the more volatile components from a cut or fraction.

Sulfuric Acid Treating: A refining process in which unfinished petroleum products such as gasoline, kerosene, and lubricating oil stocks are treated with sulfuric acid to improve their color, odor, and other characteristics.

Sweetening: Processes that either remove obnoxious sulfur compounds (primarily hydrogen sulfide, mercaptans, and thiophenes) from petroleum fractions or streams, or convert them, as in the case of mercaptans, to odorless disulfides to improve odor, color, and oxidation stability.

Synbit: Synthetic bitumen is an upgraded bitumen that may have been subjected to a hydrotreating or decoking process.

Syncrude: Unconventional crudes such as those derived from tar sands, oil shale, and coal liquefaction, and the product mixture of Fischer–Tropsch synthesis.

Tail Gas: The lightest hydrocarbon gas released from a refining process.

TAN: Total acid number, which provides a measure of the acidity of oils, especially used for crude oil assays.

Tar Sands: A mixture of 84%–88% sand and mineral-rich clays, 4% water, and 8%–12% bitumen heavy oil. Bitumen is a dense, sticky, semisolid substance that is about 83% carbon. Tar sand is also known as oil sand and heavy oils. The extensive tar sands in Canada and Venezuela have received considerable attention.

Thermal Cracking: The breaking up of heavy oil molecules into lighter fractions by the use of high temperature without the aid of catalysts.

TOC: Toxic organic compound such as dioxin.

Topping: Refers to a crude distillation unit. It is also used to describe a simple low Nelson Index refinery that is mostly a crude distillation unit.

Turnaround Maintenance: A planned complete shutdown of an entire process or section of a refinery, or of an entire refinery to perform major maintenance, overhaul, and repair operations and to inspect, test, and replace process materials and equipment.

Vacuum Distillation: The distillation of petroleum under vacuum, which reduces the boiling temperature sufficiently to prevent cracking or decomposition of the feedstock.

Vapor: The gaseous phase of a substance that is a liquid at normal temperature and pressure.

Visbreaking: Viscosity breaking is a low-temperature cracking process used to reduce the viscosity or pour point of straight-run residuum.

Viscosity: The measure of the internal friction or resistance to flow of a fluid. In determining viscosities of liquid hydrocarbon products, values are often expressed as the number of seconds in time required for a certain volume of the liquid under test to pass through a standard orifice under prescribed conditions. It is expressed either in Saybolt seconds or in mm^2/s (kinematic viscosity).

VOC: Volatile organic compound.

WAIT: Weighted average inlet temperature is used in monitoring reactor temperatures for catalytic reforming units.

Wet Chemistry: The process involving chemical reactions of reagents and soluble carrier or support in the process recovering precious metals.

Wet Gas: A gas containing a relatively high proportion of hydrocarbons that are recoverable as liquid.

WHSV: Weight hourly space velocity is defined as the mass (weight) flow rate of a feed divided by the mass (weight) of catalyst in a reactor.

WTI: West Texas Intermediate crude oil.

Index

D

E

F

G

H

I

J

K

L

M

N

O

S

T

U

V

W

Y